Gene Mapping, Discovery, and Expression

METHODS IN MOLECULAR BIOLOGY™

John M. Walker, SERIES EDITOR

METHODS IN MOLECULAR BIOLOGY™

Gene Mapping, Discovery, and Expression

Methods and Protocols

Edited by

Minou Bina

Department of Chemistry, Purdue University, West Lafayette, IN

HUMANA PRESS ✳ TOTOWA, NEW JERSEY

© 2006 Humana Press Inc.
999 Riverview Drive, Suite 208
Totowa, New Jersey 07512

www.humanapress.com

This publication is printed on acid-free paper. ∞
ANSI Z39.48-1984 (American Standards Institute)

Permanence of Paper for Printed Library Materials.

Cover illustration: Figure 2, from Chapter 4, "Quantitative DNA Fiber Mapping in Genome Research and Construction of Physical Maps," by H.-U. G. Weier and L. W. Chu

Cover design by Patricia F. Cleary.

For additional copies, pricing for bulk purchases, and/or information about other Humana titles, contact Humana at the above address or at any of the following numbers: Tel.: 973-256-1699; Fax: 973-256-8341; E-mail: orders@humanapr.com; or visit our Website: www.humanapress.com

Printed in the United States of America. 10 9 8 7 6 5 4 3 2 1

eISBN 1-59745-097-9

Library of Congress Cataloging in Publication Data
Gene mapping, discovery, and expression : methods and protocols / edited by Minou Bina.
           p. ; cm. — (Methods in molecular biology ; v. 338)
        Includes bibliographical references and index.
        ISBN 1-58829-575-3 (alk. paper)
        1. Gene mapping—Methodology. 2. Gene mapping—Data processing. 3. Genetics—Technique.
        4. Genetic expression.
        [DNLM: 1. Chromosome Mapping—methods—Laboratory Manuals. 2. Databases, Nucleic Acid—Laboratory Manuals. 3. Gene Expression Profiling—methods—Laboratory Manuals. 4. Microarray Analysis—methods—Laboratory Manuals. QU 25 G3256 2006] I. Bina, Minou. II. Series.
        QH445.2.G436 2006
        572.8'633—dc22   2005025438

Preface

Completion of the sequence of the human genome represents an unparalleled achievement in the history of biology. The project has produced nearly complete, highly accurate, and comprehensive sequences of genomes of several organisms including human, mouse, drosophila, and yeast. Furthermore, the development of high-throughput technologies has led to an explosion of projects to sequence the genomes of additional organisms including rat, chimp, dog, bee, chicken, and the list is expanding.

The nearly completed draft of genomic sequences from numerous species has opened a new era of research in biology and in biomedical sciences. In keeping with the interdisciplinary nature of the new scientific era, the chapters in *Gene Mapping, Discovery, and Expression: Methods and Protocols* recapitulate the necessity of integration of experimental and computational tools for solving important research problems. The general underlying theme of this volume is DNA sequence-based technologies. At one level, the book highlights the importance of databases, genome-browsers, and web-based tools for data access and analysis. More specifically, sequencing projects routinely deposit their data in publicly available databases including GenBank, at the National Center of Biotechnology (NCBI) in the United States; EMBL, maintained by the European Bioinformatics Institute; and DDBJ, the DNA Data Bank of Japan. Currently, several browsers offer facile access to numerous genomic DNA sequences for gene mapping and data retrieval. These include the map-view at NCBI; the genome browser at the University of California at Santa Cruz, UCSC; and the browser maintained by Ensembl. All three browsers offer sophisticated tools for gene mapping and localization on genomic DNA.

For beginners in the field, through a specific example, one chapter provides a step-by-step procedure for localization, creating a map, and a graphical representation of genes of interest using the genome browser at UCSC. Since the drafts of the genomic sequences provide primarily a reference for studies of gene organization, additional methods are needed for understanding the complexity and dynamic nature of chromosomes. Significantly, segmental duplications are a common feature of many mammalian genomes. Therefore, *Gene Mapping, Discovery, and Expression: Methods and Protocols* provides a computational protocol for identifying and mapping recent segmental and gene duplications. Another chapter offers a step-by-step procedure for identifying paralogous genes, using the genome browser at UCSC.

To examine local variations in specific regions of chromosomes experimentally, a chapter provides a novel method, Quantitative DNA Fiber Mapping, that relies on fluorescent *in situ* hybridization (FISH) to identify, delineate, and characterize selected, often small, DNA sequences along a larger piece of the human genome. In another experimental contribution, a chapter describes a sensitive and specific method, Primed *in situ* labeling, that can be used for localization of single copy genes and sequences too small for detection by conventional FISH.

Novel DNA sequence-based strategies include methods for the discovery and mapping of the functional elements and the "codes" in DNA that regulate the expression of genes. The completed sequence of the human genome and the genomic sequences of model organisms offer a rich source of data for addressing this problem. A fundamental and powerful method is based on comparing the sequences from different species to identify the conserved functional elements. A chapter in this volume describes the VISTA family of computational tools, created to assist researchers in aligning DNA sequences for locating the genomic DNA regions that are highly conserved. Another chapter aims at using sequence conservation as a guide for identifying the elements that may regulate the expression of genes. This chapter describes how to use publicly available servers (Galaxy, the UCSC Table Browser, and GALA) to find genomic sequences whose alignments show properties associated with *cis*-regulatory modules and conserved transcription factor binding sites. Furthermore, this volume describes additional versatile and web-based tools for promoter, regulatory region, and expression analyses. These tools include CORG "COmparative Regulatory Genomics" and BEARR "Batch Extraction and Analysis of *cis*-Regulatory Regions."

DNA sequence-based technologies include other strategies that could help with the identification of regulatory signals and potential protein binding elements in the regulatory regions of genes. For example, a chapter describes how a database of 9-mers from promoter regions of human protein-coding genes could be accessed via the web for the discovery of the lexical characteristics of potential regulatory motifs in human genomic DNA. These characteristics could help with predicting and classifying regulatory cis-elements according to the genes that they control.

Cis-elements can control the expression of genes in an allele-specific fashion. The analysis of allele-specific gene expression is of interest in the study of genomic imprinting. Significantly, there is growing awareness that differences in allelic expression could be widespread among autosomal non-imprinted genes. A chapter in *Gene Mapping, Discovery, and Expression: Methods and Protocols* provides protocols for in vivo analysis of allelic-specific gene expression. These include analysis of the relative allelic abundance of transcribed RNA, and of transcription factor recruitment and Pol II loading by chromatin immunoprecipitation. Another chapter describes miRNAs expression vectors containing human RNA polymerase II or III promoters for studies of the control of gene expression.

In this new scientific era, gene expression is extensively studied using microarray technologies. Two chapters describe how to use web-based tools for accessing and analyzing the microarray data. One chapter describes Gene Expression Omnibus (GEO) developed at NCBI. GEO has emerged as a leading fully public repository for gene expression data. The chapter describes how to use Web-based interfaces, applications, and graphics to effectively explore, visualize and interpret the hundreds of microarray studies and millions of gene expression patterns stored in GEO. Another chapter describes the resources at the Stanford Microarray Database (SMD). This database offers a large amount of data for public use. The chapter describes how to use the primary tools for searching, browsing, retrieving, and analyzing data available at SMD. Furthermore, researchers, educators, and students may find SMD a very useful repository of a large quantity of publicly available data that together with analysis tools, could be used for exploratory, unsupervised analysis and discovery.

Another level of sequence-based technologies depends on how best to analyze the structural organization of chromosomes, evaluate the sequence specificity of transcription factors, and isolate and identify the components of the protein complexes formed with DNA. More specifically, in cells, the chromosomal DNA is associated with proteins to form complexes referred to as chromatin. A major group of chromosomal proteins, the histones, functions in the compaction of DNA by forming nucleosomes. Another major group corresponds to transcription factors, which control the expression of genes through protein–DNA and protein–protein interactions. Evidence supports major roles for the underlying DNA sequence on the relative arrangement of proteins along the chromosomes. Two chapters in this volume provide DNA sequence-based methods for probing chromatin structure. One chapter describes a step-by-step procedure for detecting and analyzing nucleosome ladders on unique DNA sequences. Another offers a non-invasive method of assaying relative DNA accessibility in yeast chromatin without disrupting DNA–protein interactions.

The DNA sequence specificities of transcription factors are key components of the *cis* regulatory networks. However, despite their importance, the DNA binding specificities of many transcription factors remain unknown. Furthermore, methods routinely used for characterizing protein binding sites are not scalable and are time-consuming. These issues are problematic because complete, accurate, and reliable datasets of transcription factor binding elements are needed for localizing the regulatory regions of genes. This volume offers two chapters on novel DNA microarray-based technologies for rapid, high-throughput in vitro characterization of the DNA sequence specificities of transcription factors.

Lastly, several chapters in *Gene Mapping, Discovery, and Expression: Methods and Protocols* offer non-invasive technologies for the isolation of transcription factor complexes formed with specific DNA sequences used as bait. Identification of the components of large protein–DNA complexes is an important step in elucidating the mechanisms by which gene expression is controlled. Two chapters describe the use of powerful methods based on mass spectrometry for identification of proteins in the complexes formed with DNA. These methods can lead to the discovery of novel transcription factors with important roles in the control of gene expression.

Minou Bina

Contents

Contributors

CATHERINE A. BALL • *Department of Biochemistry, Stanford University Medical School, Stanford, CA*

TANYA BARRETT • *National Center for Biotechnology Information, National Institutes of Health, Bethesda, MD*

MICHAEL F. BERGER • *Biophysics Program, Harvard University, Boston, MA*

MINOU BINA • *Department of Chemistry, Purdue University, West Lafayette, IN*

EDGAR BONTE • *Department of Cell Biology, Erasmus Medical Center, Rotterdam, The Netherlands*

HARALD BRAUN • *Department of Cell Biology, Erasmus Medical Center, Rotterdam, The Netherlands*

MARTHA L. BULYK • *Department of Medicine, Division of Genetics; Department of Pathology; and Harvard–MIT Division of Health Sciences & Technology, Brighman and Women's Hospital and Harvard Medical School, Boston, MA*

JENNIFER CAMPBELL • *Department of Cell Biology, Erasmus Medical Center, Rotterdam, The Netherlands*

LISA W. CHU • *Department of Genome Biology, Life Sciences Division, Lawrence Berkeley National Laboratory, Berkeley, CA*

ALFRED CIOFFI • *Department of Biological Sciences, Purdue University, West Lafayette, IN*

ERNIE DE BOER • *Department of Cell Biology, Erasmus Medical Center, Rotterdam, The Netherlands*

CHRISTOPH DIETERICH • *Computational Molecular Biology Department, Max Planck Institute for Molecular Genetics, Berlin, Germany*

INNA DUBCHAK • *Genomics Division, Lawrence Berkeley National Laboratory, Berkeley, CA*

RON EDGAR • *National Center for Biotechnology Information, National Institutes of Health, Bethesda, MD*

LAURA ELNITSKI • *Genome Technology Branch, National Institutes of Health, Rockville, MD*

SIMON FIELD • *University of Oxford, Oxford, UK*

TOMARA J. FLEURY • *Department of Biological Sciences, Purdue University, West Lafayette, IN*

YOKO FUKUDA • *Department of Chemistry and Biotechnology, The University of Tokyo, Japan*

JEREMY GOLLUB • *Department of Biochemistry, Stanford University Medical School, Stanford, CA*

FRANK GROSVELD • *Department of Cell Biology, Erasmus Medical Center, Rotterdam, The Netherlands*

ROSS C. HARDISON • *Department of Biochemistry and Molecular Biology, Center for Comparative Genomics and Bioinformatics, The Pennsylvania State University, University Park, PA*

SCOTT A. HOOSE • *Department of Biochemistry and Biophysics, Texas A&M University, College Station, TX*

HIROAKI KAWASAKI • *Department of Chemistry and Biotechnology, The University of Tokyo, Japan*

RAZI KHAJA • *Program in Genetics and Genomic Biology, Research Institute, The Hospital for Sick Children, Toronto, ON, Canada*

DAVID KING • *Department of Biochemistry and Molecular Biology, Center for Comparative Genoics and Bioinformatics, The Pennsylvania State University, University Park, PA*

MICHAEL P. KLADDE • *Department of Biochemistry and Biophysics, Texas A&M University, College Station, TX*

JULIAN C. KNIGHT • *Wellcome Trust Centre for Human Genetics, University of Oxford, Oxford UK*

KATARZYNA E. KOLODZIEJ • *Department of Cell Biology, Erasmus Medical Center, Rotterdam, The Netherlands*

SHERYL A. LAZARUS • *Department of Chemistry, Purdue University, West Lafayette, IN*

JEFFREY R. MACDONALD • *Program in Genetics and Genomic Biology, Research Institute, The Hospital for Sick Children, Toronto, ON, Canada*

SJAAK PHILIPSEN • *Department of Cell Biology, Erasmus Medical Center, Rotterdam, The Netherlands*

JIANNIS RAGOUSSIS • *Wellcome Trust Centre for Human Genetics, University of Oxford, Oxford, UK*

PATRICK RODRIGUEZ • *Department of Cell Biology, Erasmus Medical Center, Rotterdam, The Netherlands*

DMITRIY V. RYABOY • *Genomics Division, Lawrence Berkeley National Laboratory, Berkeley, CA*

STEPHEN W. SCHERER • *Program in Genetics and Genomic Biology, Research Institute, The Hospital for Sick Children, Toronto, ON, Canada*

SYED REHAN SHAH • *Department of Chemistry Purdue University, West Lafayette, IN*

GAVIN SHERLOCK • *Department of Genetics, Stanford University Medical School, Stanford, CA*

ARNOLD STEIN • *Department of Biological Sciences, Purdue University, West Lafayette, IN*

JOHN STROUBOULIS • *Department of Cell Biology, Erasmus Medical Center, Rotterdam, The Netherlands*

KAZUNARI TAIRA • *Department of Chemistry and Biotechnology, The University of Tokyo, Japan*

PAUL TEMPST • *Molecular Biology Program, Memorial Sloan-Kettering Cancer Center, New York, NY*

AVIRACHAN T. THARAPEL • *Department of Pediatrics, University of Tennessee, Memphis, TN*

IRINA A. UDALOVA • *Kennedy Institute of Rheumatology, Imperial College, London, UK*

VINSENSIUS BERLIAN VEGA • *Genome Institute of Singapore, Singapore*

MARTIN VINGRON • *Max Planck Institute for Molecular Genetics, Germany*

STEPHEN S. WACHTEL • *Department of Obstetrics and Gynecology, University of Tennessee, Memphis, TN*

HEINZ-ULRICH G. WEIER • *Department of Genome Biology, Life Sciences Division, Lawrence Berkeley National Laboratory, Berkeley, CA*

KENNETH K. WU • *Division of Hematology, Institute of Molecular Medicine, University of Texas Health Science Center, Houston, TX*

PHILLIP WYSS • *Department of Chemistry, Purdue University, West Lafayette, IN*

MARIANA YANEVA • *Memorial Sloan-Kettering Cancer Center, New York, NY*

JUNJUN ZHANG • *The Hospital for Sick Children, Toronto, ON, Canada*

1

Use of Genome Browsers
to Locate Your Favorite Genes

Minou Bina

Summary

The completion of whole-genome sequencing projects offers the opportunity of creating high-resolution maps of specific segments in a known genomic DNA sequence. For this purpose, several genome browsers have been created. They include the map-view (http://www.ncbi.nlm.nih.gov/mapview/), the Ensembl genome browser (http://www.ensembl.org/), and the genome browser at UCSC (http://genome.ucsc.edu/). For the beginners in the field, through a specific example, this chapter provides a step-by-step procedure for creating a map using the genome browser at UCSC. The example describes mapping, in the human genome, the promoter region of the NF-IL6 gene. The procedure is applicable to creating maps of the desired regions in genomes of other species available at the genome browser at UCSC.

Key Words: The Human Genome Project; gene mapping; gene localization.

1. Introduction

The rapid advances of genome sequencing projects have offered the opportunity to map and locate genes of interest, without resorting to time-consuming and costly experimental procedures. Large sequencing projects routinely deposit their data in publicly available databases including GenBank, at the National Center for Biotechnology Information (NCBI) in the United States (*1,2*); EMBL, maintained by the European Bioinformatics Institute (*3*); and DDBJ, the DNA Data Bank of Japan (*4*).

Currently, several browsers offer facile access to numerous genomic DNA sequences for gene mapping and data retrieval. These include NCBI (*1,2*); the genome browser at the University of California at Santa Cruz (UCSC) (*5,6*); and the browser maintained by Ensembl (*7*). All three browsers offer sophisticated

From: *Methods in Molecular Biology, vol. 338: Gene Mapping, Discovery, and Expression:
Methods and Protocols*
Edited by: M. Bina © Humana Press Inc., Totowa, NJ

tools for gene mapping and localization on genomic DNA. This chapter provides an example of how to use the genome browser at UCSC *(5,6)* to obtain a map and a graphical view of a known DNA sequence.

2. Materials

The gene localization procedure was done on a PC equipped with the Windows XP operating system. The general procedure should be applicable to other computers (*see* **Note 1**).

3. Methods

The genome browser at UCSC provides numerous sophisticated tools for data access, analyses, and visualization *(5,6)*. The following sections will guide a beginner in the field through simple and general procedures for locating and mapping the positions of a known sequence on genomic DNA.

1. Use the BLAT sequence alignment program at the genome browser at UCSC *(8)*.
2. To access BLAT, go to the browser's home page (http://genome.ucsc.edu/). Click on BLAT, one of the options listed on the left side of the page. You will obtain a query box for pasting a DNA sequence for analysis by BLAT.
3. To conduct a BLAT search, you should provide the query sequence in the FASTA format. In this format, the sequence is presented as a continuous chain of nucleotides, without any numbering and blank spaces (**Fig. 1**).
4. If you know the GenBank accession number of the DNA sequence of interest, perform the following steps to obtain a FASTA formatted file:
   a. Go to NCBI (http://www.ncbi.nlm.nih.gov/).
   b. Use the pull-down menu next to the query box that contains the word All Databases.
   c. On the menu, select nucleotides.
   d. In the query box next to "for," type the known accession number. As an example, type AF350408. This accession number contains the nucleotide sequence of a cloned human DNA fragment that includes the promoter region of the NF-IL6 gene *(9)*.
   e. After typing the accession number in the NCBI query box, click on go. You will obtain a page that includes the accession number and a description of the sequence file.
   f. Above the accession number, you will find the word report, in red letters. Click on report. You will obtain a pull-down menu. On the menu, select FASTA. You will obtain the FASTA formatted version of the sequence.
5. Copy the entire sequence.
6. Paste it in the BLAT query box at the UCSC browser, described above in **step 2**.
7. Alternatively, you can scroll down the BLAT page to use the box that would allow you to upload a FASTA formatted file from your computer.

```
TCAACGGATCTTGCTTTCAATTCGTTTGGATACACACCCAGAAATAAAATTGCTGGGTCATTTGGTAATT
CTATTTTTAATTTTTTGAGGAACAGCCATACCATTTCCCACAGTAGCTACACAAGCATTCCAGTTTCTAC
ACATCCTTCCCAACGCAGCAGCCAGCCTAATGGGCGTAAGGTGGTATCTCATTGTGTTTTATTCCTTTTA
TTTTTAGTTGACACATAATTGTACATGTTCATGGGTTACAAAATGATATTTCAATACATGTATACAACGT
GTAATAATCAAATCTGGGTAATTAGGATATCCATTACCTCAAGCGTTTATCATTTCTTTGTGTCGGGAAC
ATTCAAAATCCTCTCTTCTCATTGCGTTTTGTTTGTTTTTTAGATGGAGTCTCACTCAGTTGCCCAGGCT
GGGGTGCAGTGGCGTGATCTTGGCTTGCTGCAACCTCCACCTCCCAGGTTCAAGTGATTCTCCTGCCTCA
GCCTCCCAAGTAGCTGGGACTATACGCACGTACCATCACGCCCAGCTAATTTTGTATTTTTAGTAGAGA
CGGGGTTTCACCGTGTCGGCCAGGCTGGTCTCCAACTCTTGACCTCAGGTGATCCACCAAAGTGTTGCGC
CTCATTGCAGTTTTGGTTTGTATTTCCTTAATGGTTAGTGATGTTGAGTATCTTATCATGTGATTATTAG
CCATTCACAGTTTTCTCTGGAGAAATGTTTATTCAAGTCCTTTGCCCATTTTTAATTGAGTTGTTTTTAT
TGCTGTTGAGTTGTAACAGTTCTTTATGTATTCTGGATATGAACCCCTTATCAGATATATGATTTGCAAA
TATTTTCTCCCATTCCATGTTTCCATGTTCCATTTTTTTTTTTTTTGGAGGCAGAGTCTTGCTTTGTCACC
CAGGTTGGAGTGCAGAGGCACAATTTCGGCTCACTGCAACCTCCGCCTCCCGGGTTCAAGTGATTCTCAT
GTCTCAGCCTCCTGAATAGCTGGGATTACAGATGCACGCCACCATGCCTGGCCATTTTTTTTTTTTTAAT
```

Fig. 1. Example of a FASTA formatted DNA sequence.

8. On the top of the BLAT query box, for genome, select human. Click on the pull-down menu to view the extensive list of genomic sequences offered by the browser. (You can also use the procedures described here for mapping and graphical representation of sequences from other species.)
9. Above the BLAT query box, in the box under assembly, choose the latest version (in our example, 2004). Alternatively, from the pull-down menu, select an earlier version of a genomic DNA sequence.
10. Use the pull-down menu under the Query type and select DNA.
11. For the other variables (score and output type), use the default values.
12. Finally, click on submit.
13. You will obtain a page listing the results of the BLAT search (**Fig. 2**).
14. Examine the column tagged score (**Fig. 2**). You will find the highest score (6455) for an extended region (positions 7–6477), with 100% sequence identity to the query submitted for analysis by BLAT (**Fig. 2**). In some cases, for additional extended regions, you might obtain high scores and high sequence identity to the query. These scores may represent pseudogenes or recent duplications that could be examined for further evaluation.
15. Next to each query result (Your Seq., **Fig. 2**), right-click on details to open the link in a new window. This link provides useful information (*see* **Note 2**). For example, on the top of the new window, you will find the chromosomal positions of the query sequence (in that example, chr20:48234366-48240842). Below the positions, you will find the submitted sequence with regions highlighted in different colors. Scroll down to view the results of side-by-side alignment. The quality of the alignment can guide your decision as to whether the reported matches with the query sequence are significant (*see* **Note 2**).
16. Go to the browser to obtain a map (a graphical view) of the query sequence. To do so, on the page summarizing the result of the BLAT search (**Fig. 2**), choose the top line, the line with the highest score. Right-click on the browser link on the left side, to open and view the map in a new window (**Fig. 3**).

Home   Genomes   Tables   Gene Sorter   PCR   FAQ   Help

Human BLAT Results

BLAT Search Results

ACTIONS	QUERY	SCORE	START	END	QSIZE	IDENTITY	CHRO	STRAND	START	
browser details	YourSeq	6455	7	6477	6477	100.0%	20	+	48234366	
browser details	YourSeq	331	179	1142	6477	83.2%	X	-	50368285	
browser details	YourSeq	319	424	1173	6477	85.9%	20	-	13586531	
browser details	YourSeq	310	424	1142	6477	84.3%	3	+	11893327	
browser details	YourSeq	283	377	851	6477	86.6%	9	+	138080064	1
browser details	YourSeq	263	414	1103	6477	83.9%	16	-	45428890	
browser details	YourSeq	244	211	848	6477	86.6%	2	+	158302635	1
browser details	YourSeq	228	392	842	6477	83.8%	8	+	101280868	1
browser details	YourSeq	223	195	559	6477	84.9%	3	-	109229510	1
browser details	YourSeq	222	647	1141	6477	86.6%	16	+	23564986	
browser details	YourSeq	216	386	829	6477	87.0%	21	+	32782776	
browser details	YourSeq	207	409	848	6477	86.9%	20	-	30992113	
browser details	YourSeq	206	378	753	6477	87.6%	12	+	93901634	
browser details	YourSeq	201	904	1161	6477	89.2%	2	-	181544616	1
browser details	YourSeq	200	873	1182	6477	83.9%	16	+	76416269	
browser details	YourSeq	199	647	1028	6477	85.7%	20	+	32515149	
browser details	YourSeq	194	475	855	6477	92.9%	6	-	15022785	
browser details	YourSeq	193	258	618	6477	86.7%	2	+	186017937	1
browser details	YourSeq	190	448	925	6477	89.0%	10	-	104983206	1
browser details	YourSeq	187	399	1011	6477	83.6%	5	-	137674922	1

Fig. 2. A partial listing of the result of the BLAT search.

17. Examine the page closely. The browser provides an extensive list of options from which you can choose for viewing the map *(5)*. For example, on the top of the page, you can use specific control keys (i.e., the left and right arrows) to move to and view the flanking regions in the map. You can click on zoom buttons to zoom in or out. In the example, click on the left arrow (>) twice, to move the map to include the coding region of the sequence. In that example, you will find the coding region of the human NF-IL6 gene, which is also known as C/EBPbeta (**Fig. 4**).

18. Select from the options listed below the graph (mapping and sequencing tracks), to choose what you want to include in the graph. The options are extensive. You can choose options that would allow the inclusion of additional details in the map. Each time you choose an option, or a set of options, click on the refresh button. The browser will display the selected annotations as a series of horizontal tracks *(5)*.

19. On the graph, the arrows on the tracks representing the gene provide the direction of transcription (**Fig. 4**). Click on a given track to obtain useful links and information about that track.

20. To obtain the sequence of the region shown in the graph, on the top bar (**Fig. 3**), right-click on DNA to open a new window for viewing the sequence. Follow the instructions for obtaining the desired format (for example, you can choose masking the repetitive DNA sequences to lower case letters).

21. To obtain an output of the graph, for your record or for publication, on the top bar (**Fig. 3**), right click on PDF/PS to open a new window that would provide the options to save the plot in a PDF or a postscript file (**Fig. 4**).

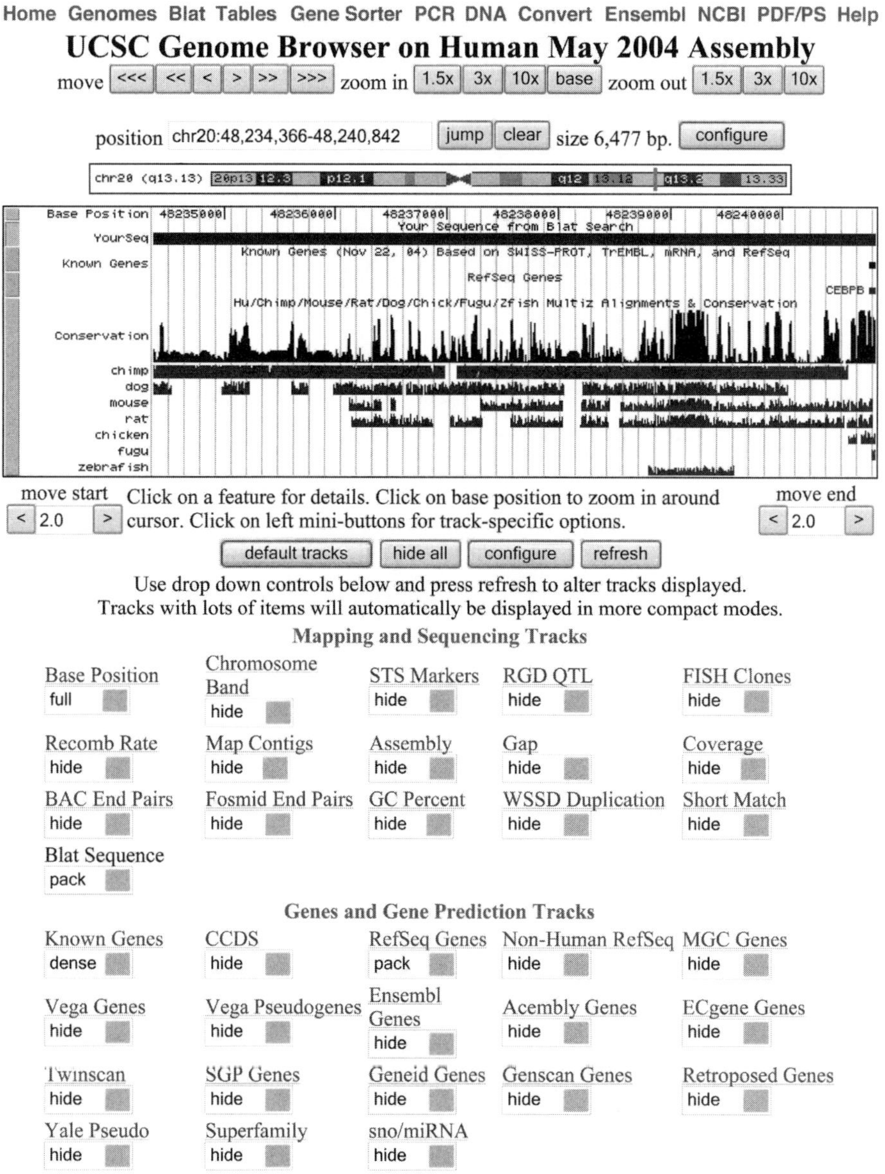

Fig. 3. Graphical representation of the promoter region of the human C/EBP (NF-IL6) gene in the genome browser at UCSC. The top of this view shows the control keys for zooming in or out, as well as keys for moving the displayed region to the left or to the right. The bottom view includes a partial listing of the control keys for adding details to and removing tracks from the map.

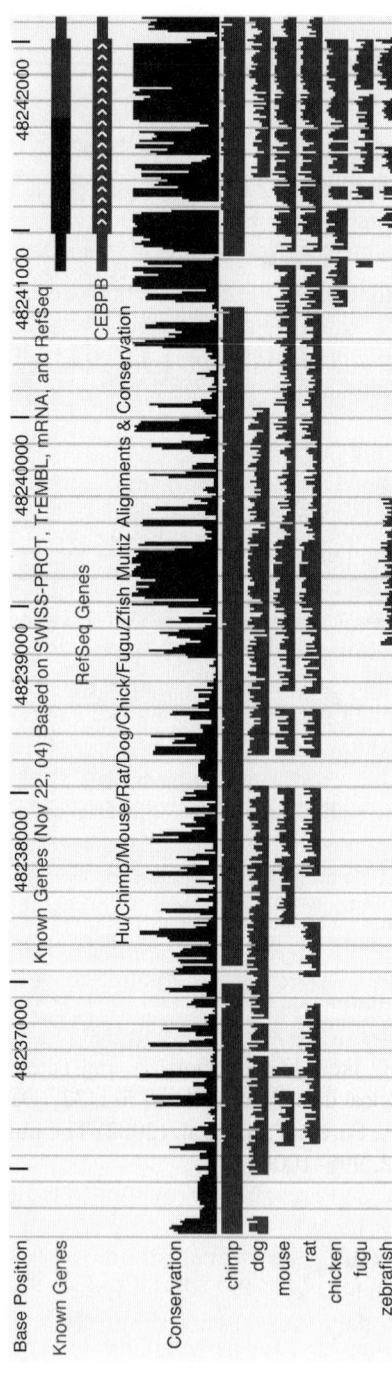

Fig. 4. Graphical representation of a region that includes both the promoter and the coding region of the human C/EBP (NF-IL6) gene. This representation was obtained by using the control key move, for including the gene in the displayed region. Subsequently, the result was saved in a PDF file. This was done by selecting the key marked PDF/PS, shown on the top of **Fig. 3**.

22. To obtain a sequence alignment of the conserved regions (**Fig. 3**), click on the area next to the track named conservation (*see* **Note 3**).
23. At the UCSC genome browser, the page that shows the map (**Fig. 3**) also provides the option of viewing that map in the Ensembl and NCBI browsers. On that page, the links are shown on the top bar (**Fig. 3**). Click on these options to view the map of the sequence of interest in these alternative browsers.

4. Notes

1. Opening a new window for each of the desired links is recommended. This would circumvent problems with losing the connection to the preceding page. The right-click option, for opening a new window to a link, is available on PCs that use Microsoft operating systems. This option might not be available on other operating systems.
2. Viewing the in-depth information can help you to evaluate whether the matches with the genomic DNA are significant.
3. Currently multispecies alignment is provided for 30,000 bases or less. Therefore, to obtain an alignment, zoom in the desired region. This works relatively well for viewing the conserved regions in the promoter regions of genes. To do so, scroll to left or right, depending on the direction of the transcript. Identify the longest cDNA by including the track for known genes. Subsequently, zoom in the 5' end of the gene, to bring the viewed region to 30,000 bases or less. Click on refresh. Then click on the track named conservation (**Fig. 3**). You will obtain alignments of the nucleotide sequences of the selected species.

References

1. Benson, D. A., Karsch-Mizrachi, I., Lipman, D. J., Ostell, J., and Wheeler, D. L. (2005) GenBank. *Nucleic Acids Res.* **33,** (Database issue) D34–38.
2. Wheeler, D. L., Barrett, T., Benson, D. A., et al. (2005) Database resources of the National Center for Biotechnology Information. *Nucleic Acids Res.* **33,** (Database issue) D39–45.
3. Kanz, C., Aldebert, P., Althorpe, N., et al. (2005) The EMBL Nucleotide Sequence Database. *Nucleic Acids Res.* **33,** (Database issue) D29–33.
4. Miyazaki, S., Sugawara, H., Ikeo, K., Gojobori, T., and Tateno, Y. (2004) DDBJ in the stream of various biological data. *Nucleic Acids Res.* **32,** (Database issue) D31–34.
5. Kent, W. J., Sugnet, C. W., Furey, T. S., et al. (2002) The human genome browser at UCSC. *Genome Res.* **12,** 996–1006.
6. Karolchik, D., Hinrichs, A. S., Furey, T. S., et al. (2004) The UCSC Table Browser data retrieval tool. *Nucleic Acids Res.* **32,** (Database issue) D493–496.
7. Hubbard, T., Andrews, D., Caccamo, M., et al. (2005) Ensembl 2005. *Nucleic Acids Res.* **33,** (Database issue) D447–453.
8. Kent, W. J. (2002) BLAT—the BLAST-Like Alignment Tool. *Genome Res.* **4,** 656–664.
9. Yang, Y., Pares-Matos, E. I., Tesmer, V. M., et al. (2002) Organization of the promoter region of the human NF-IL6 gene. *Biochim. Biophys. Acta* **1577,** 102–108.

2

Methods for Identifying and Mapping Recent Segmental and Gene Duplications in Eukaryotic Genomes

Razi Khaja, Jeffrey R. MacDonald, Junjun Zhang, and Stephen W. Scherer

Summary

The aim of this chapter is to provide instruction for analyzing and mapping recent segmental and gene duplications in eukaryotic genomes. We describe a bioinformatics-based approach utilizing computational tools to manage eukaryotic genome sequences to characterize and understand the evolutionary fates and trajectories of duplicated genes. An introduction to bioinformatics tools and programs such as BLAST, Perl, BioPerl, and the GFF specification provides the necessary background to complete this analysis for any eukaryotic genome of interest.

Key Words: Bioinformatics; BLAST/MegaBLAST; gene duplication; gene ontology; genome assembly; genomic disorder; GFF (Generic Feature Format); homology; neofunctionalization; paralogous; Perl/BioPerl; pseudogene; RefSeq; RepeatMasker; segmental duplication; sequence alignments; subfunctionalization.

1. Introduction

With the completion of the human genome sequence and the increasing availability of whole genome shotgun sequences (WGS) for numerous other eukaryotic species, we are poised to begin to understand the complexity and dynamic nature of chromosomes. Segmental duplications are nearly identical segments of DNA at two or more sites in a genome; for human they comprise about 3.5 to 5% of the total DNA content (1,2). Segmental duplications also account for 1.2 to 2% of the mouse genome (3,4) and approx 3% of the rat genome (5). Segmental duplications (also called low copy repeats [LCRs]) can be predisposition sites for increased opportunity of nonallelic homologous recombination leading to deletion, inversion, or duplication of large segments of DNA (6).

From: *Methods in Molecular Biology, vol. 338: Gene Mapping, Discovery, and Expression: Methods and Protocols*
Edited by: M. Bina © Humana Press Inc., Totowa, NJ

These structural alterations may lead to the gain or loss of dosage-sensitive genetic material and may result in a spectrum of diseases defined as genomic disorders *(7–9)*.

The presence of segmental duplications is a common feature of many mammalian genomes, and their involvement in chromosome evolution and natural variation is an area of active investigation *(10–12)*. Duplication of large segments of DNA can generate duplicate genes in whole *(13)*, or in part *(14)*, and may lead to an expanding repertoire of similar gene products. The identification of recent segmental duplication therefore gives us the ability to map the origin and fate of duplicate genes, which are a driving force in species evolution (*see* **Note 1**).

Here we define recent segmental duplications as paralogous regions of a genome having a length greater than 5000 nucleotides (nt) and having greater than 90% DNA sequence identity. We present a computational protocol for identifying and mapping recent segmental and gene duplications in eukaryotic genomes. The major procedures involved in identifying recent segmental and gene duplications include comparing genomic sequences using BLAST *(15)*, parsing and filtering BLAST alignments, and mapping genes to segmental duplications to identify gene duplicates. We note that much of our methodologies have arisen in an ongoing initiative to map segmental duplications accurately in the human *(2)*, chimpanzee, mouse *(3)*, and other mammalian genomes as displayed at publicly available websites (http://projects.tcag.ca/humandup and http://projects.tcag.ca/xenodup).

2. Materials

1. A modest-sized cluster-computer or super-computer with 4 GB of RAM per CPU running any variant of a UNIX or Linux operating system.
2. Internet connection, ftp utilities (e.g., ftp, ncftp, wget).
3. Archiving utilities (e.g., unzip).
4. An assembled genome sequence of a eukaryotic organism that is lower case masked for repetitive elements.
5. The BLAST suite of programs (particularly formatdb and MegaBLAST).
6. Perl, BioPerl.
7. Approximately 5 to 20 GB of disk space to store sequence data, blast databases, alignment data, and parsed output.

3. Methods

The methods described below outline: (1) the prerequisites and assumptions required to perform this analysis, (2) where to obtain genome assemblies of eukaryotic genomes, (3) the process for installing the BLAST suite of programs, and (4) the procedure for creating BLAST databases. To identify segmental duplications in eukaryotic genomes, the methods summarize: (5) the procedure for performing sequence alignments of all possible pairs of chromosomes using MegaBLAST, (6) how to convert MegaBLAST alignments into Generic Feature Format (GFF) format, and (7) the criteria for filtering GFF records and (8) chain alignments together. Furthermore, we describe how to identify gene duplicates by (9) mapping RefSeq genes to segmental duplications and (10) using the Gene Ontology to characterize gene duplicates by function.

3.1. Prerequisites/Assumptions

To perform segmental duplication analysis of eukaryotic genomes, the reader needs access to a modest-sized cluster-computer or super-computer with a minimum of 4 GB of RAM available to each CPU (*see* **Note 2**) running any variant of a UNIX or Linux operating system (*see* **Note 3**). Competency in using UNIX command line utilities and programming in Perl is also a necessity (*see* **Note 4**). It is also a prerequisite that the BioPerl package (*see* **Note 5**) be available in the computing environment. Furthermore, the reader should be capable of using BioPerl to convert MegaBLAST alignment files into GFF records and should be familiar with the GFF version 3 specification (*see* **Note 6**).

3.2. Download Genome of Interest

This protocol requires that the genome sequence being targeted for the identification of segmental and gene duplications be assembled and masked for repetitive elements.

Although this protocol is applicable to all eukaryotic genomes (*see* **Note 7**), the mouse genome will be used as our example. The May 2004 mouse genome assembly (referred to as mm5 by UCSC or Build 33 by NCBI) can be downloaded from UCSC (http://genome.ucsc.edu) as a zip file by executing the following command:

```
% wget http://hgdownload.cse.ucsc.edu/goldenPath/mm5/bigZips/chromFa.zip
```

This zip file contains the mouse genome assembly with one FASTA file for each chromosome. Repetitive elements within each chromosome sequence have been identified with RepeatMasker (http://www.repeatmasker.org) and are represented in lower case letters; nonrepeating DNA sequences are shown in upper case letters. Once the genome has been downloaded, the zip file is uncompressed by executing the following command:

```
% unzip chromFa.zip
```

Uncompressing this file will extract one FASTA file for each chromosome sequence. For the mouse genome, this should extract files: chr1.fa to chr19.fa, chrX.fa, chrY.fa, and chrM.fa (mitochondrial dna), as well as chr1_random.fa to chr19_random.fa, chrX_random.fa, chrY_random.fa, and chrUn_random.fa (*see* **Note 8**).

3.3. Download and Install the BLAST Suite of Programs

To perform sequence alignments for identification of segmental duplications in the genome, download and install the BLAST suite of programs on your computing environment. The BLAST suite of programs is available from the NCBI as precompiled binary distributions or as source code. The precompiled binaries are available from ftp://ftp.ncbi.nlm.nih.gov/blast/executables/LATEST/. These are compiled for many operating systems and hardware architectures (*see* **Note 9**). Installation is a simple matter of downloading and then uncompressing the distribution for your computing environment. Documentation supplied with the BLAST suite of programs describes command line options for each of the utilities. In this protocol, the formatdb and MegaBLAST *(16)* command line tools are used to identify segmental duplications in the genome. Formatdb is used to create BLAST databases, and MegaBLAST is used to perform sequence alignments.

3.4. Create BLAST Databases, One for Each Chromosome

Once the genome has been downloaded and the BLAST suite has been installed, create BLAST databases for each of the chromosome FASTA files using the formatdb command line utility. The formatdb command line utility must be used to format a FASTA file such as chr7.fa into a BLAST database before it can be searched by MegaBLAST. The following command is an example of using formatdb to create a BLAST database:

% formatdb -i chr7.fa -p F

Executing this command will create the files: chr7.fa.nhr, chr7.fa.nin, and chr7.fa.nsq, which collectively represent the BLAST database for mouse chromosome 7. This database will be searched by MegaBLAST in order to produce sequence alignments for the purpose of identifying segmental duplications in the genome. BLAST databases must be created iteratively for every FASTA file for each chromosome sequence in the genome, including the pseudo chromosomes (*see* **Note 8**).

In the example above, "–i chr7.fa" specifies the name of the input file, and "–p F" specifies that the sequence contained within the file is nucleotide. Below is a detailed description of the command line options used:

formatdb 2.2.10 arguments:
  -i   Input file(s) for formatting (this parameter must be set)
       [File In]

-p Type of file
    T - protein
    F - nucleotide [T/F] Optional
    default = T

A full description of this command and its options is included with the documentation supplied with the BLAST suite and is also available at the NCBI website (http://www.ncbi.nih.gov/BLAST/docs/formatdb.html).

3.5. Perform Sequence Alignments of All Possible Pairs of Chromosomes Using MegaBLAST

The MegaBLAST program is used to perform sequence alignments because it was designed to identify long alignments efficiently between similar sequences. Since we have defined recent segmental duplications as long stretches of DNA (>5000 nt) having greater than 90% sequence identity, MegaBLAST is ideal at identifying these paralogous regions of the genome. After creating the BLAST databases for each chromosome, MegaBLAST is used to perform sequence alignments between all possible pairs of chromosomes. In other words, each FASTA file is compared with each of the BLAST databases (*see* **Note 10**).

The following command is an example of using MegaBLAST to find sequence alignments between mouse chromosome 7 and mouse chromosome 3.

% megablast –d chr7.fa –i chr3.fa –D 2 –F 'm' –U T –o chr7.3.blast

In the example above, the "-d chr7.fa" option specifies that MegaBLAST use the mouse chromosome 7 BLAST database as the subject of this comparison and the "-i chr3.fa" option specifies mouse chromosome 3 as the query sequence. Sequence alignments are stored in the chr7.3.blast output file as specified by the option "-o chr7.3.blast" and the format of output generated is "traditional BLAST output" as specified by the "-D 2" option. Furthermore, "-U T" specifies that lower case letters in the query sequence should be recognized as a repetitive element. The "-F 'm'" option denotes that the MegaBLAST algorithm should not find word matches in the repetitive regions of the query sequence but should allow for extension of sequence alignments through these regions.

Below is a detailed description of the command line options that are required to perform sequence alignments using MegaBLAST to identify segmental duplications in a genome:

megablast 2.2.10 arguments:
  -d Database [String]
     default = nr
  -i Query File [File In]
  -D Type of output:

      0 - alignment endpoints and score
      1 - all ungapped segments endpoints
      2 - traditional BLAST output
      3 - tab-delimited one-line format [Integer]
      default = 0
  -F Filter query sequence [String]
      default = T
  -U Use lower case filtering of FASTA sequence [T/F] Optional
      default = F
  -o BLAST report Output File [File Out] Optional
      default = stdout

A full description of this command and its options is included with the documentation supplied with the BLAST suite of programs and is also available at the NCBI website (http://www.ncbi.nih.gov/BLAST/docs/megablast.html).

Sequence alignments generated by MegaBLAST between a subject database and a query sequence of the same chromosome are used to identify intrachromosomal segmental duplications (i.e., duplications that occur within the same chromosome). Sequence alignments generated by MegaBLAST between a subject database and a query sequence of different chromosomes are used to identify interchromosomal segmental duplications (i.e., duplications that occur between different chromosomes). Executing MegaBLAST on a subject database and query sequence generates many sequence alignments. Not all of these represent sequences involved in segmental duplications, so further steps are required to convert, filter, and process these alignments based on a variety of criteria. These criteria are described in the sections below.

3.6. Convert MegaBLAST Alignments Into GFF Format

In the previous step, MegaBLAST was used to generate traditional BLAST output for all pairs of chromosomes. Sequence alignments in this format are extremely informative since they visualize detailed information about homologous DNA, showing locations of nucleotide mismatches and small insertions and deletions (**Fig. 1**).

However, programmatically it is difficult to identify duplications from blast results in this format as this output is generated for visual inspection. In order to identify segmental duplications from blast results without loss of information it is necessary to transform traditional BLAST output into a tabular format. The current Generic Feature Format version 3 (GFF3) specification (http://song. sourceforge.net/gff3.shtml) is a widely accepted tabular format for describing genes and other features associated with DNA, RNA, and protein sequences. The BioPerl project (http://www.bioperl.org) supports the parsing of different output formats, including traditional BLAST output into GFF3.

```
Score = 3336 bits (1683), Expect = 0.0
Identities = 2016/2123 (94.96%), Gaps = 24/2123 (1%)
Strand = Minus / Plus


chr3 :  36201 TGAGGCAAGGGGCTCACGCTGACCTCTGTCCGCGTGGGAGGGGCCGGTGTGAGGCAAGGG 36142
               |||||| |||||| ||||||||||||||||||||  |||||||||||||||||||||||||
chr7 :  61612 TGAGGC-AGGGGGTCACGCTGACCTCTGTCCG--GGGAGGGGCCGGTGTGAGGCAAGGGG 61670


chr3 :  36141 CTCACACTGACAGCTCTCAGCGTGGGAGGGGCCGGGGTGAGGCAAGGGGCTCACGCTGAC 36082
               |||||||||||| |||||||||||||||||||| |||||||||||||||||||||||||||
chr7 :  61671 CTCACACTGACCTCTCTCAGCGTGGGAGGGGCCGGTGTGAGGCAAGGGGCTCACGCTGAC 61730


chr3 :  36081 CTCT---CGCGTGGGAGGGGCCGGTCTGAGGCAAGGGCTCACACCGACCTCTCTCAGCGT 36022
               |||||    |||||||||||||||||| |||||||||||||||| |||||||||||||||
chr7 :  61731 CTCTGTCCGCGTGGGAGGGGCCGGTGTGAGGCAAGGGCTCACACTGACCTCTCTCAGCGT 61790
```

Fig. 1. Traditional BLAST output as generated by MegaBLAST.

Using the Bio::SearchIO module that is part of the BioPerl package, it is required that BLAST alignment files for each pair of chromosomes be converted into GFF3 records. Below is an example of the result of converting the alignment shown in **Fig. 1** as a GFF3 record:

chr7 UCSC_hg17 match 61612 61790 0.0 - . Target=chr3 36022 36201;Gap=M6 I1 M25 I2 M90 D3 M53;percentId=94.96;alnLength=2123;matches=2016;gaps=24; bitScore=3336;rawScore=1683

To understand how to generate records in GFF3 format, the reader should understand the GFF3 specification. This will enable the user to apply the Bio::SearchIO module to convert BLAST alignment files to generate this output. This format allows storage of all information from the traditional BLAST output including: subject sequence start and stop coordinates, query sequence start and stop coordinates, e-value, strand, percent identity, alignment length, matched nucleotides, gaps, bit score, raw score and detailed alignment information.

3.7. Filter GFF Records Based on Many Criteria

After converting the traditional BLAST alignments into GFF format, some alignments are excluded since not all are components of recent segmental duplications. To identify sequences meeting a stringent categorization of being a "recent segmental duplication," GFF records are filtered based on the criteria described below.

3.7.1. Filter Sequence Alignments With Less Than 90 Percent Identity

Recent segmental duplications are defined as paralogous sequences that share greater than 90% sequence similarity. Remove GFF records in which the percent identity attribute does not meet this minimum percent identity cutoff. This filtering criterion is applicable to both inter- and intrachromosomal sequence alignments.

3.7.2. Filter Suboptimal Sequence Alignments

Suboptimal sequence alignments occur when one sequence alignment is redundant in the sense that the subject and query elements are completely covered or spanned by another alignment. Remove the GFF record with the smaller span, which is considered a suboptimal alignment. This filtering step is applicable to both inter- and intrachromosomal sequence alignments.

3.7.3. Filter Identical Sequence Alignments

This filtering step is only applicable to intrachromosomal sequence alignments. Exclude self-self matches, whose GFF records have subject sequence coordinates that are identical to the query sequence coordinates.

3.8. Identify Segmental Duplications by Chaining Alignments Together

To define the boundaries of segmental duplications, alignments whose coordinates are monotonically increasing are chained together to form larger contiguous alignments. This compensates for short and fragmented alignments, which have arisen because of insertion or deletion events that have modified paralogous copies of DNA. Since we defined segmental duplications as regions of the genome having length greater than 5000 nt, we need to filter chained alignments that do not meet this minimum length requirement.

1. Sort GFF records by subject and query coordinates.
2. For records of the same subject and query chromosome pair, if adjacent sequence alignments are separated by less than 3000 nt, chain the alignments together
3. Remove chained alignments that are smaller than 5000 bp.

This step concludes the identification of large regions of the genome involved in recent segmental duplications. Large segmental duplications can often contain duplicate genes and/or be implicated in genomic disease and structural rearrangements; hence they have an inherent biological interest. **Subheadings 3.9.** and **3.10.** discuss mapping genes to segmental duplications, identifying duplicate gene pairs, and characterizing gene duplications using the Gene Ontology.

3.9. Map RefSeq Genes to the Mouse Genome and to Segmental Duplications

To identify and characterize recent gene duplicates in the mouse genome, you will first need to obtain the most current curated gene data set, map the location of the gene to the genome of interest, and perform a positional colocalization of genes and duplications to detect gene paralogs.

3.9.1. Obtain RefSeq Gene Set and Mapping Location in the Mouse Genome

1. Obtain the mouse gene data set (refGene.txt.gz) from the University of California at Santa Cruz (http://hgdownload.cse.ucsc.edu/goldenPath/mm5/database/).
2. Extract the gene mapping information from the above file, and store in GFF3 format. A description of the refGene.txt table format from UCSC can be found at http://genome.ucsc.edu/goldenPath/gbdDescriptions.html#GenePredictions.

3.9.2. Identify Recent Gene Duplicates

Identifying recent gene duplications generated via a segmental duplication event can be accomplished by localizing the genes that lie within the boundaries of the duplications detected and determining the paralogous gene pair in the corresponding duplicon. The genes may be duplicated in whole or part along with the surrounding genomic DNA.

1. Identify the genes that reside completely within the defined boundary of the duplications (whole gene duplication). Compare the transcriptional start and end coordinates stored in the GFF3 file and identify those genes that fall completely within the coordinates of the duplication.
2. Identify the genes that lie partially within the defined boundary of the duplication (partial gene duplication). Compare the transcriptional start and end coordinates stored in the GFF3 file and identify those genes that overlap one or both boundaries of the duplication (as defined by either the feature start, the feature end, or those transcripts that span the entire duplication).
3. Now that you have found all RefSeq genes, which reside within or span the boundaries of segmental duplications, you will need to search for the paralogous gene pair within the related segmental duplication loci. The duplicated gene may be supported by a curated RefSeq mRNA, an unannotated full-length mRNA, or an expressed sequence tag (EST).
   a. Download EST (all_est.txt.gz) and mRNA (all_mrna.txt.gz) data sets from UCSC (http://hgdownload.cse.ucsc.edu/goldenPath/mm5/database/).
   b. Extract the EST and mRNA mapping information from the above file, and store in GFF3 format. A description of the all_est.txt and all_mrna.txt table format from UCSC can be found at http://genome.ucsc.edu/goldenPath/gbdDescriptions.html#GenePredictions.
   c. Identify the transcripts (EST and mRNA) that map completely within the defined boundary of the duplications (whole gene duplication). Compare the transcriptional start and end coordinates stored in the GFF3 file, and identify those EST or mRNA sequences that fall completely within the coordinates of the duplication.
   d. Identify the transcripts (EST and mRNA) that are located partially within the defined boundary of the duplication (partial gene duplication). Compare the transcriptional start and end coordinates stored in the GFF3 file and identify those EST or mRNA sequences that overlap one or both boundaries of the duplication (as defined by either the feature start, the feature end, or those transcripts that span the entire duplication).

4. You now have a list of all RefSeq genes and EST and mRNA sequences that reside within duplications. This data set will represent all transcribed sequences that are candidates of recent gene duplication events. To determine the relationship between duplicate genes, a pairwise comparison of all transcripts within related duplications is required.

   a. To determine whether two transcripts are related (i.e., a duplicated gene pair), you will need to BLAST pairs of transcript sequences.

   b. Based on our criteria, genes that share greater than 90% DNA sequence similarity for greater than 50% of the length of the transcript can be categorized as a duplicated gene pair.

3.10. Functional Characterization of Genes by Gene Ontology

Duplicate genes may undergo pseudogenization, subfunctionalization, or neofunctionalization *(17)*. To identify the putative function and fates of duplicate genes, an in silico analysis of gene function should be undertaken using the Gene Ontology (GO) resource *(18)*.

1. Obtain the geneID (extract the ID from the gene2refseq.gz file) for each duplicated gene from the NCBI website (ftp://ftp.ncbi.nlm.nih.gov/gene/DATA/). The geneID is a unique NCBI identifier (previously Locus Link ID) for each curated RefSeq entry. The GO database can be searched by this unique ID to extract precomputed gene ontology information. Additional information on the GO project is available at this website http://www.geneontology.org/.

2. Using the unique geneID, assign each gene to its GO annotations from each of the three GO taxonomies (biological processes, cellular component, and molecular function) by utilizing the GO Tree Machine (http://genereg.ornl.gov/gotm/). You will need to create an account. (Registration is free and will allow the user to save and retrieve analyses.)

3. Create a text file with the list of the geneIDs and save to a file.

   a. Log onto the GO Tree Machine site, and give the analysis a relevant name for future access.

   b. From the drop-down menu for "Select the ID type in your file," select Locus Link ID (same as geneID).

   c. For "What kind of analysis do you want to do?" select "single gene list" to perform a functional characterization of the duplicated genes.

   d. You will need to upload the text file with the list of geneIDs previously created and select "MAKE TREE."

   e. Alternatively, if, for **step 3c**, you select "interesting gene list vs. reference gene list" you can perform a statistical analysis of duplicated genes to detect GO terms that are relatively enriched compared with the full RefSeq data set. You will need to choose the "MOUSE" reference list.

4. Notes

1. Gene duplication allows for relaxed selection owing to redundancy, and this may allow for processes such as subfunctionalization, neofunctionalization, and pseu-

dogenization. Subfunctionalization occurs when two gene copies specialize to perform complementary functions. Neofunctionalization involves gene duplication whereby one of the genes acquires a new biochemical function. Furthermore, pseudogenization occurs when one of the duplicated genes acquires mutations rendering it nonfunctional.

2. Since chromosome sequence FASTA files are quite large and range in size from 50 to 250 Mb, a significant amount of computational power and memory is required to perform the sequence alignments using MegaBLAST.

3. We will explain how to perform this analysis in a serial manner. It is up to the reader to understand the nuances of their particular cluster or supercomputing installation in order to parallelize the algorithm and achieve the desired results in less time. This means understanding whether using MPI or forking and executing processes is suitable.

4. This protocol can be written in any programming language such as Perl, Java, Python, Ruby, C, or C++. However, typically in bioinformatic applications, algorithms are written in Perl.

5. The BioPerl package is available from http://www.bioperl.org/.

6. The current Generic Feature Format version 3 (GFF3) specification is available at http://song.sourceforge.net/gff3.shtml.

7. Assembled genomes of several species such as: human, rat, chimpanzee, dog, chicken, and others are available from the download page of the University of California at Santa Cruz (UCSC), http://hgdownload.cse.ucsc.edu/downloads.html.

8. The main chromosome sequence assemblies are found in the chrN.fa files, where N is the name of the chromosome. The chrN_random.fa files are pseudo chromosomes containing sequences that are not yet finished or cannot be localized with certainty at any particular place in the chromosome assembly. The chrUn_random.fa file is another pseudo chromosome containing clones that have not been localized to a particular chromosome in the genome. These pseudo chromosomes should not be overlooked since they can often contain sequences that are involved in segmental duplications and have not been included in the main genome assembly perhaps because of their duplicated nature.

9. If the precompiled binaries do not match your computing environment, source code is available from NCBI at ftp://ftp.ncbi.nlm.nih.gov/toolbox/ncbi_tools/ncbi.tar.gz. The instructions detail how to compile and install this suite of tools for your particular computing environment.

10. A total of N^2 sequence alignments are performed for all sequence files where N is the number of files in the genome (i.e., chr1.fa vs chr2 BLAST database and chr2.fa vs chr1 BLAST database). Sequence comparisons are required for all chromosomes in the genome including the pseudo chromosomes.

References

1. Bailey, J. A., Gu, Z., Clark, R. A., et al. (2002) Recent segmental duplications in the human genome. *Science* **297,** 1003–1007.

2. Cheung, J., Estivill, X., Khaja, R., et al. (2003) Genome-wide detection of segmental duplications and potential assembly errors in the human genome sequence. *Genome Biol.* **4,** R25.

3. Cheung, J., Wilson, M. D., Zhang, J., et al. (2003) Recent segmental and gene duplications in the mouse genome. *Genome Biol.* **4,** R47.

4. Bailey, J. A., Church, D. M., Ventura, M., Rocchi, M., and Eichler, E. E. (2004) Analysis of segmental duplications and genome assembly in the mouse. *Genome Res.* **14,** 789–801.

5. Tuzun, E., Bailey, J. A., and Eichler, E. E. (2004) Recent segmental duplications in the working draft assembly of the brown Norway rat. *Genome Res.* **14,** 493–506.

6. Lupski, J. R. (1998) Genomic disorders: structural features of the genome can lead to DNA rearrangements and human disease traits. *Trends Genet.* **14,** 417–422.

7. Stankiewicz, P. and Lupski, J. R. (2002) Genome architecture, rearrangements and genomic disorders. *Trends Genet.* **18,** 74–82.

8. Eichler, E. E. (2001) Recent duplication, domain accretion and the dynamic mutation of the human genome. *Trends Genet.* **17,** 661–669.

9. Ji, Y., Eichler, E. E., Schwartz, S., and Nicholls, R. D. (2000) Structure of chromosomal duplicons, and their role in mediating human genomic disorders. *Genome Res.* **10,** 597–610.

10. Iafrate, A. J., Feuk, L., Rivera, M. N., et al. (2004) Detection of large-scale variation in the human genome. *Nat. Genet.* **36,** 949–951.

11. Armengol, L., Pujana, M. A., Cheung, J., Scherer, S. W., and Estivill, X. (2003) Enrichment of segmental duplications in regions of breaks of synteny between the human and mouse genomes suggest their involvement in evolutionary rearrangements. *Hum. Mol. Genet.* **12,** 2201–2208.

12. Bailey, J. A., Baertsch, R., Kent, W. J., Haussler, D., and Eichler, E. E. (2004) Hotspots of mammalian chromosomal evolution. *Genome Biol.* **5,** R23.

13. Ohno, S. (1970) *Evolution by Gene Duplication.* Springer, New York, NY.

14. Buiting, K., Korner, C., Ulrich, B., Wahle, E., and Horsthemke, B. (1999) The human gene for the poly(A)-specific ribonuclease (PARN) maps to 16p13 and has a truncated copy in the Prader-Willi/Angelman syndrome region on 15q11→q13. *Cytogenet. Cell Genet.* **87,** 125–131.

15. Altschul, S. F., Gish, W., Miller, W., Myers, E. W., and Lipman, D. J. (1990) Basic local alignment search tool. *J. Mol. Biol.* **215,** 403–410.

16. Zhang, Z., Schwartz, S., Wagner, L., and Miller, W. (2000) A greedy algorithm for aligning DNA sequences. *J. Comput. Biol.* **7,** 203–214.

17. Prince, V. E. and Pickett, F. B. (2002) Splitting pairs: the diverging fates of duplicated genes. *Nat. Rev. Genet.* **3,** 827–837.

18. Ashburner, M., Ball, C. A., Blake, J. A., et al. (2000) Gene ontology: tool for the unification of biology. *Nat. Genet.* **25,** 25–29.

3

Identification and Mapping of Paralogous Genes on a Known Genomic DNA Sequence

Minou Bina

Summary

The completion of whole genome sequencing projects offers the opportunity to examine the organization of genes and the discovery of evolutionarily related genes in a given species. For the beginners in the field, through a specific example, this chapter provides a step-by-step procedure for identifying paralogous genes, using the genome browser at UCSC (http://genome.ucsc.edu/). The example describes identification and mapping in the human genome, the paralogs of TCF12/HTF4. The example identifies TCF3 and TCF4 as paralogs of the TCF12/HTF4 gene. The example also identifies a related sequence, corresponding to a pseudogene, in one of the introns of the JAK2 gene. The procedure described should be applicable to the discovery and creation of maps of paralogous genes in the genomic DNA sequences that are available at the genome browser at UCSC.

Key Words: The Human Genome Project; mapping of gene families; gene discovery.

1. Introduction

Paralogs refer to genes that appear in more than one copy in the genome of a given organism *(1)*. Paralogs arise from gene duplication events. If it is advantageous, duplicated genes evolve independently to produce distinct but related proteins. This process often involves specialization of paralogous genes into specific functions *(1)*. The evolution of paralogous genes can generate developmental and physiological novelties by changing the patterns of regulation of these genes, by changing the functions of the proteins they encode, or by both *(1)*.

From the complete genomic sequence of a given species, it is possible to identify the paralogous genes in that species. This chapter describes an example of how to map and obtain a graphical representation of paralogous genes in a genomic DNA. The example uses the genome browser at the University of California

From: *Methods in Molecular Biology, vol. 338: Gene Mapping, Discovery, and Expression: Methods and Protocols*
Edited by: M. Bina © Humana Press Inc., Totowa, NJ

at Santa Cruz (UCSC) *(2,3)*. In the analysis paralogy is defined on the basis of significant scores obtained for global alignments of amino acid sequences. This can be contrasted with local alignments, which are often utilized for the discovery of conserved motifs in the amino acid sequences of proteins (*see*, for example, **ref. 4**). The example given provides a relatively simple case, a good starting point for a beginner in the field. More complex cases would require additional tool sets. As an example, see the publication that describes how to explore relationships and mine data with the browser at UCSC *(5)*.

2. Materials

The gene localization procedure was done on a PC equipped with the Windows XP operating system. The general procedure should be applicable to other computers (*see* **Note 1**).

3. Methods

The following sections will guide a beginner in the field through simple and general procedures for the localization and mapping of paralogous genes in a known genomic DNA sequence. The approach is based on searching conceptual translations of a genomic DNA for blocks of highly conserved amino acid sequences that occur in more than one location *(6)*. For identification and mapping of paralogous genes, the browser at the UCSC is used here since it provides numerous tools for data access and visualization *(2,3)*.

3.1. Using the Amino Acid Sequence of a Protein as Query to Identify Potential Paralogous Genes

1. In the browser at UCSC (http://genome.ucsc.edu/), use the BLAT sequence alignment tool *(7)* for locating genes that might be paralogous to the gene of interest. To access BLAT, in the genome browser click on BLAT, one of the options listed on the left side of the page. You will obtain a query box for pasting the amino acid sequence of a protein.
2. If you want to analyze a predicted protein sequence that was compiled in your lab, you should convert it to a FASTA format. In this format, the sequence is presented as a continuous chain of amino acids, without any numbering, blank spaces, or annotation (**Fig. 1**).
3. If you know the accession number for a DNA or a protein sequence of interest, perform the following steps to obtain a FASTA formatted file from GenBank *(8,9)*. In the example shown below, the accession number of a DNA file is used to obtain a FASTA formatted file for the corresponding protein.
   a. Go to NCBI (http://www.ncbi.nlm.nih.gov/).
   b. Use the pull-down menu next to the query box that contains the word All Databases.
   c. On the menu, select nucleotides.

```
MNPQQQRMAAIGTDKELSDLLDFSAMFSPPVNSGKTRPTTLGSSQFSGSGIDERGGTTSWGTSGQPSPSY
DSSRGFTDSPHYSDHLNDSRLGAHEGLSPTPFMNSNLMGKTSERGSFSLYSRDTGLPGCQSSLLRQDLGL
GSPAQLSSSGKPGTAYYSFSATSSRRRPLHDSAALDPLQAKKVRKVPPGLPSSVYAPSPNSDDFNRESPS
YPSPKPPTSMFASTFFMQDGTHNSSDLWSSSNGMSQPGFGGILGTSTSHMSQSSSYGNLHSHDRLSYPPH
SVSPTDINTSLPPMSSFHRGSTSSSPYVAASHTPPINGSDSILGTRGNAAGSSQTGDALGKALASIYSPD
HTSSSFPSNPSTPVGSPSPLTGTSQWPRPGGQAPSSPSYENSLHSLKNRVEQQLHEHLQDAMSFLKDVCE
QSRMEDRLDRLDDAIHVLRNHAVGPSTSLPAGHSDIHSLLGPSHNAPIGSLNSNYGGSSLVASSRSASMV
GTHREDSVSLNGNHSVLSSTVTTSSTDLNHKTQENYRGGLQSQSGTVVTTEIKTENKEKDENLHEPPSSD
DMKSDDESSQKDIKVSSRGRTSSTNEDEDLNPEQKIEREKERRMANNARERLRVRDINEAFKELGRMCQL
HLKSEKPQTKLLILHQAVAVILSLEQQVRERNLNPKAACLKRREEEKVSAVSAEPPTTLPGTHPGLSETT
NPMGHM
```

Fig. 1. Example of a FASTA formatted protein sequence.

d. In the query box next to for, type the known accession number for the DNA sequence of interest. As an example, type BK001049. This accession number contains the nucleotide sequence of HTF4c, one of the spliced transcripts of the human TCF12/HTF4 gene *(10)*. After typing the accession number in the query box, click on go. You will obtain a page that includes the accession number and a description of the DNA sequence file.

e. On the right side of the accession number click on link. On the pull-down menu select protein.

f. Above the accession number of the retrieved protein sequence file, you will find the word "report," in red letters. On the pull-down menu, click on FASTA. You will obtain the FASTA format of the protein sequence.

g. Copy the entire sequence.

h. Paste the sequence in the BLAT query box at the UCSC browser, described above in **step 1**.

i. Alternatively, you can scroll down the BLAT page to use the box that would allow you to upload a FASTA formatted protein sequence from your computer.

4. On the top of the BLAT query box, for genome, select human. Click on the pull-down menu to view the extensive list of the genomic sequences that are offered by the browser. Therefore, you can also use the procedures that are described here for mapping and graphical representation of sequences from other species.

5. Above the BLAT query box, in the box under assembly, select the latest version (in our example, 2004). Alternatively, from the pull-down menu, choose an earlier version of the genomic DNA.

6. Go to the pull-down menu under the Query type and select protein.

7. For the other variables (score and output type), use the default values.

8. Finally, click on submit.

9. Upon completion of the BLAT search, you will receive a table listing the results (**Fig. 2**).

10. Examine the column tagged score. You will find the highest score (2100) for an extended region (positions 1–706), with 100% sequence identity to the query sequence analyzed by BLAT. The second and third highest scores (512 and 316) also reflect global alignments corresponding to positions 5 and 679 and to positions 4 to 681,

Home   Genomes   Tables   Gene Sorter   PCR   FAQ   Help

Human BLAT Results

BLAT Search Results

```
ACTIONS          QUERY           SCORE START  END QSIZE IDENTITY CHRO STRAND  START
-----------------------------------------------------------------------------------
browser details YourSeq          2100     1   706   706 100.0%    15   ++    54999404
browser details YourSeq           512     5   679   706  86.7%    18   +-    51046532
browser details YourSeq           316     4   681   706  89.8%    19   +-     1562784
browser details YourSeq           163   182   675   706  80.4%     9   ++     5101422
browser details YourSeq           123   587   661   706  77.4%    19   +-     1563201
browser details YourSeq            28   146   175   706  96.7%    20   ++    23754587
browser details YourSeq            12   237   240   706 100.0%     9   ++     5101587
browser details YourSeq            12   131   134   706 100.0%    20   ++    23754542
```

Fig. 2. A partial listing of the result of the BLAT search.

respectively (**Fig. 2**). The fourth score might or might not be significant; therefore it should also be analyzed. The other scores correspond to relatively short local alignments and therefore do not appear to be significant. This conclusion can be deduced by examining the information provided in detail (**Fig. 2**). Therefore, next to each sequence, right-click on details to open a new window for the link to obtain useful information (*see* **Note 2**).

11. First examine the details for the sequence with the highest score, the first line in **Fig. 2** (*see* **Note 1**). In the details for that line, you will obtain the position of the submitted sequence on human chromosome 15. Also, you will obtain the submitted amino acid sequence with regions highlighted in different colors. These regions might correspond to spliced sites in the DNA. On that page, scroll down to view the nucleotide positions of the exons and the splice sites in the genomic DNA. Scroll further down to examine the predicted amino acid sequence encoded by the exons of the gene.

12. Next, examine the details for the sequence with the second highest score, the second line in **Fig. 2**. You will obtain the position of a genomic DNA region, with a predicted amino acid sequence that shows similarity to the sequence analyzed by BLAT. You will find that the genomic DNA is on human chromosome 18. Also, you will obtain the similarity of the predicted sequence to the submitted amino acid sequence. The regions exhibiting sequence similarity are highlighted in different colors. The result indicates global similarities over an extended region.

13. Scroll down the page to view the genomic positions of the exons in the gene on chromosome 14. Scroll down to view the results of side-by-side alignments.

14. Next examine the details for the sequence with the third highest score (the third line in **Fig. 2**). As detailed above in **step 12**, you will obtain the position of a genomic DNA region with a predicted amino acid sequence that shows similarity to the sequence analyzed by BLAT. The genomic DNA is on human chromosome 19. As described above in **step 12**, you can view the blocks that show similarity to the query sequence. These blocks are highlighted in different colors. Again, the result indicates global similarities over an extended region. As detailed above in

step 13, by scrolling down, you can obtain genomic positions of the exons of the gene in the genomic DNA (in this case chromosome 19) and view the results of side-by-side alignments.

15. Finally, examine the details for the other sequences listed in **Fig. 2**. The details for the sequence on the fourth line indicate a global alignment that might be significant. However, as shown in **Subheading 3.2., steps 10** and **11**, you will find that the sequence corresponds to a pseudogene. The details for the sequence on the fifth line identify the same genomic region obtained from the details for chromosomes 19, the third line. The details describing the sequence on the sixth line reveal relatively short local alignments with a protein sequence predicted for a gene on chromosome 20. The sequence matches with low scores are unlikely to correspond to paralogous genes.

3.2. Mapping and Viewing the Chromosomal Positions of the Candidate Paralogous Genes

1. The BLAT report (**Fig. 2**) includes links for viewing the genomic locations of candidate paralogous genes in the browser at UCSC.
2. First, obtain a map of the query sequence on the genomic DNA. To do so, in the BLAT results (**Fig. 2**) right-click on the browser link, on the left side of the first line, to open this link in a new window. The first line contains the highest score and indicates 100% sequence identity to the query.
3. Examine the browser page closely (*see* **Note 3**). You will find that the map provides the genomic position of the gene encoding TCF12/HTF4, on human chromosome 15 *(11)*. Note that the browser offers an extensive list of options from which you can chose for viewing and analyzing the map. For example, on the top, you can use the left and right arrows to move to the flanking regions in the map. You can use the zoom buttons (on the top right of the page) to zoom in, to obtain an expanded view, or zoom out, to include additional sequences in the map (**Fig. 3A**).
4. Explore the options that are listed below the graph (below the bar indicating mapping and sequencing tracks), to choose what you want to include in the graph. The options are extensive. You can chose options to create tracks for viewing additional details *(2)*. The options include creating a track for reference sequences (RefSeq). Each time you chose an option, or a set of options, click on the refresh button.
5. On the map displayed, the arrows on the tracks corresponding to known genes provide the direction of transcription (**Fig. 3**). Click on one of these tracks to obtain information about that track and useful links that could help with data analysis and evaluation (*see* **Note 3**). For example, click one of the tracks labeled TCF12. You will obtain a page that includes the accession number for the transcript that corresponds to that track. Scroll down the page to obtain additional information about the gene.
6. To obtain an output of the graph showing the map, for your record or for publication, click on PDF/PS file. In that link, you will be able to save the plot in a PDF file or a postscript file. **Figure 3** displays examples of outputs obtained from PDF files.

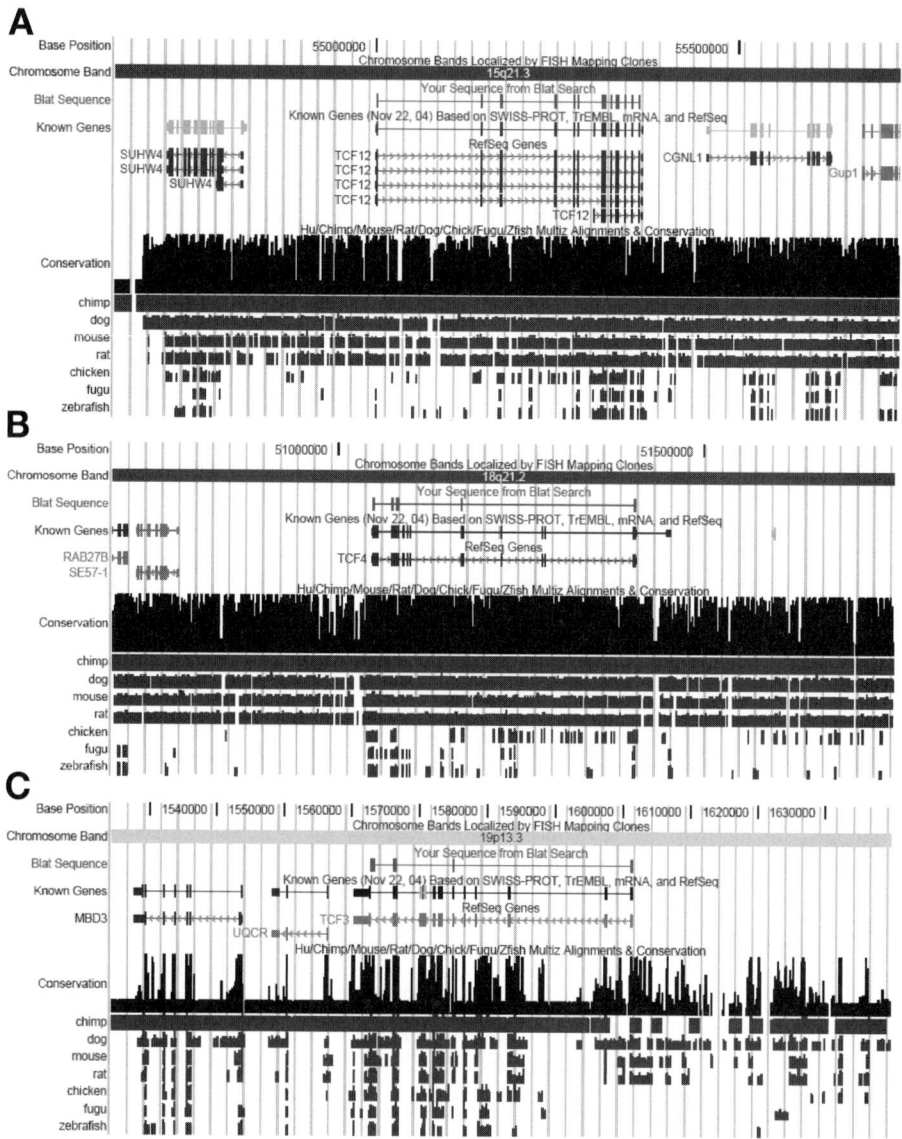

Fig. 3. Composite maps of three paralogous transcription factor genes. (**A**) The TCF12 gene on human chromosome 15q21. (**B**) The TCF4 gene on human chromosome 18q21.1. (**C**) The TCF3 gene on human chromosome 19p13.3. The paralogs of TCF12 were identified from the result of the BLAT search (**Fig. 2**). The map for each gene was obtained via the corresponding link (browser), in the result of BLAT (**Fig. 2**). The control key zoom out (x3) was selected in order to include and view the flanking genes in each map. The resulting displays were saved in separated PDF files. This can be done by using the control key PDF/PS in the genome browser. Subsequently, the composite map was created using tools offered by Adobe Acrobat.

7. Next, obtain a map locating the position of the first candidate paralogous genes. To do so, as described above in **Subheading 3.1., step 12**, in the BLAT results (**Fig. 2**) right-click on the browser link on the left side of the second line to open the link in a new window. This line contains the second highest score. The details indicate a global alignment exhibiting 80% sequence identity to the query.

8. As done for the previous graph, closely examine the map and the position of the gene on the browser. By exploring the listed options, you will find that the match with the query corresponds to the TCF4 gene. This gene encodes E2-2, also known as ITF2 and SEF2 *(12,13)*. Previous studies have shown that TCF4 is a paralog of TCF12/HTF4 (for details, *see* **ref.** *10* and the references therein). In the genome browser, click on a track representing one of the transcripts of the gene to obtain additional details and descriptions (*see* **Notes 1** and **3**).

9. Next, follow similar procedures to obtain a map of the second candidate paralogous gene (the third line in **Fig. 2**). You will find that the gene corresponds to TCF3 (**Fig. 3C**). This gene is also known as E2A *(14)* and has been shown to be a paralog of TCF12/HTF4 (for details, *see* **ref.** *10* and the references therein).

10. Lastly, follow similar procedures to obtain a map of the gene on chromosome 9 (the fourth line in **Fig. 2**). You may find the result surprising since the sequence maps within an intron in the JAK2 gene (**Fig. 4**). From this result, one could suspect that the sequence might correspond to a pseudogene.

11. You can explore this possibility by adding to the map a track defining the position of pseudogenes. To do so, from the options listed under the bar named Mapping and Sequencing tracks, turn on two of the control keys. One key is named Retrotposed genes and the other Yale Pseudo. For both keys, select the dense option and subsequently click on the refresh button. In the resulting map you will find that the gene on chromosome 9 is not a candidate paralog but corresponds to a pseudogene (**Fig. 4**).

4. Notes

1. Opening a new window for each of the desired links is recommended. This would circumvent problems with losing the connection to the preceding page. The option of opening a new window for a link is available on PCs that use Microsoft operating systems. This option might not be available on other operating systems.

2. Viewing and analyzing the information provided in detail can help you to evaluate whether or not the matches with the submitted sequence are significant.

3. The default view of the browser may vary. However, you can experiment with the options to create the desired view.

References

1. True, J. R. and Carroll, S. B. (2002) Gene co-option in physiological and morphological evolution. *Annu. Rev. Cell Dev. Biol.* **18**, 53–80.
2. Kent, W. J., Sugnet, C. W., Furey, T. S., et al. (2002) The human genome browser at UCSC. *Genome Res.* **12**, 996–1006.

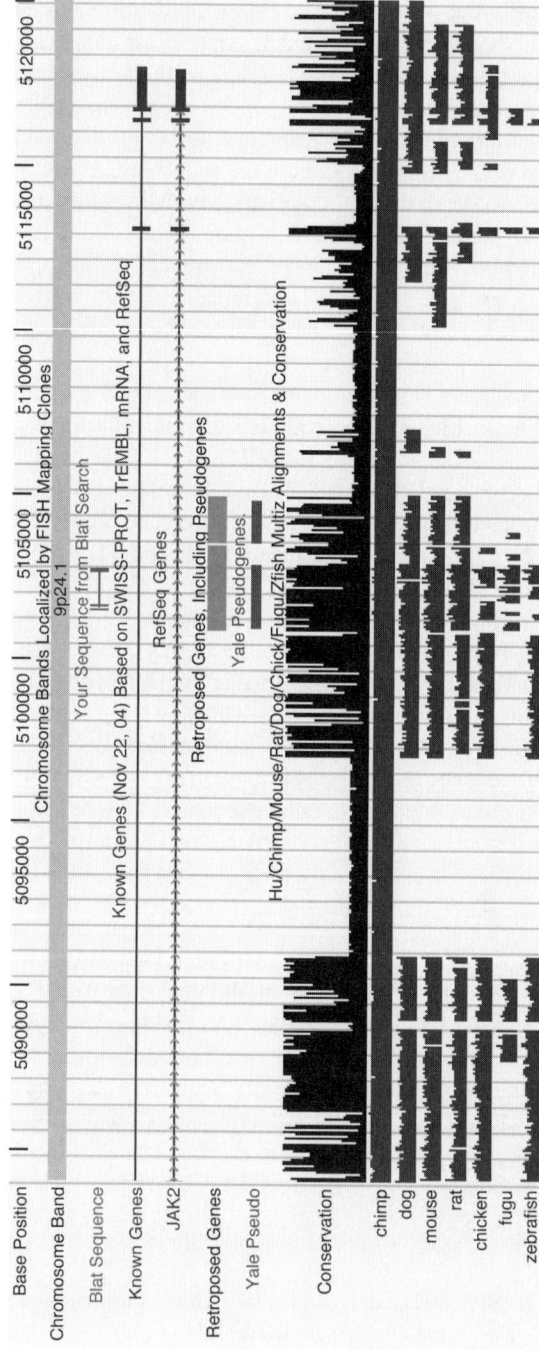

Fig. 4. Localization of a pseudogene on human chromosome 9p24.1. The result of the BLAT search (**Fig. 2**) also identified a region that showed sequence similarity to the products of the three paralogous transcription factor genes (**Fig. 3**). As seen in this map, the sequence identified by BLAT (Track 2) is within an intron of the JAK2 gene (Track 4). To examine whether this sequence corresponds to a pseudogene, we activated two control keys (Retroposed genes and Yale Pseudo) in the genome browser displaying the map (producing Track 5 and Track 6). From the result displayed in the map, we can infer a correlation between the position of a pseudogene and the position of the BLAT Sequence in the intron of the JAK2 gene.

28

3. Karolchik, D., Hinrichs, A. S., Furey, T. S., et al. (2004) The UCSC Table Browser data retrieval tool. *Nucleic Acids Res.* **32,** (Database issue) D493–496.
4. Xie, J., Li, K., and Bina, M. (2004) A Bayesian insertion/deletion algorithm for distant protein motif searching via entropy filtering. *JASA* **99,** 409–420.
5. Kent, W. J., Hsu, F., Karolchik, D., et al. (2005) Exploring relationships and mining data with the UCSC Gene Sorter. *Genome Res.* **15,** 737–741.
6. Venter, J. C., Adams, M. D., Myers, E. W., et al. (2001) The sequence of the human genome. *Science* **291,** 1304–1351.
7. Kent, W. J. (2002) BLAT—the BLAST-Like Alignment Tool. *Genome Res.* **12,** 656–664.
8. Benson, D. A., Karsch-Mizrachi, I., Lipman, D. J., Ostell, J., and Wheeler, D. L. (2005) GenBank. *Nucleic Acids Res.* **33,** (Database issue) D34–38.
9. Wheeler, D. L., Barrett, T., Benson, D. A., et al. (2005) Database resources of the National Center for Biotechnology Information. *Nucleic Acids Res.* **33,** (Database issue) D39–45.
10. Gan, T. I., Rowen, L., Nesbitt, R., et al. (2002) Genomic organization of human TCF12 gene and spliced mRNA variants producing isoforms of transcription factor HTF4. *Cytogenet. Genome Res.* **98,** 245–248.
11. Zhang, Y., Flejter, W. L., Barcroft, C. L., et al. (1995) Localization of the human HTF4 transcription factors 4 gene (TCF12) to chromosome 15q21. *Cytogenet. Cell Genet.* **68,** 235–238.
12. Henthorn, P., McCarrick-Walmsley, R., and Kadesch, T. (1990) Sequence of the cDNA encoding ITF-2, a positive-acting transcription factor. *Nucleic Acids Res.* **18,** 678.
13. Corneliussen, B., Thornell, A., Hallberg, B., and Grundstrom, T. (1991) Helix-loop-helix transcriptional activators bind to a sequence in glucocorticoid response elements of retrovirus enhancers. *J. Virol.* **65,** 6084–6093.
14. Murre, C., McCaw, P. S., and Baltimore, D. (1989) A new DNA binding and dimerization motif in immunoglobulin enhancer binding, daughterless, MyoD, and myc proteins. *Cell* **56,** 777–873.

4

Quantitative DNA Fiber Mapping in Genome Research and Construction of Physical Maps

Heinz-Ulrich G. Weier and Lisa W. Chu

Summary

Efforts to prepare a first draft of the human DNA genomic sequence forced multidisciplinary teams of researchers to face unique challenges. At the same time, these unprecedented obstacles stimulated the development of many highly innovative approaches to biomedical problem solving, robotics, and bioinformatics. High-resolution physical maps are required for ordering individual segments of information for the construction of a comprehensive map of the entire genome. This chapter describes a novel way to identify, delineate, and characterize selected, often small DNA sequences along a larger piece of the human genome. The technology is based on immobilization of high molecular weight DNA molecules on a solid substrate (such as a glass slide) followed by uniform stretching of the DNA molecule by the force of a receding meniscus. The hydrodynamic force stretches the DNA molecules homogeneously to approximately 2.3 kb/μm, so that distances measured after probe binding in μm can be converted directly into kb distances. Out of a large number of applications, this article focuses on mapping of genomic sequences relative to one another, the assembly of physical maps with near kb resolution, and, finally, quality control during physical map assembly and sequencing.

Key Words: Genome research; physical map assembly; DNA molecules; DNA fibers; fluorescence in situ hybridization (FISH); multicolor analysis; digital image analysis.

1. Introduction

High-resolution physical maps are indispensable for large-scale, cost-effective DNA sequencing and disease gene discovery. Thus, the construction of high-resolution physical maps of the human genome was one of the major goals of the human genome project in an effort to assemble a first draft of the human genome sequence *(1)*. Prevalent physical mapping strategies for most organisms studied

From: *Methods in Molecular Biology, vol. 338: Gene Mapping, Discovery, and Expression: Methods and Protocols*
Edited by: M. Bina © Humana Press Inc., Totowa, NJ

in recent years implemented a bottom-up approach for organizing individual pieces of DNA sequence information (typically provided in the form of recombinant DNA clones from large genomic libraries) into a high-resolution map of contiguous overlapping fragments (contigs). Large-scale sequencing, which is under way for various organisms, will greatly benefit from further progress in the creation of contig maps in a form suitable for DNA sequencing *(2,3)*. A particular challenge to physical map assembly is regions rich in DNA repeats such as the centromeres of various vertebrates including human *(4)*.

The cost-efficient sequencing of complex genomes also requires innovative approaches toward reducing the redundancy of sequencing templates, which can further reduce the cost of the overall project. Template redundancy is caused by extensive overlap between sequencing templates (clones) and the presence of cloning vector sequence in these clones. This redundancy can be minimized by construction of physical maps of the highest possible resolution and depth, definition of tiling paths comprised of minimally overlapping clones, knowledge of the extent of overlap between paths, and characterization of sequencing templates prior to sequencing to reject clones that contain only a cloning vector sequence or that are truly chimeric.

The recent progress in cloning large, megabasepair (Mbp) size genomic DNA fragments in yeast artificial chromosomes (YACs) *(5–7)* made it possible to rapidly construct low-resolution framework maps based on overlapping YAC clones. Fluorescence *in situ* hybridization (FISH) has proved indispensable for identification of nonchimeric YAC clones and for physical mapping of individual YAC clones onto metaphase chromosomes *(8)*. To generate physical maps with a resolution of 400 kb or better, the YAC clones were ordered by combining different complementing analytic techniques including pulsed-field gel electrophoresis (PFGE) *(9)*, FISH with interphase cell nuclei or metaphase spreads *(10–13)*, sequence-tagged sites (STS) content mapping *(5,14–17)*, optical mapping *(18,19)*, and/or DNA repeat fingerprinting *(20–23)*. These approaches have been successfully used to assemble YAC contigs covering most of the human genome *(24,25)*.

High-resolution maps providing ordered sets of cloned DNA fragments at the 100-kb level of resolution are assembled with smaller, more manageable DNA fragments isolated from other libraries. Most mapping and sequencing groups prefer cloning of the genomic DNA in vectors that maintain relatively large DNA fragments without rearrangements, are nonchimeric, and allow easy DNA purification. In general, high-resolution maps are comprised of overlapping cosmids *(26–30)*, P1/PAC clones *(31,32)*, or bacterial artificial chromosomes (BACs) *(33)*.

Assembly of high-resolution maps requires identification of cloned DNA sequences that contain overlapping regions of the genome. To minimize the effort invested in contiging a clone set, overlaps need to be determined quickly

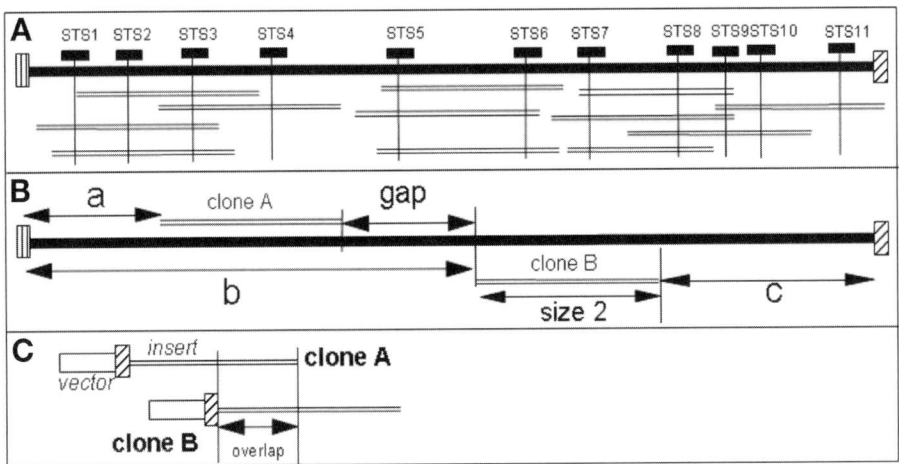

Fig. 1. Strategies for assembly of high-resolution physical maps. (**A**) STS are generated along the YAC clone (solid black line) and used to screen P1, PAC, or BAC libraries by PCR. This allows one to order the clones and group them into contigs. (**B–D**) Physical map assembly using QDFM. (**B**) Mapping of small clones such as clones A and B shown here onto larger DNA molecules (i.e., BACs, PACs, YACs) allows measurement of clone size (size 2), distance of the hybridization domains from either end of the larger molecule (a, b, c) and between clones (gap). (**C**) Mapping of one clone relative to another allows determination of overlap between clones and their relative orientation. Hatched boxes indicate vector DNA-specific sequences that can be tagged with specific DNA probes.

and accurately. This has been accomplished by various forms of clone fingerprinting (i.e., by identification of common restriction fragment or inter-Alu polymerase chain reaction (PCR) patterns *(29,34–38)*, by hybridization clone arrays bound to filters *(21,39,40)* or oligonucleotide arrays *(41)*, and by identification of overlapping STS *(42–44)*. The development of radiation hybrid (RH) maps *(45)* and the availability of large numbers of STS markers, together with extensive bacterial clone resources, provide additional means to accelerate the process of mapping a chromosome and preparing clone contigs ready for sequencing (**Fig. 1**) *(46)*. These techniques, although effectively used by the genome community, are limited because they do not readily yield information about contig orientation, extent of deletions or rearrangements in clones, overlap of contig elements, or their chimerism status nor do they provide information about the extent of gaps in the maps.

At Lawrence Berkeley National Library (LBNL), physical maps have been constructed for specific regions of the human genome. The typical procedure for physical map assembly at LBNL involved construction of a low-resolution framework map based on YAC clones selected from the CEPH YAC library *(6)* using publicly accessible STS content mapping and fingerprint data (**Fig. 1A**) *(3)*. High-resolution physical maps were then built with P1/PAC or BAC clones. The clones were isolated initially by screening filters containing library copies with inter-Alu PCR products prepared from nonchimeric YAC clones *(4)* and later by PCR-based library screening *(3)*. Next, more STS were generated by clone end sequencing and used to screen the libraries for additional clones. Finally, physical maps were assembled based on clone overlap as indicated by STS content (**Fig. 1A**).

Techniques to rapidly identify minimally overlapping clones and to determine the extent as well as the orientation of overlap will expedite the construction of minimal tiling paths, facilitate the sequence assembly process, and lower the overall cost of the project. FISH can provide important information for high resolution physical map assembly. For example, FISH to interphase nuclei allows probes to be ordered with several 100-kb resolution *(10–12)*, and FISH to preparations of decondensed nuclear *(47,48)* or isolated cloned DNA *(49)* allows visualization of probe overlap and provides some information about the existence and size of gaps in the map. However, none of these techniques provides quantitative information about the extent of clone overlap or about the separation between map elements because the chromatin onto which clones are mapped is condensed to varying degrees from site to site in these preparations.

The work of A. Bensimon and colleagues *(50)* showed that the extent of DNA condensation can be controlled by using a process termed *molecular combing*. In molecular combing, a solution of purified DNA molecules is placed on a flat surface prepared so that the DNA molecules slowly attach at one or both ends. The DNA solution is then spread over a larger area by placing a cover slip on top. DNA molecules are allowed to bind to the surface. During drying, the molecules are straightened and uniformly stretched by the hydrodynamic action of the receding meniscus (**Fig. 2**). Molecules prepared in this manner are stretched remarkably homogeneously to approx 2.3 kb/µm, i.e., approx 30% over the length predicted for a double-stranded DNA molecule of the same size *(50–53)*. We showed previously that cloned DNA fragments can readily be mapped by FISH onto DNA molecules prepared by molecular combing; in reference to its quantitative nature, we termed our technique *quantitative DNA fiber mapping* (QDFM) *(51)*. We also showed that QDFM can be applied to DNA molecules larger than 1 Mbp, which allowed us to map probes with near kilobase resolution onto whole yeast chromosomes and large YAC clones from the CEPH/Genethon library *(52)*. Because the DNA fibers are easily accessible to probes and detection reagents, hybridization efficiencies are typically high and allow the routine detection of DNA targets as small as 500 to 1000 bp *(51,54–55)*. In addi-

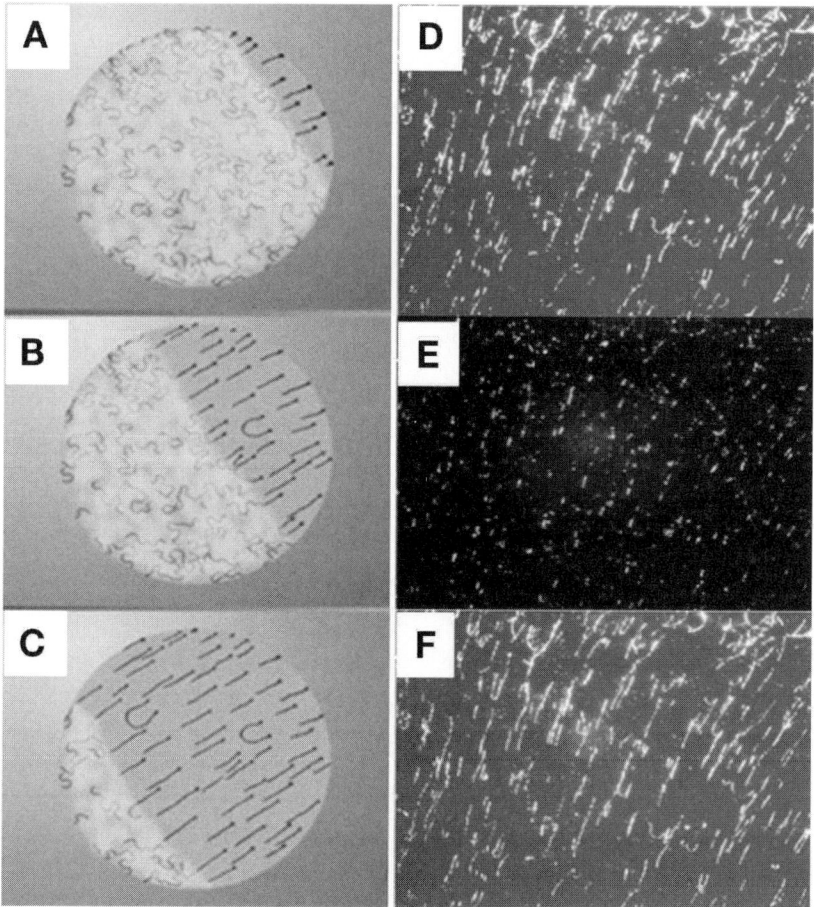

Fig. 2. Orientation and stretching of DNA molecules using the surface tension of a receding meniscus. (**A–C**) Schematic diagram showing DNA molecules bound at one or both ends and being pulled in the direction of the receding menicus during drying. (**D–F**) High density of λ DNA molecules after "molecular combing" and hybridization with DNA probes specific for individual *Hind*III fragments. These gray-scale images show either the red or green signals after detection of biotinylated or digoxigenin-labeled probes (**F,E**) or the superposition of both (**D**).

tion to the construction of high-resolution physical maps *(56,57)*, QDFM has proved useful in studies of DNA replication *(58)*.

The present chapter provides an in-depth description of the experimental procedures that will allow FISH experts as well as the novice to apply QDFM successfully to their research projects. Three practical examples demonstrate the application of QDFM in genome research.

2. Materials

2.1. Cell Culture and DNA Isolation

1. AHC medium: add 36.7 g of AHC powder (BIO 101, Vista, CA) per liter of water (*see* **Note 1**), and autoclave at 121°C for 15 min.
2. AHC agar (BIO 101): add 53.7 g of AHC agar medium per liter of purified water. Autoclave at 121°C for 15 min and cool to 50°C. Then mix well and pour plates. Store plates upside down in plastic bags at 4°C.
3. Tris-HCl (tris(hydroxymethyl)aminomethane) (*see* **Note 2**): prepare 1 M stock solutions in 500-mL bottles, adjust pH to 7.5, 8.0, or 8.3, and autoclave.
4. Lysozyme (Sigma, St. Louis, MO): prepare stock solution (50 mg/mL in 10 mM Tris-HCl, pH 7.5), and store in 100 to 150-µL aliquots at –20°C. Once thawed, do not refreeze.
5. Zymolase (70,000 U/g): prepare 10 mg/mL stock in 50 mM potassium phosphate (KH$_2$PO$_4$) buffer, pH 7.8, 50% glycerol. Store at –20°C.
6. Luria broth (LB): dissolve 10 g of Luria-Bertani powder per 400 mL of purified water in a 500-mL bottle. Autoclave at 121°C for 15 min and allow to cool for storage and use.
7. SCE buffer: 1 M sorbitol, 0.1 M Na citrate, 10 mM EDTA, pH 7.8.
8. ES buffer: 0.5 M EDTA (pH 8.0; Invitrogen, Gaithersburg, MD), 1% sarcosyl.
9. TE (Tris/EDTA) buffer (1X): 10 mM Tris-HCl, 1 mM EDTA, pH 7.4, 7.5, or 8.0.
10. TE50 buffer: 10 mM Tris-HCl, 50 mM EDTA, pH 7.8.
11. 10X Tris/borate/EDTA (TBE) buffer: 890 mM Tris base, 890 mM boric acid, 20 mM EDTA.
12. Alkaline lysis (AL) solution I: 50 mM glucose, 10 mM EDTA, 25 mM Tris-HCl, pH 8.0. Add 4 mL of 0.5 M glucose, 0.8 mL of 0.5 M EDTA, and 1 mL of 1 M Tris-HCl, pH 8.0, to 34.2 mL water. Store at 4°C. The amount is sufficient for 12 isolations at a level of 20 mL cell culture (*see* **Note 3**).
13. AL solution II: 0.2 N NaOH, 1% sodium dodecyl sulfate (SDS). Add 1.4 mL of 10 N NaOH, 7 mL of 10% SDS to 61.6 mL water. The amount is sufficient for 12 isolations at a level of 20 mL cell culture.
14. AL solution III: 3 M NaOAc, pH 4.8, in water.
15. Plasmid and cosmid DNAs: isolated using commercially available kits such as the Geneclean II kit (BIO 101).
16. β-Agarose (New England Biolabs).

2.2. Functionalization of Glass Surfaces and Molecular Combing

1. 0.1% Aminopropyltriethoxysilane (APS): prepare in 95% ethanol just prior to use.
2. YOYO-1 (Invitrogen): prepare stock solution of 1 mM in dimethyl sulfoxide (DMSO). Dilute 1:1000 with water prior to use. Store at –20°C and discard diluted dye after 1 wk.

2.3. Preparation of DNA Probes

1. Gel loading dye (6X): 1% bromophenol blue in 30% glycerol.

2. Thermus aquaticus (Taq) DNA polymerase buffer (10X): 500 m*M* KCl, 100 m*M* Tris-HCl (made from 1 *M* Tris-HCl, pH 8.3), 15 m*M* MgCl$_2$.
3. Modified nucleotide mix (10X): combine 5 μL each of 100 m*M* dATP, 100 m*M* dGTP, and 100 m*M* dCTP (Amersham) with 2.5 μL of 1 *M* Tris-HCl, pH 7.5, 0.5 μL 0.5 *M* EDTA, pH 8.0 (Invitrogen), and 232 μL water for a total of 250 μL. The final concentration of nucleoside triphosphates is 2 m*M* each. Store at –20°C. This solution for labeling is used in combination with 1 m*M* digoxigenin-11-dUTP, fluorescein isothiocyanate (FITC)-12-dUTP, or other dTTP analogs.

2.4. FISH

1. 20X SSC stock buffer: 3 *M* NaCl, 0.3 *M* Na$_3$·citrate, pH 7.0.
2. Denaturing solution: 70% formamide (FA; Invitrogen), 2X standard saline citrate (SSC), pH 7.0. Store at 4°C.
3. Hybridization master mix (MM2.1): 14.3% (w/v) dextran sulfate, 78.6% FA, 2.9X SSC, pH 7.0. For 10 mL MM2.1, mix 1.45 mL of 20X SSC with 0.7 mL water, dissolve 1.43 g dextran sulfate (Calbiochem, San Diego, CA), incubate overnight, and then add 7.86 mL FA (Invitrogen). Aliquot in 1.5-mL microcentrifuge tubes and store at –20°C.
4. Maleic acid buffer: 100 m*M* maleic acid, 150 m*M* NaCl; adjust to pH 7.5 with concentrated NaOH.
5. Blocking solutions.
   a. Blocking stock solution: dissolve blocking reagent (Roche Molecular Biochemicals) in 10% (w/v) maleic acid buffer with shaking and heating. Autoclave stock solution and store in aliquots at 4°C.
   b. Slide blocking solution: 5X SSC containing 2% blocking reagent and 0.1% *N*-lauroyl sarcosine. Combine 0.05 g *N*-lauroyl sarcosine (Na salt) and 1 g blocking reagent with 12.5 mL of 20X SSC, pH 7.0, and add 30 mL water. Heat to 60°C while stirring and bring the final volume to 50 mL with water, when the blocking reagent is dissolved. Aliquot into 1.5-mL tubes, spin at 300*g* for 10 min, and store at 4°C.
6. PN buffer: 0.1 *M* sodium phosphate, pH 8.0, 0.1% Nonidet-P40 in water. This buffer is prepared by dissolving 26.8 g of Na$_2$HPO$_4$ (dibasic) in 1 L water and 8.2794 g of Na$_2$HPO$_4$ (monobasic) in 600 mL water. Note the amount of monobasic solution added. Tirate the dibasic solution (pH > 9) by adding small volumes of the monobasic solution (pH ~4.5) until a pH of 8.0 is reached. Add Nonidet-P40 to 0.05% (v/v).
7. PNM buffer: dissolve 5 g of nonfat dry milk (Carnation, Wilkes-Barre, PA) in 100 mL of PN buffer, incubate at 50°C overnight, add 1/50 vol sodium azide, spin at 1000*g* for 30 min, aliquot clear supernatant into 1.5-mL tubes, and store at 4°C. Spin at 2000*g* for 30 s prior to use of the clear supernatant.
8. Rhodamine-conjugated antibodies against digoxigenin, made in sheep (Roche Molecular Biochemicals), mouse-derived antibodies against FITC (DAKO, Carpintera, CA), FITC-conjugated antimouse antibodies made in horse (Vector, Burlingame, CA), and biotinylated anti-avidin antibodies made in goat (Vector) are prepared

as stock solutions (1 mg/mL) in PNM, stored at 4°C, and diluted 1:50 with PNM just prior to use. Store at 4°C. Avidin conjugated to AMCA or FITC (Vector): stock solution is 2 mg/mL in PNM, diluted 1:500 prior to use. Store at 4°C.

9. Antifade solution: 1% *p*-phenylenediamine, 15 mM NaCl, 1 mM H$_2$PO$_4$, pH 8.0, 90% glycerol. Store in 1-mL aliquots at −80°C, and keep one aliquot at −20°C for everyday use (*see* **Note 4**).

2.5. Image Acquisition and Analysis and Construction of High-Resolution Physical Maps

1. Quantitative DNA fiber mapping experiments require only standard laboratory equipment and access to a fluorescence microscope equipped with a film or electronic camera.
2. Filters capable of excitation in single bands centered around 360, 405, 490, 555, and 637 nm and visualization in multiple bands in the vicinities of 460 nm (blue), 520 nm (green), 600 nm (red), and 680 nm (infrared) are desirable. Single-color images are collected using a CCD camera (Xilix, Hamamatsu, Vosskuehler, Photometrics, or similar) connected to a computer workstation *(51)*. When images are recorded on film, standard film with a sensitivity of ASA 400 is sufficient.

3. Methods

In QDFM, a small volume of isolated DNA molecules in aquous solution is placed on glass *(51,53)* or freshly cleaved sheets of mica *(54)* surface functionalized so that most DNA molecules attach at one or both ends. The DNA solution is then spread over a larger area by placing a cover slip on top; additional DNA molecules bind to the surface. During the subsequent drying step, molecules tethered to the support via either one or both ends (if linear) or via intermediate nicks in the double-stranded molecule (if the molecule is circular) are straightened and uniformly stretched by the hydrodynamic action of the receding meniscus, i.e., the surface tension at the water-air interface.

3.1. Isolation of DNA for QDFM

1. DNAs are isolated from plasmid, cosmid, P1/PAC, and BAC clones using either a commercially available purification kit or an alkaline lysis protocol (*see* **Subheading 3.3.1.** below). For the larger inserts, inserts are sized by PFGE. Digestion of DNA with a rare cutting restriction enzyme produces linear high molecular weight DNA molecules, but the alkaline lysis procedure typically provides sufficient amounts of nicked circular or randomly broken DNA suitable for QDFM *(52)*.
2. In general, the DNA is loaded on a 1.0% low melting point agarose (Bio-Rad, Hercules, CA) gel and electrophoresed for about 15 h. To efficiently separate DNA molecules of several hundred kb, we use a PFGE system (Bio-Rad).

3.1.1. Pulsed Field Gel Electrophoresis

1. The agarose plug preparation and PFGE using a PFGE system (Bio-Rad) follow standard protocols. Typically, a diluted solution of YACs in AHC medium is plated on AHC plates, and 5 to 15 individual YAC colonies are tested to account for deletions. In most cases, the largest clone carries the least deletion(s).
2. Preparation of gel plugs containing YACs (Invitrogen; stored at −80°C): spin down cells grown in 5 mL AHC media at 30g for 6 min. Resuspend cells in 0.5 mL of 0.125 M EDTA, pH 7.8. Spin again and resuspend the cell pellet in 500 μL of SCE. Mix with an equal volume of 1.5% low melting point (LMP) agarose preheated to 43°C. Quickly pipet up and down, and then vortex gently for 1 to 2 s to mix. Pipet into plug molds (Bio-Rad) and allow to solidify at room temperature or on ice.
3. Remove plugs from molds, incubate samples in 2 mL SCE containing 100 μL of zymolase, and shake at 150 rpm at 30°C for 2.5 h to overnight. Replace SCE buffer with 2 mL of ES containing 100 μL of proteinase K (20 mg/mL; Roche). Shake for 5 h to overnight at 50°C, and rinse plugs rinse 5 times with 6 mL of TE50 for 30 min each rinse. Store the plugs at 4°C until use.
4. PFGE running conditions for separation of YACs from yeast chromosomes: voltage gradient, 6 V/cm; switching interval, 79 s forward, 94 s reverse; running time, 38 h; agarose concentration, 1.0% LMP agarose; running temperature, 14°C; running buffer, 0.5X TBE.
5. PFGE running conditions for separation of full-length P1/PAC/BAC clones from debris: voltage gradient, 6 V/cm; switching interval, 2 s forward, 12 s reverse; running time, 18 h; agarose concentration, 1.0% LMP agarose; running temperature, 14°C; running buffer, 0.5X TBE (*see* **Note 5**).
6. For probe production and determination of optimal PFGE conditions: stain the gel with ethidium bromide (EB; 0.5 μg/mL in water), cut out a gel slice containing the target DNA band, and transfer slice to a 14-mL polystyrene tube (cat. no. AS-2264, Applied Scientific). Wash slice with water for 30 min, and then wash with 1X agarose buffer for 30 min.
7. For isolation of high molecular weight DNAs: run duplicate samples on the right and left side of the gel, respectively. After a predetermined run time, cut gel in half, and stain one half with EB. Measure the migrated distance on a UV transilluminator, cut out a gel slice at the corresponding position from the unstained half, and proceed as described in **step 5**.

3.1.2. Recovery of High Molecular Weight DNA From LMP Agarose Gel Slices

1. The DNA is recovered from the low melting point agarose slab gel by excising the appropriate bands using a knife or razor blade. High molecular weight DNA is then isolated by β-agarose digestion of the gel slices. Equilibrate gel slice in agarose buffer.
2. Melt the gel completely by incubating it for 10 min at 85°C, and then transfer the molten agarose to a 43°C water bath.

3. Add 1 μL β-agarose for every 25 μL of molten agarose, and incubate at 43°C for 2 h.
4. Add an equal volume of 200 m*M* NaCl, and store the DNA samples at 4°C until use (*see* **Note 5**).

3.2. Pretreatment of Microscope Slides and Preparation of DNA Fibers on Glass

1. The derivatization of glass substrates is among the most critical steps of the procedure. The slides should have the capacity to bind DNA molecules at one or both ends but should allow the molecules to stretch during the subsequent drying.
2. Solid substrates for QDFM are prepared in batches of 20 to 50 by derivatization of glass microscope slides (*see* **Note 7**), cover slips, or sheets of mica with APS, resulting in primary amino groups on the glass surface (*51,53*; *see* **Note 8**).
3. Clean glass slides mechanically by repeated rubbing with wet cheesecloth to remove dust and glass particles.
4. Rinse several times with water, immerse slides in boiling water for 10 min, and air-dry.
5. Immerse slides in 18 *M* sulfuric acid for at least 30 min to remove organic residues, followed by immersion in boiling water for 2 min.
6. Immerse precleaned dry slides in a solution of 0.1% APS in 95% ethanol for 10 min.
7. Remove slides from the silane solution, rinse several times with water, and immerse in water for 2 min.
8. Dehydrate by immersing in absolute ethanol and dry slides upright for 10 min at 65°C on a hot plate.
9. Store slides for 2 to 6 wk at 4°C in a sealed box under nitrogen prior to use.

3.2.1. Molecular Combing

1. In a typical experiment, 1 to 2 μL of clonal DNA are mixed with an equal amount of YOYO-1 (1 or 0.1 μ*M*) and 8 μL water. Then 1 or 2 μL of this diluted DNA is applied to an untreated cover slip, which is then placed DNA side down on the APS-derivatized slide.
2. The DNA concentration can be estimated in the fluorescence microscope using a filter set for FITC and adjusted as needed (*see* **Notes 9** and **10**). As early as after 2 min of incubation at room temperature, the untreated cover slip can be removed slowly from one end, allowing the receding meniscus to stretch the bound DNA molecules ("fibers") in one direction (*53,59*) (*see* **Note 11**).
3. Alternatively, the slide or cover slip sandwich can be allowed to dry overnight at room temperature, after which the untreated cover slip is removed by lifting it on one side with a razor blade. Slides carrying DNA fibers are rinsed briefly with water, drained, allowed to dry at room temperature, and "aged" in ambient air at 20°C for 1 wk before hybridization. Extra slides are stored at 4°C.

3.3. Preparation of DNA Probes

A typical QDFM experiment uses several different probes simultaneously. One probe is needed to counterstain the DNA fibers. This probe is usually prepared by labeling DNA from the same batch that was used to prepare the fibers. Probes for sequences to be mapped along the DNA fibers are made such that they can be detected in a different color. Furthermore, it is recommended to include landmark probes that provide reference points by binding specifically to the vector part or the ends of DNA molecules *(60)*.

3.3.1. Alkaline Lysis Protocol and Purification of DNA from P1, PAC, or BAC Clones

1. The P1/PAC/BAC clones typically show far fewer deletions than YACs, so that it often suffices to pick two to three colonies from a plate, grow them overnight in AHC, and extract the DNA using an alkaline lysis protocol. The DNA can then be loaded directly onto the PFGE gel using a common gel loading dye. This protocol describes the isolation of DNA from approx 20-mL overnight cultures using 40-mL Oak Ridge centrifugation tubes (Nalgene). The protocol can be scaled down to accommodate smaller volumes.
2. Grow cultures overnight in 25 to 30 mL LB medium containing the recommended amount of antibiotic.
3. Prepare Oak Ridge tubes. Write the clone ID on a small piece of tape stuck to the cap. Spin 18.5 mL of culture at 2000g for 10 min at 4°C and discard the supernatant.
4. Resuspend the pellet in 2340 µL of AL solution I, then add 100 µL of lysozyme stock to each tube. Incubate tubes for 5 min at room temperature. Then place the tubes on ice.
5. Add 5.2 mL of AL solution II. The mixture should now become clear. Mix gently by inverting the tubes several times. Incubate for 5 min on ice.
6. Add 3.8 mL of AL solution III and mix gently by inverting the tubes several times. Incubate for 10 min on ice.
7. Spin for 15 min at 14,000g.
8. Transfer 10.4 mL of supernatant into a new Oak Ridge tube, add 5.8 mL of isopropanol, and mix gently by inverting tubes several times. Use the old cap (with the ID sticker) on the new tube.
9. Spin for 5 min at approx 10,000g and discard the supernatant. Watch the pellet!
10. Wash the pellet in cold 70% ethanol. Let the pellets dry briefly, i.e., at approx 20 to 40 min at 20°C to 37°C.
11. Resuspend the pellet in 0.8 mL of TE buffer and split the volume into two 1.5-mL microcentrifuge tubes.
12. Add 400 µL phenol/chloroform/isoamyl alcohol (Invitrogen) to each tube. All centrifugations during the following phenol/chloroform extraction are done at 12,000g.
13. Vortex for 15 s and spin down for 3 min.
14. Remove most of the bottom layer and spin again for 3 min.
15. Transfer the top layer to new microcentrifuge tubes and add 400 µL chloroform/isoamyl alcohol (24:1, v/v; Invitrogen).
16. Vortex well for 15 s, spin down for 3 min, and remove most of the bottom layer followed by a second centrifugation for 3 min.

17. Transfer top layer to a new microcentrifuge tube, add 2.5 vol, i.e., 1 mL of 100% ethanol, and let the DNA precipitate for 30 min at −20°C.
18. Spin down for 15 min, discard the supernatant, wash the pellet in ice-cold 70% ethanol, spin again briefly, remove supernatant, and air-dry the pellet.
19. Resuspend the pellet in 20 to 40 μL TE, pH 7.4, containing 10 μg/mL RNAse (Roche) made DNase-free by boiling at 100°C for 10 min and store in aliquots at −20°C.
20. Incubate for 30 min at 37°C (in water bath) and store at −20°C until use.

3.3.2. Preparation of DNA From Yeast Artificial Chromosome Clones

1. Retrieve the desired yeast clone containing the YAC from the library and grow it on AHC agar for 2 to 3 d at 30°C. Pick colonies from the plate and culture the clones in up to 35 mL AHC media at 30°C for 2 to 3 d.
2. Centrifuge cells in AHC media at 2000*g* at 4°C for 5 min.
3. Decant the supernatant and resuspend cells in 3 mL total of 0.9 *M* sorbitol, 0.1 *M* EDTA, pH 7.5, containing 4 μL β-mercaptoethanol, followed by addition of 100 μL of zymolase (2.5 mg/mL), and incubate at 37°C for 60 min.
4. Pellet the cells at 2000*g* and 4°C for 5 min and decant supernatant.
5. Resuspend pellet in 5 mL of 50 m*M* Tris-HCl, pH 7.4, 20 m*M* EDTA. Add 0.5 mL of 10% SDS and mix gently. Incubate at 65°C for 30 min.
6. Add 1.5 mL of 5 *M* potassium acetate and place on ice for 60 min.
7. Spin at 12,000*g* for 15 min at 4°C, and transfer the supernatant to a new tube.
8. Mix the supernatant gently with 2 vol of 100% ethanol by inverting the tube a few times. Spin at 2000*g* for 15 min at room temperature.
9. Prepare sets of four 1.5-mL microcentrifuge tubes for each clone.
10. Decant supernatant and air-dry the pellet. Resuspend pellet in 3 mL of 1X TE, pH 7.5.
11. Transfer 750 μL of the DNA solution to each of the four 1.5-mL microcentrifuge tubes.
12. Add an equal volume of phenol/chloroform/isoamyl alcohol (25:24:1, pH 8.0), vortex well, and spin at 10,000*g* for 3 min.
13. Transfer the top layer to new 1.5-mL microcentrifuge tubes and add an equal volume of chloroform/isoamyl alcohol (24:1). Vortex well and centrifuge at high speed (10,000*g*) for 3 min.
14. Transfer the top layer to new 1.5-mL microcentrifuge tubes. Add 40 μL of RNAse (1 mg/mL, DNAse free) to each of the four tubes and incubate at 37°C for 30 min.
15. Add 1 vol of isopropanol and gently mix by inversion. Centrifuge at high speed (10,000*g*) for 20 min.
16. Decant supernatant, wash pellet with 1 vol of cold 70% ethanol, and centrifuge at 10,000*g* for 3 min.
17. Decant the 70% ethanol, air-dry the pellet, and resuspend the pellet in 30 μL 1X TE.
18. The DNA concentration is measured after the pellet is completely dissolved.

3.3.3. Generation of Probes by In Vitro DNA Amplification

1. In vitro DNA amplification using PCR is a very efficient method to synthesize probe DNA. It can be applied to amplify a particular DNA sequence, such as a part of the

cloning vector *(60)*, or with mixed-base primers to perform arbitrary amplification of virtually any sequence of interest *(51,59,61)*. As illustrated in the following paragraphs, the former amplification can be applied to prepare DNA landmark probes, whereas the latter allows the preparation of probes to counterstain the fibers.

2. The generation of P1/PAC-, BAC-, and YAC-vector probe DNA takes advantage of the access to published vector sequences. PCR primers are typically designed to amplify fragments of 1100 to 1400 bp of vector sequence *(51,52,60)*. Various oligonucleotide pairs have been designed in several laboratories including ours and are used in either single pairs or combinations *(51,52,56,60)*. The PCR usually follows standard conditions, i.e., a Tris-HCl buffer containing 1.5 mM MgCl$_2$ and 1 U Taq DNA polymerase per 50 µL reaction is used, and annealing temperatures range from 50°C to 60°C.

3. The YAC cloning vectors pJs97 and pJs98, cloned in plasmid vectors (Invitrogen), can be used to prepare probes that are useful to determine the orientation of the YAC insert *(56)*. For this purpose, plasmid DNA is extracted using the alkaline lysis protocol in **Subheading 3.3.1.** or a commercial kit and labeled by random priming as described below in **Subheading 3.3.4.**

4. The DNA probes for counterstaining the YAC DNA fibers are generated by mixed-base oligonucleotide-primed PCR (also referred to as degenerate oligonucleotide-primed PCR [DOP-PCR] *(61,62)*. An aliquot of the high molecular weight DNA obtained by PFGE for fiber preparation is PCR amplified for a total of 42 cycles with oligonucleotide primers that anneal about every 200 to 800 nucleotides. In our preferred scheme, we use two different DNA amplification programs *(62)*. Initially we perform a few manual PCR cycles using T7 DNA polymerase to extend the oligonucleotide primers at a relatively low temperature. Next, DNA copies prepared in these first cycles are amplified using the thermostable Taq DNA polymerase and a rapid thermal cycling scheme.

5. In the first amplification stage, T7 DNA polymerase (Sequenase II, Amersham) is used in five to seven cycles to extend the mixed-base primer JUN1 (5'-CCAAGCT TGCATGCGAATTCNNNNCAGG-3'; N = ACGT) that is annealed at low temperature. Briefly, 2 to 3 µL of high molecular weight DNA solution (after PFGE purification) are removed from the bottom of each tube and PCR amplified using the following conditions: denaturation at 92°C for 3 min, primer annealing at 20°C for 2 min, and extension at 37°C for 6 min. Sequenase II enzyme must be added after each denaturation.

6. In the second amplification stage, 10 µL of the reaction products are resuspended in a 200-µL Taq DNA amplification buffer and amplified with primer JUN15 (5'-CCCAAGCTTGCATGCGAATTC-3') with the following PCR conditions: denaturation at 94°C for 1 min, primer annealing at 50°C for 1 min, and extension at 72° for 2 min, repeated for 30 cycles. After precipitation of the PCR products in 1.2 vol of isopropanol, the products are resuspended in 30 µL of TE buffer. Subsequently, 1.5 µL of this solution is labeled in a 25-µL random priming reaction incorporating digoxigenin-11-dUTP or FITC-12-dUTP.

7. DNA amplification is confirmed by electrophoresing a 5-µL aliquot on a 3% agarose gel in TBE buffer containing 0.5 µg/mL ethidium bromide.

3.3.4. Probe Labeling via Random Priming and Hybridization

1. Labeling of DNA by random priming is a reliable method and, in our laboratory, is applied routinely to label DNA fragments from 100 bp to several hundred kb. The procedure involves an initial thermal denaturation of the DNA to allow the random oligonucleotides ("primers") to anneal. Thus, restriction or hydrolysis of large molecules is not necessary. Several companies now offer kits for random priming reactions. Slight differences exist with regard to enzyme activity, amount of random primers, and cost per reaction.
2. The concentration of PCR products can be estimated from the agarose gels run to confirm target amplification. If a sufficient amount of clonal or genomic DNA is available, 1 or 2 µL can be used to accurately determine the concentration with Hoechst 33258 fluorometry using a TK100 fluorometer (Pharmacia).
3. Add 250 ng of DNA to water to a final volume of 7 µL in a 0.5-mL microcentrifuge tube.
4. Boil DNA at 100°C for 5 min, and then quickly chill on ice.
5. For labeling with either digoxigenin-dUTP or FITC-dUTP, add:

   2.5 µL   10X modified nucleotide mixture
   3.25 µL   1 mM dTTP
   1.75 µL   digoxigenin-11-dUTP or FITC-12-dUTP (1 mM each, Roche)
   10 µL   2.5X random primers (BioPrime kit, Invitrogen)
6. For labeling the DNA with biotin, add 2.5 µL 10X dNTP mix provided with the BioPrime kit (containing biotin-14-dCTP), 5 µL water, and 10 µL 2.5X random primers).
7. Mix well, add 0.5 µL DNA polymerase I Klenow fragment (40 U/µL, Invitrogen) and incubate in a water bath at 37°C for 120 min.
8. Add 2.5 µL of 10X stop buffer (Invitrogen, part of the BioPrime kit).
9. Store probe at –20°C until use.

3.4. FISH

1. The hybridization procedure is very similar to protocols to used with metaphase spreads. In the hybridization mix, combine 1 µL of each probe, 1 µL of human COT1™ DNA (1 µg/µL, Invitrogen; optional), 1 µL of salmon sperm DNA (5 Prime-3 Prime, Boulder, CO), and 7 µL of MM2.1.
2. Fiber hybridizations include a comparatively low concentration of a biotin- or FITC-labeled DNA probe prepared from the high molecular weight DNA that is used to prepare the fibers. This counterstain highlights the otherwise invisible DNA fibers and allows competitive displacement by the probes to be mapped along the DNA fiber *(51,56)*. Additionally, one or several cloning vector-specific probes are included to allow determination of the orientation of the insert *(60)*.

3. Apply the hybridization mixture to the slides and cover-slip. Avoid bubbles; if bubbles occur, try to squeeze them out gently with fine-tip forceps, avoiding movement of the cover slip.
4. Transfer the slides to a dry bath (or "hot plate") and denature the DNA at 88°C for 90 s.
5. Transfer the slides to a moisture chamber (a plastic or stainless steel box with a wet paper towel at the bottom and support such as cut disposable plastic pipetors to raise the slide) and incubate it overnight at 37°C.
6. The wash and detection steps are not much different from protocols used for FISH to interphase and metaphase cells and have been described in detail *(51,54)*. After hybridization, the slides are washed three times in 2X SSC at 20°C for 10 min each and then incubated with 100 µL PNM buffer or blocking stock solution under a plastic cover slip at 20°C for 5 min (*see* **Note 12**). The slides are then incubated at room temperature for 30 min with 100 µL PNM buffer containing AMCA-avidin (Pharmacia), anti-digoxigenin-rhodamine (Roche), and a mouse antibody against FITC (DAKO) (*see* **Note 13**).
7. The slide is then washed two to three times in 2X SSC for 15 min each at 20°C with constant motion on a shaking platform.
8. If necessary, signals are amplified using a biotinylated antibody against avidin raised in goat (Vector) followed by another layer of AMCA-avidin, a Texas Red-labeled antibody against sheep raised in rabbit, and a horse-anti-mouse antibody conjugated to FITC (Vector) *(52)*.
9. The slide is mounted in 8 µL of antifade solution and covered by a 22 × 22-mm cover slip.

3.5. Digital Image Acquisition and Analysis and Map Assembly

1. Although not a prerequisite for QDFM, digital image acquisition and computer-assisted analysis greatly facilitate the quantitative analysis of hybridization images. Since QDFM is based on simple measurements of distances between probe hybridization domains, the analysis can alternatively be performed on images recorded on film and either printed or projected on a screen.
2. Images are acquired using a standard fluorescence microscope (Zeiss Axioskop or similar) equipped with 63X, 1.25 N.A. and 40X, 1.2 N.A. objectives and a filter set for excitation and simultaneous observation of DAPI, Texas Red/rhodamine, FITC, and CY5 fluorescence, respectively (ChromaTechnology, Brattleboro, VT) (*see* **Notes 14** and **15**).
3. For determination of map positions, interactive software is available for either Apple Macintosh, IBM/PC, or SUN computers that allows the user to trace DNA fibers by drawing a straight or segmented line and then calculates the length of the line in pixels *(52,56)*. The pixel spacing for the camera and the microscope objective used in the experiment (a 63X objective is used for molecules up to 100 kb and a 40X objective for larger molecules) is known and is converted into µm or into kb using the factor of 2.3 kb/µm *(51)*. After measuring all relevant distances

along the DNA fibers in triplicate, the results in the form of lists are then imported into Microsoft Excel spreadsheets and used to calculate average values for each fiber and mean values and standard deviations for individual experiments.

3.5.1. Construction of High-Resolution Physical Maps

1. QDFM can facilitate the construction of high-resolution physical maps comprised of any combination of cosmid, P1, PAC, or BAC clones in two ways:
   a. If a low-resolution map is available, for example, in the form of a YAC contig, individual clones can be mapped directly onto DNA fibers prepared from the larger clones (*3,51*).
   b. Alternatively, a high-resolution map can be constructed by measuring the extent and orientation of overlap between individual clones by hybridizing one clone onto another (**Fig. 1**). In most experiments, the approach taken will depend on the sources of the clones and might combine both schemes.
2. **Figure 3** shows the mapping of a P1 clone (approx 81-kb insert, red) onto a colinear YAC clone (green). Precise localization of the region of overlap and measurement of distances from the ends of the YAC (distances A and B) are facilitated by probes that mark specifically the ends of the YAC molecules (red, arrows).
3. The DNA fiber-based mapping of two P1 clones (1107 and 1143) onto the colinear YAC clone 141G6 (approx 490 kb) (*5*) and determination of the size of the gap between these P1 clones by QDFM has been described (*51*) (**Fig. 4**). Briefly, the degree and uniformity of stretching achieved for the YAC molecules was assessed by measuring the lengths of the domains produced by hybridization with DNA from the approx 81-kb P1 1143 along 10 YAC fibers. The length of 1143 along the YAC fibers was 34.5 ± 2.55 µm, corresponding to a stretching of 2.3 kb/µm, almost identical to that achieved for λ phage (*51*) (**Fig. 4C**). This suggested that the degree of stretching is highly reproducible and independent of the length of the combed molecule.
4. The location of the P1 clone 1143 along the YAC was determined by measuring the distance of its hybridization signal from the marked end of the YAC (**Fig. 4B** and **C**). The hybridization domain of clone 1143 began at 49.2 µm or 114 kb (± 5.7 kb, $n = 10$) from the proximal end of the YAC and extended 81 kb, assuming a conversion factor of 2.3 kb/µm. Measurement of the hybridization domain of clone 1107 suggested a mean size of 89.9 kb (± 8.2 kb), which agrees well with the size of 88 kb obtained by PFGE analysis. This conversion also allowed us to estimate the size of the YAC as 496 kb (SD 37 kb, $n = 4$). This was in good agreement with published values ranging from 430 to 495 kb (*5,29*).
5. The extent of the gap between clones 1143 and 1107 (**Fig. 4A**) was found to be 10.9 µm or 25.4 kb (± 1.1 kb) by measuring the physical distances between the P1 hybridization signals on 10 YAC fibers (*51*). Partial fibers showing hybridization signals along the gap region and part of the flanking P1s were sufficient for determining the size of the gap region since these all appeared to be equally stretched.

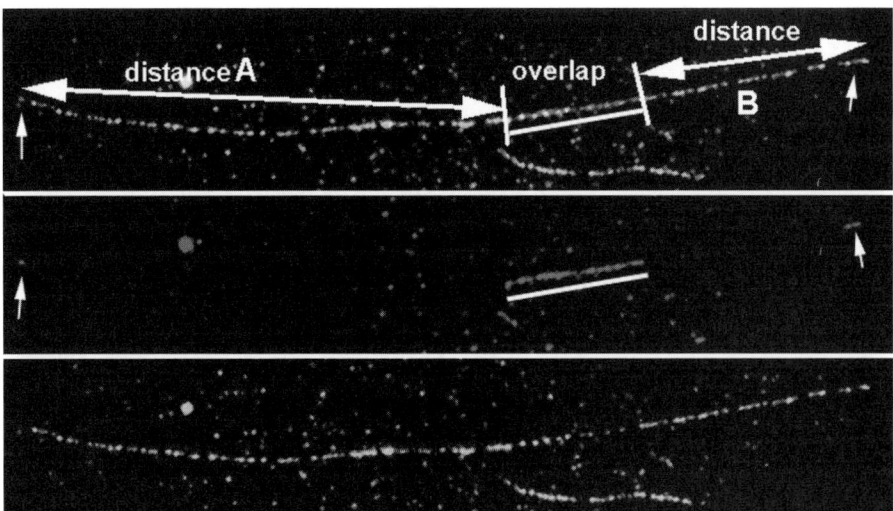

Fig. 3. Application of QDFM for the precise localization of a genomic interval represented by a P1 clone along a larger YAC DNA molecule. (**A–C**) This dual-color FISH experiment allows one to measure the physical distances from the ends of the YAC (distances A, B) as well as the overlap between these clones (overlap). The image in (**A**) shows the superposition of red (**B**) and green (**C**) images recorded from digoxigenin-labeled and biotinylated probes, respectively.

3.5.2. Quality Control of Individual DNA Sequencing Templates

1. The P1 clone #39 maps to the long arm of chromosome 20 band q13 *(63)*. Sequencing templates were prepared by cloning size-selected approx 3-kb fragments from sonicated P1 #39 DNA into the plasmid vector pOT2 *(63)*.
2. Digoxigenin-labeled probes prepared by random priming of plasmid DNA from eight sequencing templates (plasmid clones) were mapped onto DNA fibers prepared from *Not*I-linearized molecules of P1 #39. The hybridization mixture contained the plasmid probe, 100 ng/μL human COT1 DNA (Invitrogen) to block hybridization of DNA repeat sequences, biotinylated P1 #39 DNA (here visualized in blue), and an FITC-labeled probe for the P1 vector (green). The bound plasmid DNA was visualized in red (**Fig. 5**). Plasmid probes that showed hybridization to vector sequences were also mapped by hybridization to DNA fibers prepared from an empty recombinant P1 (Genome Systems, St. Louis, MO). **Figure 5** shows representative images of hybridized DNA fibers using six different plasmid probes. The map position of the approx 3-kb plasmids was readily visible and is indicated in the black-and-white reproduction by an arrow (**Fig. 5**).
3. Map positions relative to the *Sal*I site (near the Sp6 promotor) of the P1 vector pAd10SacBII are obtained from QDFM images. Plasmid clones 9-d4, 2-a2, 4-c11, 4-h3, and 10-h8 were found to map exclusively to the insert of P1 #39, whereas

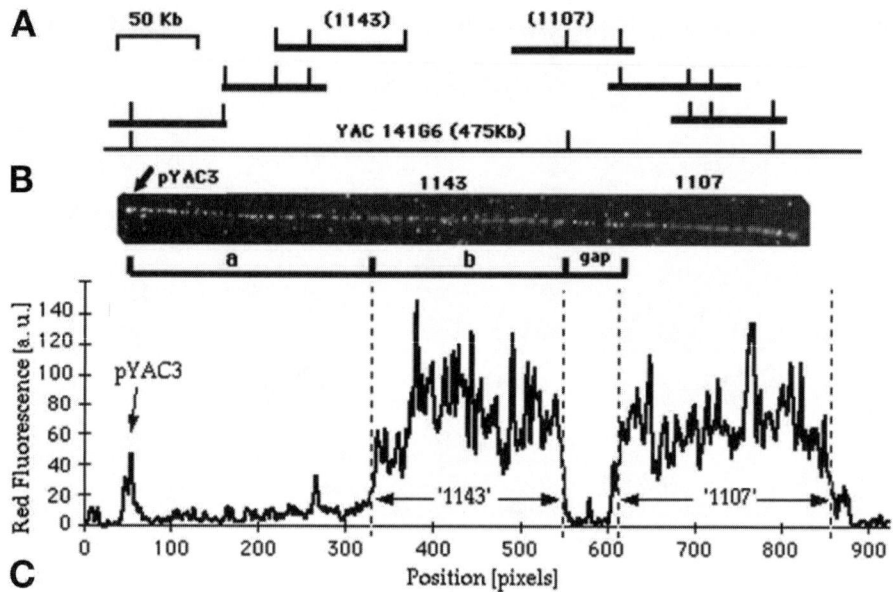

Fig. 4. Physical mapping of P1 clones and estimation of the size of a gap. (**A**) The P1 clones 1143 and 1107 represent the ends of two contigs of P1 clones that map to the long arm of chromosome 21. The YAC clone 141G6 carries an insert of about 475 kb that contains the same genomic region. (**B**) Hybridization of two digoxigenin (dig)-labeled P1 probes (detected with anti-dig-rhodamine, red) and a biotinylated DNA probe prepared from YAC DNA (detected with avidin-FITC, green) allows measurement of the positions of the P1 probes in the larger genomic interval defined by the YAC insert. The same QDFM experiment also provides an accurate measurement of the distance between the two P1 clones (gap). (**C**) The extent of the gap between clones 1143 and 1107 was determined to be 10.9 μm or 25.4 kb (±1.1 kb) by measuring the physical distances between the P1 hybridization signals on 10 YAC fibers.

clones 1c7 and 4-h5 hybridized to the insert as well as vector regions (**Fig. 5**). Gel electrophoretic analysis indicated that clones 1-c7 and 4-h5 were significantly larger than the other clones (**Table 1**). Clone 1-c7 mapped to the insert adjacent to the unique *Not*I (T7) site at approximately the same position as clone 1-h3 but showed an additional hybridization signal adjacent to the *Sal*I site (not shown). Clone 4-h5 hybridized to the insert approximately 52 ± 4 kb from the *Sal*I site and showed a second hybridization signal of approximately 3 kb centered between positions 4500 and 7500 of the 16-kb pAd10SacBII vector (**Fig. 5**, arrowhead) (Genbank accession # U09128; **Table 1**).

3.5.3. Localizing Contigs Comprised of Approx 3 kb Plasmid Clones

1. QDFM can map and orient contigs comprised of approx 3-kb sequencing templates onto P1 or BAC molecules with a precision of a few kb.

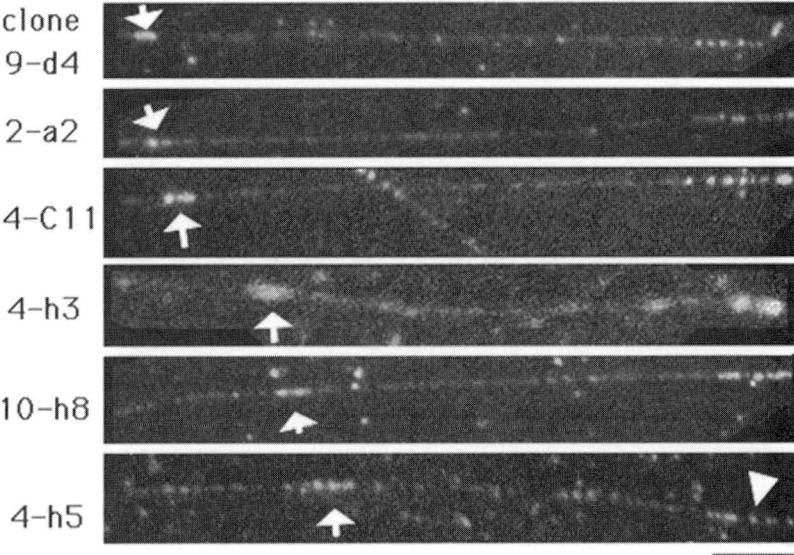

clone
9-d4

2-a2

4-C11

4-h3

10-h8

4-h5

Fig. 5. High-resolution mapping of sequencing templates (approx 3 kb plasmid clones) along a human P1 clone. Six plasmid clones were hybridized onto DNA fibers prepared from *Not*1-digested DNA of clone P1 #39. The position of the plasmid-specific hybridization signals is indicated by arrows. Two signals, in the insert (arrow) and in the vector part (arrowhead), were detected when hybridizing a probe prepared from plasmid clone 4-h5. The horizontal bar indicates the part of the DNA fiber representing the P1 vector.

2. The P1 clone #30 maps in roughly the same region of chromosome 20q13 as P1 #39 *(3)*. **Figure 6A** shows the mapping of a five-member contig (contig3) onto an *Sfi*I-linearized P1 DNA molecule (clone #30). While this P1 clone was being prepared to be sequenced at the LBNL Human Genome Center, we mapped onto it three plasmid contigs. Each contig was comprised of five to seven plasmid clones of approx 3 kb in length (sequencing templates). QDFM let us measure the contig map position and extent, as well as the overlap between contigs with an accuracy of approx 1 kb. The results (summarized in **Fig. 6B**) showed that contigs 2 and 3 overlap about 11 kb, but neither of them overlap with contig 1. Thus, either of these two contigs can be sequenced in parallel with the nonoverlapping contig 1.
3. The information regarding contig position will also facilitate the preparation of sequencing templates comprised of large-insert DNA clones. In genomic regions, for which only shallow coverage exists, a minimal tiling path typically contains clones with extensive overlap. QDFM allows elimination of approx 3-kb plasmid contigs that fall in overlapping regions. These investigations can be performed rapidly on either circular, enzymatically restricted, or randomly broken DNA molecules.

Table 1
Mapping Approx 3-kb Sequencing Templates on DNA Fibers Prepared From P1 #39

Clone ID	Proximal position (kb)	Distal position (kb)	Plasmid insert size (kb)[a]	Homology with vector	Chimeric clone	Vector map position
1-h3	80 ± 1.6	83 ± 1.3	2.6	No	No	—
9-d4	77.3 ± 1.4	79.6 ± 1.3	2.9	No	No	—
2-a2	76.3 ± 1.3	79.5 ± 1.3	3.2	No	No	—
4-c11	74 ± 2.2	77.5 ± 1.9	3.4	No	No	—
10-h8	55.1 ± 1.7	58.8 ± 1.9	3.6	No	No	—
4-h3	54.3 ± 3.2	58.8 ± 3.2	4.5	No	No	—
4-h5	52.3 ± 3.8	56.4 ± 4.2	6.4 {2.3, 4.1}	Yes	Yes	4500–7500
1-c7 signal1	0	3.4 ± 0.6	8.0	Yes	Yes	Sal1 side
1-c7 signal2	82.8 ± 4.1	85.1 ± 4.2		No		Not1 side

[a]Measured by agarose gel electrophoresis.

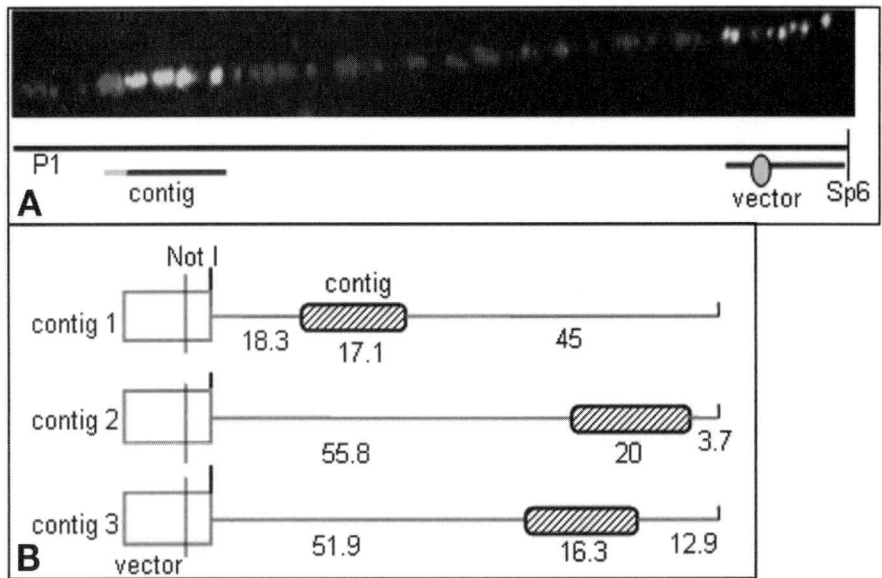

Fig. 6. High-resolution mapping of contigs allows selection of nonoverlapping clones for parallel sequencing. (**A**) Mapping on a contig comprised of five plasmid clones onto linearized P1 #30 fibers. (**B**) The absolute positions of three plasmid contigs along P1 #30. The numbers indicate mean values of distances or contig sizes in kb.

4. Notes

1. Unless stated otherwise, all solutions should be prepared with purified water having a restistivity of 18.2 MΩ-cm and total organic content of less than five parts per billion. This standard is referred to as "water" in this protocol.
2. Unless noted otherwise, all chemicals were purchased from Sigma.
3. The AL solution set (solutions I–III) is for isolation of P1, PAC, or BAC DNA.
4. Allow the antifade solution to warm up for only a short period, remove an aliquot, and then place it back into the freezer. Fresh antifade solution is colorless. Discard aliquot when solution has turned dark.
5. DNA isolated from P1, PAC, or BAC clones can be loaded directly into the wells of the PFGE gel. Use 1 µL of loading buffer per 5 µL of DNA solution.
6. The integrity of DNA molecules can be assessed by microscopic inspection of aliquots of DNA stained with 0.5 µM YOYO-1, before high molecular weight DNAs are used for DNA fiber preparation or stored at 4°C in 100 mM NaCl.
7. Slides from different manufacturers or the same brand but different batches may produce different qualities of DNA fibers. Purchase a sufficiently large batch of slides from one manufacturer such as Erie Scientific (Portsmouth, NH), a supplier of glass slides that are sold under Fisher or BD labels. Avoid slides that are painted

on one end, since the paint might dissolve during pretreatment. Slides that have a sandblasted or etched area at one end are preferable.

8. Cover slip silanation is performed like the procedure described for slides. Briefly, cover slips are rinsed with distilled water and dehydrated in 100% ethanol. Cover slips are derivatized with a 0.05 to 0.1% solution of APS in 95% ethanol for 2 min. Cover slips are then rinsed and dried as described in **Subheading 3.2.**

9. Binding of DNA molecule ends to the substrate and the stretching effect can be monitored in the fluorescence microscope by staining the DNA with 0.2 to 0.5 μM YOYO-1 prior to deposition. This allows exclusion of batches of slides that bind DNA too tightly. Dilute the YOYO-1 in water, since the antifade solution prevents the DNA molecules from binding.

10. The density of DNA molecules after DNA fiber stretching can be adjusted by altering the concentration of the DNA molecules prior to binding. **Figure 2D** to **F** shows the typical density of hybridized λ DNA molecules. In experiments depositing λ, P1, PAC, or BAC DNA molecules, the fraction of intact DNA molecules can reach approx 50 to 70%. Mapping can utilize both linear and circular DNA molecules. Although binding of DNA molecules in their circular form helps to maintain their integrity, it interferes with DNA fiber stretching, and the molecules are found to be stretched to varying degrees. Mapping onto circular molecules can thus be used for rough estimation of overlap and mapping on linear fibers for high-precision measurements. This can be done in a single experiment, because some circular DNA molecules are sheared during deposition, thus providing randomly broken linear DNA molecules. (*see* **ref. 53** for a more detailed description).

11. Different procedures have been described to stretch DNA molecules *(47–51)*. In our hands, stretching involving a hydrodynamic force (meniscus) at 20°C or 4°C has proved most reproducible. There is, however, no need to wait until the preparation has dried to completion. Once the DNA molecules have bound to the substrate, the cover slip can be lifted to exert the hydrodynamic stretching force *(53)*.

12. We found that even short periods of drying out of slides or parts of them during the immunocytochemical signal amplification lead to unacceptable levels of background fluorescence owing to nonspecific binding of detection reagents. It is important to just drain the liquids from the slides and then rapidly apply the next solution, a blocking solution, or antibodies.

13. If only two labels are used, i.e., biotin and digoxigenin, bound probes are detected with avidin-FITC DCS (Vector) and antidigoxigenin-rhodamine, respectively.

14. Most fluorochromes fade quickly. Under the microscope, we always minimize the exposure of slides to the excitation light. The key factor in producing good images is to use the DNA counterstain or one fiber probe to quickly localize areas showing a sufficiently high density of well-stretched molecules and then switch to the image acquisition mode.

15. Always measure additional segments of the molecule such as the vector segment since these might provide additional information about the extent and homogeneity of DNA stretching.

Acknowledgments

This work was supported by a grant from the Director, Office of Energy Research, Office of Health and Environmental Research, U.S. Department of Energy, under contract DE-AC-03-76SF00098, the LBNL/UCSF Training Program in Genome Research sponsored by the University of California Systemwide Biotechnology Research and Education Program, and NIH grants CA80792 and HD45736.

Disclaimer

This document was prepared as an account of work sponsored by the United States Government. While this document is believed to contain correct information, neither the United States Government nor any agency thereof, nor The Regents of the University of California, nor any of their employees, makes any warranty, express or implied, or assumes any legal responsibility for the accuracy, completeness, or usefulness of any information, apparatus, product, or process disclosed, or represents that its use would not infringe privately owned rights. Reference herein to any specific commercial product, process, or service by its trade name, trademark, manufacturer, or otherwise, does not necessarily constitute or imply its endorsement, recommendation, or favoring by the United States Government or any agency thereof, or The Regents of the University of California. The views and opinions of authors expressed herein do not necessarily state or reflect those of the United States Government or any agency thereof, or The Regents of the University of California.

References

1. Collins, F. and Galas, D. (1993) A new five-year plan for the U.S. human genome project. *Science* **262,** 43–46.
2. Boguski, M., Chakravarti, A., Gibbs, R., Green, E., and Myers, R. M. (1996) The end of the beginning: the race to begin human genome sequencing. *Genome Res.* **6,** 771–772.
3. Cheng, J.-F. and Weier, H.-U. G. (1997) Approaches to high resolution physical mapping of the human genome, in *Biotechnology International* (Fox, C. F. and Connor, T. H., eds.), Universal Medical Press, San Francisco, pp. 149–157.
4. Martin, J., Han, C., Gordon, L. A., et al. (2004) The sequence and analysis of duplication-rich human chromosome 16. *Nature* **4,** 988–994.
5. Chumakov, I., Rigault, P., Guillou, S., et al. (1992) Continuum of overlapping clones spanning the entire human chromosome-21q. *Nature* **359,** 380–387.
6. Cohen, D., Chumakov, I., and Weissenbach, J. (1993) A first-generation physical map of the human genome. *Nature* **366,** 698–701.
7. Olson, M. V. (1993) The human genome project. *Proc. Natl. Acad. Sci. USA* **90,** 4338–4344.

8. Selleri, L., Eubanks, J. H., Giovannini, M., et al. (1992) Detection and characterization of "chimeric" yeast artificial chromosome clones by fluorescent in situ suppression hybridization. *Genomics* **1**, 536–541.

9. Vetrie, D., Bobrow, M., and Harris, A. (1993) Construction of a 5.2-megabase physical map of the human X chromosome at Xq22 using pulsed-field gel electrophoresis and yeast artificial chromosomes. *Genomics* **15**, 631–642.

10. Trask, B., Pinkel, D., and van den Engh, G. (1989) The proximity of DNA sequences in interphase cell nuclei is correlated to genomic distance and permits ordering of cosmids spanning 250 kilobase pairs. *Genomics* **5**, 710–717.

11. Brandriff, B. F., Gordon, L. A., Tynan, K. T., et al. (1992) Order and genomic distances among members of the carcinoembryonic antigen (CEA) gene family determined by fluorescence in situ hybridization. *Genomics* **12**, 773–779.

12. Warrington, J. A. and Bengtsson, U. (1994) High-resolution physical mapping of human 5q31-q33 using three methods: radiation hybrid mapping, interphase fluorescence in situ hybridization, and pulse field gel electrophoresis. *Genomics* **24**, 395–398.

13. Lu-Kuo, J. M., Le Paslier, D., Weissenbach, J., Chumakov, I., Cohen, D., and Ward, D. C. (1994) Construction of a YAC contig and a STS map spanning at least seven megabasepairs in chromosome 5q34-35. *Hum. Mol. Genet.* **3**, 99–106.

14. Green, E. D. and Olson, M. V. (1990) Systematic screening of yeast artificial-chromosome libraries by use of the polymerase chain reaction. *Proc. Natl. Acad. Sci. USA* **87**, 1213–1217.

15. Coffey, A. J., Roberts, R. G., Green, E. D., et al. (1992) Construction of a 2.6-Mb contig in yeast artificial chromosomes spanning the human dystrophin gene using an STS-based approach. *Genomics* **12**, 474–484.

16. Weissenbach, J., Gyapay, G., Dib, C., et al. (1992) A second-generation linkage map of the human genome. *Nature* **359**, 794–801.

17. Locke, J., Rairdan, G., McDermid, H., et al. (1996) Cross-screening: a new method to assemble clones rapidly and unambiguously into contigs. *Genome Res.* **6**, 155–165.

18. Cai, W., Aburatani, H., Stanton, V. P., Housman, D. E., Wang, Y. K., and Schwartz, D. C. (1995) Ordered restriction endonuclease maps of yeast artificial chromosomes created by optical mapping on surfaces. *Proc. Natl. Acad. Sci. USA* **92**, 5164–5168.

19. Samad, A., Huff, E. J., Cai, W., and Schwartz, D. C. (1995) Optical mapping: a novel, single-molecule approach to genomic analysis. *Genome Res.* **5**, 1–4.

20. Waterston, R. and Sulston, J. (1995) The genome of *Caenorhabditis elegans*. *Proc. Natl. Acad. Sci. USA* **92**, 10836–10840.

21. Bellanné-Chantelot, C., Lacroix, B., Ougen, P., et al. (1992) Mapping the whole human genome by fingerprinting yeast artificial chromosomes. *Cell* **70**, 1059–1068.

22. Zucchi, I. and Schlessinger, D. (1992) Distribution of moderately repetitive sequences pTR5 and LF1 in Xq24-q28 human DNA and their use in assembling YAC contigs. *Genomics* **12**, 264–275.

23. Porta, G., Zucchi, I., Hillier, L., et al. (1993) Alu and L1 sequence distributions in Xq24-q28 and their comparative utility in YAC contig assembly and verification. *Genomics* **16**, 417–425.

24. Bell, C., Budarf, M. L., Nieuwenhuijsen, B. W., Barnoski, B. L., and Buetow, K. H. (1995) Integration of physical, breakpoint and genetic maps of chromosome 22. Localization of 587 yeast artificial chromosomes with 238 mapped markers. *Hum. Mol. Genet.* **4**, 59–69.
25. Foote, S., Vollrath, D., Hilton, A., and Page, D. C. (1992) The human Y chromosome: overlapping DNA clones spanning the euchromatic region. *Science* **258**, 60–66.
26. Stallings, R. L., Doggett, N. A., Callen, D., et al. (1992) Evaluation of a cosmid contig physical map of human chromosome 16. *Genomics* **13**, 1031–1039.
27. Tynan, K., Olsen, A., Trask, B., et al. (1992) Assembly and analysis of cosmid contigs in the CEA-gene family region of human chromosome 19. *Nucleic Acids Res.* **20**, 1629–1636.
28. Nizetic, D., Gellen, L., Hamvas, R. M., et al. (1994) An integrated YAC-overlap and 'cosmid-pocket' map of the human chromosome 21. *Hum. Mol. Genet.* **3**, 759–770.
29. Patil, N., Peterson, A., Rothman, A., DeJong, P. J., Myers, R. M., and Cox, D. R. (1994) A high resolution physical map of 2.5 Mbp of the Down syndrome region on chromosome 21. *Hum. Mol. Genet.* **3**, 1811–1817.
30. Cherry, J. M., Ball, C., Weng, S., et al. (1997) Genetic and physical maps of *Saccharomyces cerevisiae*. *Nature* **387(Suppl)**, 67–73.
31. Pierce, J. C., Sauer, B., and Sternberg, N. (1992) A positive selection vector for cloning high molecular weight DNA by the bacteriophage-P1 system—improved cloning efficacy. *Proc. Natl. Acad. Sci. USA* **89**, 2056–2060.
32. Ioannou, P. A., Amemiya, C. T., Garnes, J., et al. (1994) A new bacteriophage P1-derived vector for the propagation of large human DNA fragments. *Nat. Genet.* **6**, 84–89.
33. Shizuya, H., Birren, B., Kim, U. J., et al. (1992) Cloning and stable maintenance of 300-kilobase-pair fragments of human DNA in *Escherichia coli* using an F-factor-based vector. *Proc. Natl. Acad. Sci. USA* **89**, 8794–8797.
34. Branscomb, E., Slezak, T., Pae, R., Galas, D., Carrano, A. V., and Waterman, M. (1990) Optimizing restriction fragment fingerprinting methods for ordering large genomic libraries. *Genomics* **8**, 351–366.
35. Nelson, D. L. (1991) Applications of polymerase chain reaction methods in genome mapping. *Curr. Opin. Genet. Dev.* **1**, 62–68.
36. Chang, E., Welch, S., Luna, J., Giacalone, J., and Francke, U. (1993) Generation of a human chromosome 18-specific YAC clone collection and mapping of 55 unique YACs by FISH and fingerprinting. *Genomics* **17**, 393–402.
37. Riles, L., Dutchik, J. E., Baktha, A., et al. (1993) Physical maps of the six smallest chromosomes of *Saccharomyces cerevisiae* at a resolution of 2.6-kilobase pairs. *Genetics* **134**, 81–150.
38. Gillett, W., Hanks, L., Wong, G. K., Yu, J., Lim, R., and Olson, M. (1996) Assembly of high-resolution restriction maps based on multiple complete digests of a redundant set of overlapping clones. *Genomics* **33**, 389–408.
39. Hoheisel, J. D. and Lehrach, H. (1993) Use of reference libraries and hybridisation fingerprinting for relational genome analysis. *FEBS Lett.* **325**, 118–122.

40. Locke, J., Raidan, G., McDermid, H., et al. (1996) Cross-screening: a new method to assemble clones rapidly and unambiguously into contigs. *Gen. Res.* **6,** 155–165.
41. Sapolsky, R. J. and Lipshutz, R. J. (1996) Mapping genomic libray clones using oligonucleotide arrays. *Genomics* **33,** 445–456.
42. Green, E. D., Mohr, R. M., Idol, J. R., et al. (1991) Systematic generation of sequence-tagged sites for physical mapping of human chromosomes: application to the mapping of human chromosome 7 using yeast artificial chromosomes. *Genomics* **11,** 548–564.
43. Arveiler, B. (1994) Yeast artificial chromosome recombinants in a global strategy for chromosome mapping. Amplification of internal and terminal fragments by PCR, and generation of fingerprints. *Methods Mol. Biol.* **29,** 403–423.
44. Aburatani, H., Stanton, V. P., and Housman, D. E. (1996) High-resolution physical mapping by combined Alu-hybridization/PCR screening: contruction of a yeast artificial chromosome map covering 31 centimorgans in 3p21-p14. *Proc. Natl. Acad. Sci. USA* **93,** 4474–4479.
45. Cox, D. R., Burmeister, M., Price, E. R., Kim, S., and Myers, R. M. (1990) Radiation hybrid mapping: a somatic cell genetic method for constructing high-resolution maps of mammalian chromosomes. *Science* **250,** 245–250.
46. Mungall, A. J., Edwards, C. A., Ranby, S. A., et al. (1996) Physical mapping of chromosome 6: a strategy for the rapid generation of sequence-ready contigs. *DNA Sequence* **7,** 47–49.
47. Haaf, T. and Ward, D. C. (1994) High resolution ordering of YAC contigs using extended chromatin and chromosomes. *Hum. Mol. Genet.* **3,** 629–633.
48. Florijn, R. J., Bonden, L. A. J., Vrolijk, H., et al. (1995) High-resolution DNA Fiber-FISH for genomic DNA mapping and colour bar-coding of large genes. *Hum. Mol. Genet.* **4,** 831–836.
49. Heiskanen, M., Karhu, R., Hellsten, E., Peltonen, L., Kallioniemi, O. P., and Palotie, A. (1994) High resolution mapping using fluorescence in situ hybridization to extended DNA fibers prepared from agarose-embedded cells. *Biotechniques* **17,** 928–933.
50. Bensimon, A., Simon, A., Chiffaudel, A., Croquette, V., Heslot, F., and Bensimon, D. (1994) Alignment and sensitive detection of DNA by a moving interface. *Science* **265,** 2096–2098.
51. Weier, H.-U. G., Wang, M., Mullikin, J. C., et al. (1995) Quantitative DNA fiber mapping. *Hum. Mol. Genet.* **4,** 1903–1910.
52. Wang, M., Duell, T., Gray, J. W., and Weier, H.-U. G. (1996) High sensitivity, high resolution physical mapping by fluorescence in situ hybridization on to individual straightened DNA molecules. *Bioimaging* **4,** 1–11.
53. Hu, J., Wang, M., Weier, H.-U. G., et al. (1996) Imaging of single extended DNA molecules on flat (aminopropyl)triethoxysilane-mica by atomic force microscopy. *Langmuir* **12,** 1697–1700.
54. Weier, H.-U. G. (2001) Quantitative DNA fiber mapping, in *Methods in Cell Biology*, vol. 64, part B, *Cytometry*, 3rd ed. (Darzynkiewicz, Z., Chrissman, H. A., and Robinson, J. P., eds.), Academic Press, San Diego, pp. 37–56.

55. Weier, H.-U. G. (2001) DNA fiber mapping techniques for the assembly of high-resolution physical maps. *J. Histochem. Cytochem.* **49,** 939–948.

56. Duell, T., Wang, M., Wu, J., Kim, U.-J., and Weier, H.-U. G. (1997) High resolution physical map of the immunoglobulin lambda variant gene cluster assembled by quantitative DNA fiber mapping. *Genomics* **45,** 479–486.

57. Duell, T., Nielsen, L. B., Jones, A., Young, S. G., and Weier, H.-U. G. (1998) Con-struction of two near-kilobase resolution restriction maps of the 5' regulatory region of the human apolipoprotein B gene by quantitative DNA fiber mapping (QDFM). *Cytogenet. Cell. Genet.* **79,** 64–70.

58. Breier, A. M., Weier, H.-U. G., and Cozzarelli, N. R. (2005) Independence of replisomes in *Escherichia coli* chromosomal replication. *Proc. Natl. Acad. Sci. USA* **102,** 3942–3947.

59. Weier, H.-U. G. (2002) Quantitative DNA fiber mapping, in *FISH Technology* (Rautenstrauss, B. and Liehr, T., eds.), Springer Verlag, Heidelberg, pp. 226–253.

60. Hsieh, H. B., Wang, M., Lersch, R. A., Kim, U.-J., and Weier, H.-U. G. (2000) Rational design of landmark probes for quantitative DNA fiber mapping (QDFM). *Nucleic Acids Res.* **28,** e30.

61. Cassel, M. J., Munné, S., Fung, J., and Weier, H.-U. G. (1997) Carrier-specific breakpoint-spanning DNA probes for pre-implantation genetic diagnosis [PGD] in interphase cells. *Hum. Reprod.* **12,** 101–109.

62. Fung, J. L., Munné, S., Garcia, J., Kim, U.-J., and Weier, H.-U. G. (1999) Reciprocal translocations and infertility: molecular cloning of breakpoints in a case of constitutional translocation t(11;22)(q23;q11) and preparation of probes for pre-implantation genetic diagnosis (PGD). *Reprod. Fertil. Dev.* **11,** 17–23.

63. Collins, C., Rommens, J. M., Kowbel, D., et al. (1998) Positional cloning of ZNF217 and NABC1: genes amplified at 20q13.2 and overexpressed in breast carcinoma. *Proc. Natl. Acad. Sci. USA* **95,** 8703–8708.

5

PRINS for Mapping Single-Copy Genes

Avirachan T. Tharapel and Stephen S. Wachtel

Summary

Primed *in situ* labeling (PRINS) is a sensitive and specific method that can be used for the localization of single copy genes and sequences too small for detection by conventional fluorescence *in situ* hybridization. By the use of PRINS, the human *SRY* gene was localized to Yp11.31-p11.32 and the *SOX3* gene to Xq26-q27. In other studies, we localized specific deletions of *RBM* and *DAZ*, candidate genes for *AZF* (azoospermia factor) to proximal Yq11.2, the *AZF* region, in an infertile male. Locus-specific oligonucleotide probes (PRINS primers) were annealed to chromosomal DNA *in situ* and extended on preparations fixed on glass slides in the presence of dATP, dCTP, dGTP, dTTP, biotin-16-dUTP, and Taq DNA polymerase. After addition of avidin-conjugated fluorophore, signals were visualized by fluorescence microscopy in metaphase spreads from patients and controls. With further development, the PRINS method may prove useful for localization of single-copy genes, in general, and for the detection of gene deletions.

Key Words: PRINS; gene localization; single-copy genes; oligonucleotide probe; primer; FISH.

1. Introduction

Chromosome abnormalities produce a variety of clinical phenotypes including malformation and mental retardation. This is a significant factor in human mortality and morbidity because some 0.7% of live-born children have congenital malformations owing to chromosome abnormalities. Fine study of the chromosomes is thus crucial in the analysis of most congenital diseases, with implications for clinical management and counseling.

For nearly 20 years, fluorescence *in situ* hybridization (FISH) has been an important and useful adjunct to the cytogenetic evaluation of chromosome rearrangements and gene deletions. FISH allows resolution of DNA sequences separated by only a few megabases in metaphase or prophase chromosomes; the low-end

From: *Methods in Molecular Biology, vol. 338: Gene Mapping, Discovery, and Expression:*
Methods and Protocols
Edited by: M. Bina © Humana Press Inc., Totowa, NJ

resolution is approximately 20 kb. However, production of locus-specific FISH probes is complicated, expensive, and time consuming, and many of the probes are not readily available.

We are concerned here with a powerful new method—primed *in situ* labeling (PRINS)—that can be used as an alternative to FISH *(1,2)*. With PRINS, simple oligonucleotide primers are employed instead of cloned DNA probes for the *in situ* labeling of chromosomal loci. According to the PRINS method, unlabeled oligonucleotide primers are annealed *in situ* to complementary target sequences on chromosomes and extended in the presence of labeled nucleotides and DNA polymerase. Chromosomal DNA acts as template for the extension and the labeled nucleotides as substrate for the DNA polymerase. The labeled extended target is visualized by fluorescence microscopy.

Although most early applications of the method dealt with repetitive sequences, PRINS could also be used to identify single-copy genes. In fact, Cinti et al. *(3)* used PRINS to identify the X-linked *factor IX* gene more than 10 years ago. That was the first published example of identification of a single-copy gene by PRINS, and it represents a significant biologic milestone, because every known gene sequence is a potential primer source. It follows that any gene or exon could be located in chromosome preparations *in situ*, provided that a unique nucleotide sequence is available. Further development and adaptation of the PRINS method could have a fundamental impact on medical genetics. The physical mapping of individual loci would now be possible in most laboratories concerned with the study of human chromosomes and the diagnosis of genetic diseases.

Here we describe the PRINS method with special emphasis on the detection of unique sequences and single-copy genes such as *SRY* and *SOX3 (4)*.

2. Materials

2.1. Lymphocyte Culture

1. RPMI 1640 culture medium (Gibco Invitrogen, Grand Island, NY) supplemented with 20% fetal bovine serum (FBS; Gibco Invitrogen). Stored at 4°C.
2. Gentamycin sulfate (Irvine Scientific, Santa Ana, CA) as a supplement to RPMI 1640 medium. Stored at 4°C and added to medium at 10 mg/mL.
3. Na heparin (Elkins-Sinn, Cherry Hill, NJ) as a supplement to RPMI 1640. Stored at 4°C and added to medium at 1000 U/mL.
4. L-Glutamine (Gibco Invitrogen) as a supplement to RPMI 1640. Stored at −20°C and added to medium at room temperature (4 mL/500 mL medium).
5. Phytohemagglutinin (PHA), lyophilized (Gibco Invitrogen). Stored at 4°C and reconstituted in distilled water at room temperature.
6. Potassium chloride and sodium citrate (Fisher Scientific, Pittsburgh, PA) for KCl-Na citrate hypotonic solution.

2.2. Collection of Cells and Preparation of Slides

1. Colcemid (Gibco Invitrogen). Purchased in lyophilized form and stored at 4°C. Used at room temperature after reconstitution in distilled water.
2. Methanol (Fisher Scientific).
3. Glacial acetic acid (Fisher Scientific).
4. Glass slides and cover glasses (Fisher Scientific).

2.3. Primed In Situ Hybridization

1. Programmable thermal cycler equipped with a flat plate for holding slides (MISHA, Shandon Lipshaw, Pittsburgh, PA) (*see* **Note 1**).
2. Dimethyl sulfoxide (DMSO; Sigma Genosys, St. Louis, MO) molecular biology or high-performance liquid chromatography (HPLC) grade.
3. Ethanol (Fisher Scientific).
4. PRINS primers (Research Genetics, Huntsville AL). The following oligonucleotide probes for the *SRY* gene *(4)* were HPLC purified and stored at −20°C (*see* **Note 2**):

   5'-GCAGGGCAAGTAGTCAACGTT-3'
   5'-AAGCGACCCATGAACGCATTC-3'
   5'-AGAAGTGAGCCTGCCTATGTT-3'
   5'-GCCGACTACCCAGATTATGGA-3'

   The following probes for the *SOX3 (SRY*-related HMG box 3) gene *(4)* were HPLC purified and stored at −20°C:

   5'-CGCATCACTGCGAACCTGTCA-3'
   5'-ACGGCCCATGAACGCCTTCAT-3'
   5'-CTGCAGTACAGCCCAATGATG-3'
   5'-AAGTCAGGAGCAGCGAAAATGG-3'

5. Trinucleotides (dATP, and so on; TSA™ kit; Tyramide Signal Amplification, for Chromogenic and Fluorescence *In Situ* Hybridization and Immunohistochemistry, NEN Life Science Products, Boston, MA; *see* **step 13**).
6. Biotin-16-dUTP (Roche Molecular Systems, Alameda, CA; formerly Boehringer-Mannheim).
7. Potassium chloride (KCl).
8. Magnesium chloride ($MgCl_2$).
9. Tris-HCl buffer (Sigma Genosys).
10. Bovine serum albumin (BSA; Invitrogen Life Technologies, Carlsbad, CA).
11. Taq DNA polymerase (Amplitaq, Perkin-Elmer, Foster City, CA) with TaqStart antibody (Clontech, Palo Alto, CA).
12. Formamide (Sigma Genosys) with NaCl and Na citrate: 20X SSC: 3 *M* NaCl, 0.3 *M* Na citrate, pH 7.0 (Invitrogen Life Technologies) to make formamide-SSC solution.
13. Tyramide Signal Amplification System (TSA™ Biotin System, NEN Life Science Products). The kit should be kept at 4°C until used, although the blocking reagent may be kept at room temperature. With proper storage, the kit is useful for 6 mo, after which the contents may become unstable.

14. Tween®20 (Sigma Genosys).
15. Triton X-100 (Sigma Genosys).

2.4. Fluorescence Microscopy: Visualization and Scoring

1. Fluorochrome-conjugated avidin (fluorescein isothiocyanate [FITC] or Texas red; Vysis, Downers Grove, IL).
2. Anti-fluorescein horseradish peroxidase (HRP; NEN).
3. Streptavidin-Texas red (NEN).
4. Counterstain: 4',6-diamino-phenylindole (Vysis).
5. Microscope equipped for fluorescence microscopy and an image capture system. An Olympus B-Max, U-M510 with triple bandpass filters, Red, Green and DAPI Bandpass, and Triple Bandpass Filter Set, DAPI/Green/Red (Vysis) was used here.

3. Methods

The basic method consists of four steps: (1) DNA denaturation, (2) annealing of primers, (3) chain extension in the presence of labeled nucleotides and Taq DNA polymerase, and (4) detection of the newly synthesized labeled DNA by use of antibodies complexed to fluorescent dyes.

3.1. An Overview of the General Procedures

1. A typical PRINS reaction mixture has a volume of 50 µL and contains: 200 to 500 pmol of primer, 0.2 mM dATP, dCTP, and dGTP, 0.02 mM dTTP, 0.02 mM bio-16-dUTP, digoxigenin-11-dUTP, or fluorescein dUTP, 50 mM Tris-HCl (pH 8.3), 1.5 mM MgCl$_2$, 50 mM KCl, 0.01% BSA, and 2 Us Taq polymerase (*see* **Note 3**).
2. The reaction mixture is placed on the denatured chromosome preparation.
3. Denaturation is accomplished by immersing slides in 70% formamide-2X SSC for 2 to 3 min at 72°C and then dehydrating them through a series of cold ethanol washes. After the slides are air-dried, the reaction mixtures are layered onto the slides and covered with cover glasses that are sealed in place with rubber cement.
4. The slides are put on a flat plate programmable thermal cycler and maintained at annealing temperatures for 5 to 10 min followed by extension for 20 to 30 min at 70 to 72°C. The reaction is stopped by immersing the slides in 500 mM NaCl/50 mM EDTA, pH 8, for 5 min at 72°C.
5. Labeled sequences are detected by immunocytochemistry and fluorescence microscopy.
6. Biotin-complexed fragments can be visualized by the use of fluoresceinated avidin and digoxigenin by the use of fluoresceinated digoxigenin antibody. The chromosome preparations are counterstained with 4,6-diamidino-2-phenylindole (DAPI) or propidium iodide and scored under ultraviolet (UV) light.
7. Blood lymphocytes, a ready source of DNA for PRINS testing, can be induced to divide in culture by addition of PHA. This allows preparation of mitotic spreads on glass microscope slides for evaluation by PRINS and other procedures, such as FISH). For these procedures, whole venous blood is drawn into heparinized glass

tubes. The detailed procedure for lymphocyte culture in advance of PRINS is given in **Subheadings 3.2** to **3.6.** below.

3.2. Preparing the Culture Medium

1. Add 100 mL FBS to 400 ml RPMI 1640 in a separate container. Mix well. Store at 4°C and prepare fresh mixture every 30 d.
2. Into each bottle of medium add: 3 mL of Na heparin, 2 mL of penicillin-streptomycin, 2 mL of gentamycin sulfate, and 4 mL of L-glutamine.

3.3. Preparing the Phytohemagglutinin

1. Reconstitute one bottle of PHA with 10 mL of distilled water.
2. The PHA should be prepared fresh every 30 d and stored at 4°C.
3. Preparation of the KCl-Na citrate hypotonic mixture:
   a. Add 0.560 g of KCl to 100 mL deionized or distilled water to make 0.075 M KCl solution.
   b. Add 0.8 g Na citrate to 100 mL deionized or distilled water to make 0.8% solution, and mix the two solutions just before use: 2:1 (v:v), KCl/Na-citrate.
   c. Store at room temperature and prepare a fresh mixture every 7 d.

3.4. Initiation of Lymphocyte Culture

1. Two cultures are initiated for each subject.
2. For each culture, add 10 mL of culture medium (*see* **Subheading 3.1.** above) to two 15-mL centrifuge tubes.
3. Using a Pasteur pipet, inoculate each centrifuge tube with 12 to 15 drops of whole blood. Each centrifuge tube represents a separate lymphocyte culture. Label the tubes.
4. Add 0.15 mL of PHA to each tube and mix well.
5. Incubate the cultures at 37°C for 72 h. (In general, the cultures may remain undisturbed for the full 3-d period, although some groups gently agitate the cultures several times each day.)

3.5. Harvesting the Lymphocytes

1. Using an 18-gage syringe needle, add 5 drops of reconstituted colcemid (10 μg/mL) to each culture. The colcemid, which prevents formation of the mitotic spindle, thereby "freezing" the cells in metaphase, should be discarded after 1 mo of use.
2. After mixing well, incubate the cultures for an additional 60 to 120 min at 37°C in the colcemid.
3. Centrifuge at 250g for 10 min.
4. Remove the supernatant and break up the pellet by agitation with a vortex mixer.
5. Add 10 mL of hypotonic KCl-Na citrate mixture to each culture. Carefully resuspend the cells by gently inverting the centrifuge tubes manually. Allow the tubes to stand at room temperature for 30 min.

6. Gently resuspend the cells again by inverting the tubes. Then add 2 mL of freshly prepared (3:1) methanol/glacial acetic acid fixative directly to the contents of each tube. Mix the contents by inverting the tubes, and centrifuge the tubes again at 250*g* for 10 min.
7. Remove all the supernatant from each culture. Gently thump the tubes to break up the cellular pellets.
8. Add 10 mL of fresh fixative to each culture. Resuspend the cell pellets by inverting tubes, and allow the cultures to stand at room temperature for 10 min, and centrifuge.
9. Repeat the preceding step.
10. The next steps are performed in a "harvesting room" with a humidifier. For maximal chromosome spreading, adjust the humidity to 55 to 65% (*see* **Note 4**).
11. Set a hot plate to 65°C, and check the temperature with a surface thermometer.
12. Centrifuge the cell suspension at 250*g* for 10 min.
13. Remove the supernatant from the centrifuge tubes, and add methanol/glacial acetic acid fixative to the pellet drop by drop until the suspension becomes semiclear. The amount of fixative required will depend on the size of the pellet. The final cell concentration may have to be adjusted after evaluation of the first slide.

3.6. Preparing Slides for PRINS

High-quality chromosome preparations are a key factor for obtaining satisfactory results after hybridization on glass slides.

1. Clean microscope slides are placed in a slide tray containing deionized water. The slides are chilled in a refrigerator.
2. One cold slide at a time is removed from the slide tray and maintained in a slanting position on a stand or a device (at about a 10°–20° angle).
3. Drop or place 40 to 50 μL of cell suspension on the slide. Allow the cell suspension to roll down the slide. Wipe the back of each slide and shake off excess fixative and water. Label the slides, and place them on a hot plate at 65°C for 2 to 3 min.
4. Store slides at room temperature for 24 h after which they are ready for the PRINS procedure.

3.7. Primed In Situ Hybridization: Standard Protocol

1. Immerse slides in 0.02 *N* HCl for 20 min (*see* **Note 5**).
2. Now immerse slides in 70% formamide/SSC, pH 7.0, for 2 min at 72°C, to denature chromosomal DNA.
3. Dehydrate slides, by passage through a cold ethanol series, 70%-85%-100% EtOH, 5 min each. Air-dry.
4. Prepare reaction mixture in a final volume of 40 μL containing: 50 pmol of each primer (*see* **Note 6**), 0.2 m*M* each dATP, dCTP, and dGTP, 0.02 m*M* dTTP, 0.02 m*M* biotin/16-dUTP, 50 m*M* KCl, 10 m*M* Tris-HCl, pH 9.0, 2 m*M* MgCl$_2$, 0.01%

BSA, and I U *Taq* DNA polymerase with TaqStart antibody (*see* **Note 7**). Pipet 40 µL reaction mixture onto the freshly prepared slide (*see* **Note 8**).

5. Cover working area completely with a cover glass. Seal the ends of the cover glass in place with a thinly applied layer of rubber cement.

6. Incubate slides. Our incubations are carried out on a programmable thermal cycler equipped with a flat plate for slides (MISHA, Shandon Lipshaw; *see* **Note 1**). The program consists of one cycle of 10 min at annealing temperature (55–75°C) with an additional 30 min at 72°C for extension (*see* **Note 9** for computation of annealing temperature).

7. After extension, the slides are removed from the cycler, the cover glasses are removed, and the slides are washed in 0.4X SSC at 72°C for 2 min to stop the reaction (*see* **Note 10**).

8. In our studies, biotin-labeled nucleotides are detected with the TSA™ Biotin System (*see* **Subheading 3.8.**).

3.8. Signal Amplification

1. Prepare Biotinyl Tyramide Stock Solution. Reconstitute biotinyl tyramide (amplification reagent) by addition of 0.3 to 1.2 mL DMSO (amount of DMSO will depend on the particular NEN kit that is used: 0.3 mL for 50–150 slides; 1.2 mL for 200–600 slides). Since DMSO freezes at 4°C, it is necessary to thaw the stock solution after removal from refrigerator.

2. For each test, dilute the stock solution 1:50 with 1X Amplification Diluent to prepare the working solution; 100 to 300 µL of working solution are needed for each slide.

3. Prepare TNT washing buffer: 0.1 *M* Tris-HCl, pH 7.5, 0.15 *M* NaCl, and 0.05% Tween 20. The manufacturer advises that PBS may be used as an alternative buffer and that 0.3% Triton X-100 may be used instead of 0.05% Tween 20.

4. Prepare TNB blocking buffer: 0.1 *M* Tris-HCl, pH 7.5, 0.15 *M*. NaCl, and 0.5% Blocking Reagent (in kit). According to the manufacturer's protocol, the blocking reagent should be added slowly to the buffer in small amounts while stirring. The mixture should be heated to 60°C with continuous stirring (*see* **Note 11**). The blocking reagent should be dissolved completely (this may take several hours). After preparation, store at −20°C.

5. After hybridization, block slides by incubation with 100 to 300 µL of TNB buffer. This may be done for 30 min in the water-filled chamber of the temperature cycler with biotin-labeled probes: 100 to 300 µL SA-HRP (streptavidin-horseradish peroxidase: from the TSA kit) diluted 1:100 in TNB buffer. Fluorescein-labeled probes may be substituted: 100 to 300 µL antifluorescein-HRP diluted 1:250 in TNB buffer. Optimal concentrations of HRP-labeled reagents should be determined for individual laboratories.

6. Wash slides in TNT buffer with agitation, 3 times for 5 min at room temperature.

7. Using a pipet, place 100 to 300 µL working solution onto each slide. Maintain slides at room temperature for 5 to 10 min.

8. Repeat washing step (**step 6**).

3.9. Fluorescence Microscopy and Visualization

1. To each slide, add 100 to 300 µL streptavidin-fluorophore conjugate diluted in TNB buffer; use dilution recommended by manufacturer (streptavidin-Texas red is used at 1:500).
2. Place slides in a humidified chamber for 30 min at room temperature.
3. Wash slides in TNT buffer with agitation, 3 times for 5 min at room temperature.
4. For background staining of chromosomes, place 2 drops DAPI II on slide and mount for microscopy. Blot excess DAPI II, add cover glass, gently seal ends of cover glass with rubber cement, and refrigerate at 4°C for 30 to 60 min (*see* **Note 12**).

4. Notes

1. Annealing and extension may be accomplished on thermoblocks, or even in metal containers suspended in hot water baths (*5*), but temperature is critical and must be carefully controlled.
2. In the authors' experience, signal is increased by use of multiple primers for a single locus and by single-step annealing and extension.
3. An unamplified slide without TSAG reagents and an amplified slide without primer should be included as controls for each hybridization.
4. Slides should be kept moist during the PRINS procedure. If a humidified chamber is not available, cover slides with a damp paper towel in a closed box. If a humidifier is available, maintain humidity at 55 to 65% for optimal chromosome spreading (*see* **Subheading 3.5., step 10**).
5. Treatment of slides with 0.02 N HCl removes loosely bound protein, thereby rendering DNA more accessible to the primers.
6. For each study, the primer concentration should be optimized. New England Nuclear recommends a 10-fold reduction in "probe" (primer) concentration as optimal. This is a critically important step in PRINS, as improper concentration of probe can obviate development of the hybridization signal.
7. TaqStart monoclonal antibody binds Taq DNA polymerase, thereby minimizing nonspecific amplification and formation of primer dimers.
8. Reagents should completely cover cells or metaphase spreads on slides.
9. After counting the A, C, G, and T nucleotide residues of the primers, annealing temperatures are computed by use of either of the following formulas:

$$T_M = 69.3 + 0.41 \ (\% \ G + C) - 650 / L$$

where L = the length of the primer, i.e., the total number of nucleotides in the primer.

$$T_M = 4 \ (G + C) + 2 \ (A + T)$$

When different temperatures are obtained, the results may be averaged. In general, good results are obtained with annealing temperatures between 55°C and 75°C. Higher temperatures increase specificity of annealing.

10. Background staining is minimized by stringent washing of slides in SSC.
11. Innis and Gelfand (*6*) note that at 20°C, Tris-HCl buffer has a pK_a of 8.3, and a ΔpK_a of –0.021/°C. So the actual pH of Tris-HCl may vary during thermal cycling.

12. Low signal may be corrected by titration of HRP conjugate to optimize concentration, by increasing concentration of amplification reagent (or incubation time), or by addition of a step to optimize penetration of reagents *(7)*. Excessive background staining may be reduced by decreasing concentration of HRP conjugate or primers, by increasing endogenous peroxide quenching, by filtration of buffers, or by increasing the number or length of washes *(7)*.

References

1. Pellestor, F., Girardet, A., Andréo, B., and Charlieu, J.-P. (1994) A polymorphic alpha satellite sequence specific for human chromosome 13 detected by oligonucleotide primed *in situ* labelling (PRINS). *Hum. Genet.* **94,** 346–348.
2. Pellestor, F., Girardet, A., Lefort, G., Andréo, B., and Charlieu, J. P. (1995) Use of the primed in situ labelling (PRINS) technique for a rapid detection of chromosomes 13, 16, 18, 21, X and Y. *Hum. Genet.* **95,** 12–17.
3. Cinti, C., Santi, S., and Maraldi, N. M. (1993) Localization of single copy gene by PRINS technique. *Nucleic Acids Res.* **21,** 5799–5800.
4. Kadandale, J. S., Tunca, Y., and Tharapel, A. T. (2000) Chromosomal localization of single copy genes SRY and SOX3 by primed *in situ* labeling (PRINS). *Microb. Compar. Genom.* **5,** 71–74.
5. Koch, J., Hindkjær, J., Kølvraa, S., and Blund, L. (1995) Construction of a panel of chromosome-specific oligonucleotide probes (PRINS-primers) useful for the identification of individual human chromosomes in situ. *Cytogenet. Cell Genet.* **71,** 142–147.
6. Innis, M. A. and Gelfand, D. H. (1990) Optimization of PCRs, in *PCR Protocols, A Guide to Methods and Applications* (Innis, M. A., Gelfand, D. H., Sninsky, J. J., and White, T. J., eds.), Academic Press, San Diego, CA, pp. 3–12.
7. NEN® Life Science Products, *RENAISSANCE* ® "TSA™ Biotin System" Laboratory Manual.

6

VISTA Family of Computational Tools for Comparative Analysis of DNA Sequences and Whole Genomes

Inna Dubchak and Dmitriy V. Ryaboy

Summary

Comparative analysis of DNA sequences is becoming one of the major methods for discovery of functionally important genomic intervals. Presented here the VISTA family of computational tools was built to help researchers in this undertaking. These tools allow the researcher to align DNA sequences, quickly visualize conservation levels between them, identify highly conserved regions, and analyze sequences of interest through one of the following approaches:

- Browse precomputed whole-genome alignments of vertebrates and other groups of organisms.
- Submit sequences to GenomeVISTA to align them to whole genomes.
- Submit two or more sequences to mVISTA to align them with each other (a variety of alignment programs with several distinct capabilities are made available).
- Submit sequences to Regulatory VISTA (rVISTA) to perform transcription factor binding site predictions based on conservation within sequence alignments.
- Use stand-alone alignment and visualization programs to run comparative sequence analysis locally.

All VISTA tools use standard algorithms for visualization and conservation analysis to make comparison of results from different programs more straightforward. The web page http://genome.lbl.gov/vista/ serves as a portal for access to all VISTA tools. Our support group can be reached by email at vista@lbl.gov.

Key Words: VISTA; comparative genomics; global DNA alignment; visualization; genome browser; sequence analysis.

1. Introduction

Comparison of the DNA sequences of different species is a powerful method for decoding genomic information, since functional sequences tend to evolve at a slower rate than nonfunctional sequences. Analysis of conservation helps to identify coding sequences *(1–4)* and conserved noncoding elements with regulatory

From: *Methods in Molecular Biology, vol. 338: Gene Mapping, Discovery, and Expression: Methods and Protocols*
Edited by: M. Bina © Humana Press Inc., Totowa, NJ

functions *(5,6)*, and also to determine which sequences are unique for a given species. Several groups have aligned entire vertebrate genome assemblies, such as human, mouse, dog, chicken, and others, thus allowing for comprehensive statistical data on the patterns of DNA conservation among these species *(7–10)*. Recently published reviews on comparative sequence analysis *(11–15)* describe this fast growing field and present computational resources available for a wide range of biological investigations.

All VISTA tools *(16–19)* utilize global (or global/local) alignment strategy that assumes monotonous end-to-end correspondence of DNA intervals. The AVID *(20)* and LAGAN *(21)* programs allow for fast global alignment of up to megabase-long sequences. Alignment is visualized as a continuous curve showing the level of conservation in a moving window of a pre-defined length. A global alignment strategy together with an additional, mapping component as a first step was also used in the pairwise and three-way alignment of whole genome assemblies performed in our group *(22,23)*. A different web-accessible software package for comparative genomics, PipMaker *(24–26)*, is based on a local alignment approach. Comparison of PipMaker and VISTA (14) as well as comparative assessment of several alignment methods and programs *(27)* were published elsewhere.

The web page http://genome.lbl.gov/vista serves as a portal for access to the suite of the VISTA tools. VISTA Browser gives access to precomputed pairwise and multiple alignments of whole-genome assemblies of different groups of organisms. The three main VISTA servers (GenomeVISTA, mVISTA, and rVISTA) offer a range of options for comparative analysis of submitted by a user sequences. GenomeVISTA aligns and compares a single sequence (draft or finished) with whole genomes. mVISTA is designed for comparison of orthologous sequences from different species. rVISTA *(19)* takes into consideration conservation among species to improve prediction of transcription factor binding sites (TFBS). VISTA pages offer extensive help on selecting a type of analysis, finding optimal parameters for a particular project, and navigating the web site.

The VISTA web site also provides access to the results of comparative analysis of specific sets of genes, as well as other relevant internal and external resources. In addition to a description of these services, widely used by the biological community, a short overview of recently developed tools and techniques is given. The capabilities of our programs are illustrated by analyzing an arbitrarily selected disease-associated gene, *RUNX1*, located on human chromosome 21.

2. Using the VISTA Browser

The VISTA Browser is a Java applet that allows the researcher to interactively visualize results of VISTA comparative sequence analysis. It is used by

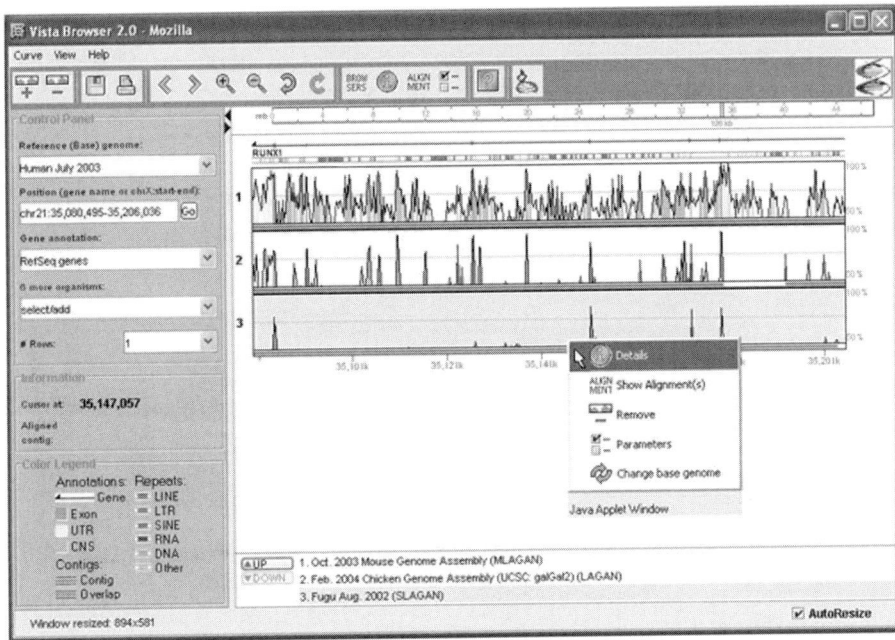

Fig. 1. VISTA Browser is a tool that allows the researcher to visualize alignments of long sequences in an intuitive way. Conserved regions are colored according to the annotation, which is shown above the conservation curve. Whole genome alignments can be navigated quickly and easily. Multiple pairwise alignments can be viewed at the same time to assist in analysis.

the VISTA servers described in **Subheadings 3** to **5** and also displays a large and constantly increasing number of precomputed whole-genome alignments. This section describes browser navigation and the kinds of information a researcher may extract from it. The browser can be launched through links provided in mVISTA and GenomeVISTA results or by going to the Berkeley Genome Pipeline gateway at http://pipeline.lbl.gov (also accessible through the VISTA home page) and selecting a genome of interest to examine whole genome alignments. **Figure 1** shows a sample screenshot of the VISTA Browser. The following discussion explains how to interpret the display, navigate the alignments, and extract the results of the conservation analysis.

2.1. Understanding the Display

2.1.1. VISTA Graph Display

The central window of the browser contains several important blocks of information. The most visible of these are the VISTA graph panels, which show the so-called peaks and valleys graphs. These graphs represent percent conservation between aligned sequences at a given coordinate on the base sequence (*see* **Note 1** for how the curve is calculated). Multiple pairwise alignments that share the base sequence can be displayed simultaneously, one under another. The percent identity bounds are shown to the right of every row. These bounds can be adjusted by clicking on the curve to select it, and then clicking the "Parameters" button. The graphs are numbered, you can identify each graph in the list underneath the VISTA panel.

2.1.2. Annotation

The browser shows the base genome annotation directly above the curves. Arrows signifying genes are drawn above the graphs, pointing in the direction of the gene. Exons and UTRs are marked on the gene as colored blocks. Gene names appear underneath the arrows. Repeats are shown directly above the plot, colored according to the legend in the lower left-hand corner of the display. The annotation can be changed in the control panel, as described in **Subheading 2.2.1.**

2.1.3. Conserved Regions

Regions of high conservation are colored according to the annotation as exons (dark blue), UTRs (light blue), or noncoding (pink). The thresholds of conservation that determine what gets colored, as well as minimum and maximum y-axis values, can be easily adjusted by right-clicking on the curve and selecting the "Parameters" option (*see* **Note 2**).

2.1.4. Contigs

The thick gray or red lines under the plot show where the second species was aligned to the base sequence. To find out the name of the contig or chromosome that was aligned, one can hover the mouse cursor over the graph—the name will be displayed in the information panel on the left. Red lines indicate that multiple regions from the second sequence were aligned to this base sequence location, and several alignments are overlapped here (*see* **Note 3**). Gray lines indicate an unambiguous alignment.

2.1.5 Location on the Base Sequence

At the very top of the central window there is a long bar representing the whole contig or chromosome that one is browsing. A moveable red rectangle on top of that bar represents the location of the region one is looking at on that contig or chromosome. The rectangle can be stretched or moved to change the location being browsed.

2.2. Navigating the Base Sequence

2.2.1. Using the Control Panel

The green Control Panel on the left features a drop-down box called "Reference (Base) genome," which lists all available base genomes. Changing the base genome in this box will switch over the whole browser; the current curves and annotations will disappear, and the browser will go to the newly selected genome.

The next field specifies the genome segment displayed by the browser. To change positions, one can enter specific coordinates on the genome, such as chr9:102,923,121-103,070,274 or contig1080:1152-7781, a gene name, or a contig name. The browser will go directly to the specified location if it is unambiguous, or display a dialog box if there are multiple matches. Partial gene names are accepted.

Some genomes may have multiple annotation sets available. To switch between them, use the drop-down "Gene Annotation" menu in the control panel. Curves will be instantly recomputed using the new annotation data.

2.2.2. Zooming In and Out

To zoom in on an interesting region, position the mouse cursor over the start of the area you wish to see closer, and then click and drag, selecting a section of the curve, just as you select text in a word processor. The browser will zoom in once you let go of the mouse button. The "Escape" key cancels the selection. One can also zoom in and out using the buttons on the toolbar at the top of the screen.

2.2.3. Adding and Removing Curves

To add or remove curves, you can use the "+" and "–" buttons on the toolbar, or the drop-down menu in the control panel. You will be able to specify visualization and conservation analysis parameters when you add the curve, but you can also come back and change these parameters later by selecting the curve of interest and clicking the "settings" button.

2.2.4. Switching Base Sequences

You can change which of the aligned sequences is used as base by right-clicking on a curve and selecting the "change base genome" option. If only one alignment is displayed, you will be immediately taken to the corresponding location on the second sequence. Otherwise, you will be presented with a list of locations that mapped to the region you are looking at, and you will be able to select the one in which you are interested.

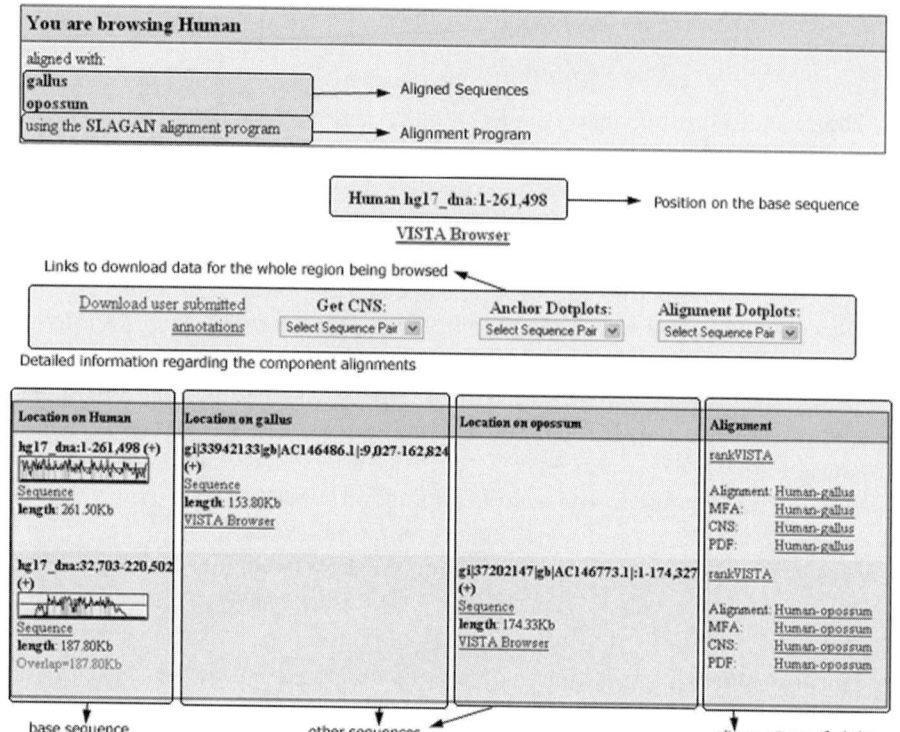

Fig. 2. The detailed information display ("Text Browser") provides access to the data underlying a VISTA graph. From this page, one can see all the component alignments that compose the curve seen in the VISTA browser, get the specific coordinates of the aligned segments on original sequences, and download the alignments, annotations, conserved regions, and other pertinent information.

2.3. Extracting Detailed Information

Once you have found regions you are particularly interested in, you may want to see the coordinates of conserved regions, their sequences, their underlying alignments and annotations, and other detailed information. All these data can be easily accessed by right-clicking on a curve and selecting the "Details" option or by clicking the "i" button on the toolbar. A new browser window will open (*see* **Note 4**), which shows detailed information about all the underlying alignments that form the curve you are looking at (**Fig. 2**). The detailed view is also referred to as Text Browser.

At the top of the Text Browser page there is a banner that displays the names of the aligned organisms. The sequence listed in the darker header area is acting as base. This banner also lists the program used to align your sequences. In some cases, a pairwise alignment might be part of a multiple alignment, in which case the rest of the organisms will be present here as well.

Underneath the top banner there is the navigation area, which shows the co-ordinates of the currently displayed region and offers a link back to the VISTA Browser and a link to a list of all conserved regions found. When available, there will also be a link to something called "VISTA Tracks," which are VISTA curves, like the ones you see in the browser, integrated into the University of California at Santa Cruz (UCSC) Genome Browser. This tool can be very use-ful, since the UCSC browser has many additional annotation tracks and fea-tures. Links to other outside browsers may also appear here. In addition, if Shuffle-Lagan (*see* the description in **Subheading 4**) was used as the align-ment program, there will be a link to download dot-plots of the alignments produced.

The main table follows, which lists each alignment that was generated for the base organism. Each row is a separate alignment. Each column, except the last one, refers to the sequences that participate in the alignment. The last column contains information pertaining to the alignment as a whole. The first cell of each row also contains a preview of the VISTA plot of this particular alignment, which allows one to quickly evaluate the quality of this alignment and to see alignment overlaps.

By looking at a row in this table, you can see which genomic interval of each organism aligned to which. The "Sequence" links will return a FASTA-format-ted segment that participates in the alignment. Clicking on the "VISTA Browser" links will launch the VISTA browser with the associated species as the base.

The last column provides links to the alignments in human-readable and MFA (multi-FASTA alignment) formats, a list of conserved regions from this align-ment alone, and links to pdf plots of this alignment alone. If the region being examined is 20 Kbp or shorter, rVISTA analysis can be performed, and a link to rVISTA will also be displayed here. You will notice that when the conserved regions are displayed, their lengths are actually web links. Clicking on them will pull up the conserved sequences from both of the participating organisms.

3. Using GenomeVISTA

GenomeVISTA is a web utility that allows the researcher to align a sequence of interest to a user-selected complete genome assembly.

3.1. Submission Instructions

The submission process consists of several simple steps.

1. Prepare a simple FASTA-formatted text file (*see* **Note 5**) that contains the sequence, or find its GenBank accession number.

2. Go to the VISTA website, http://genome.lbl.gov/vista and click on the GenomeVISTA link.

3. Enter the sequence (you can paste it in as FASTA-formatted text, upload a FASTA file, or enter the GenBank number).

4. Choose the base genome from the drop-down list.

5. (Optional) If you wish, you can enter your e-mail in the optional section. The results of the analysis will be sent to you as soon as the computation is finished. We highly recommend doing so, since that way it will be easy to refer back to your results without having to rerun the analysis. You can also enter an identifying name for the sequence here—instead of the generic "Sequence 1," you can assign something meaningful, such as "GibbonMRNA."

6. Click the "Submit" button. A link to the results, available once the computation is complete, will be displayed on screen—do not lose it if you did not enter your email!

3.2. Results

After your query is processed by the GenomeVISTA server, you will receive an e-mail with a link to the results. (If you did not enter the e-mail, it's the same link as the one displayed after a successful submission.) The page pointed to by this link lists every base genome location to which your sequence is aligned and provides two ways of viewing the resulting alignments—through the VISTA Browser, described in **Subheading 2.**, and directly through the Text Browser, described in **Subheading 2.3.** VISTA Browser can display available precomputed alignments for different genomes in parallel with the alignment obtained by GenomeVISTA. When available, a link to the VISTA Track (VISTA image on top of the UCSC Genome Browser) is also provided.

3.3. Sample Analysis

In this example, we use GenomeVISTA to align and analyze a draft opossum sequence from GenBank, filed under accession number AC146773. This sequence is composed of five ordered pieces. We align this sequence to the May 2004 release of the Human Genome assembly using the procedure described above. After GenomeVISTA analysis is performed on the sequence, the results page shows that all the pieces aligned to one general location on the human genome, namely, chr21:35,139,670-35,268,797. Click on the VISTA Browser link to view the plot of the alignment (**Fig. 3**). In addition to the opossum-human alignment, a precomputed human-mouse alignment is automatically shown to assist in the analysis. One can see that three of the opossum contigs had significant similarity to the human gene *RUNX1*, a disease-related gene that encodes for the α-subunit of core binding factor (CBF), a transcription factor that binds to the core element of many enhancers and promoters *(28,29)*.

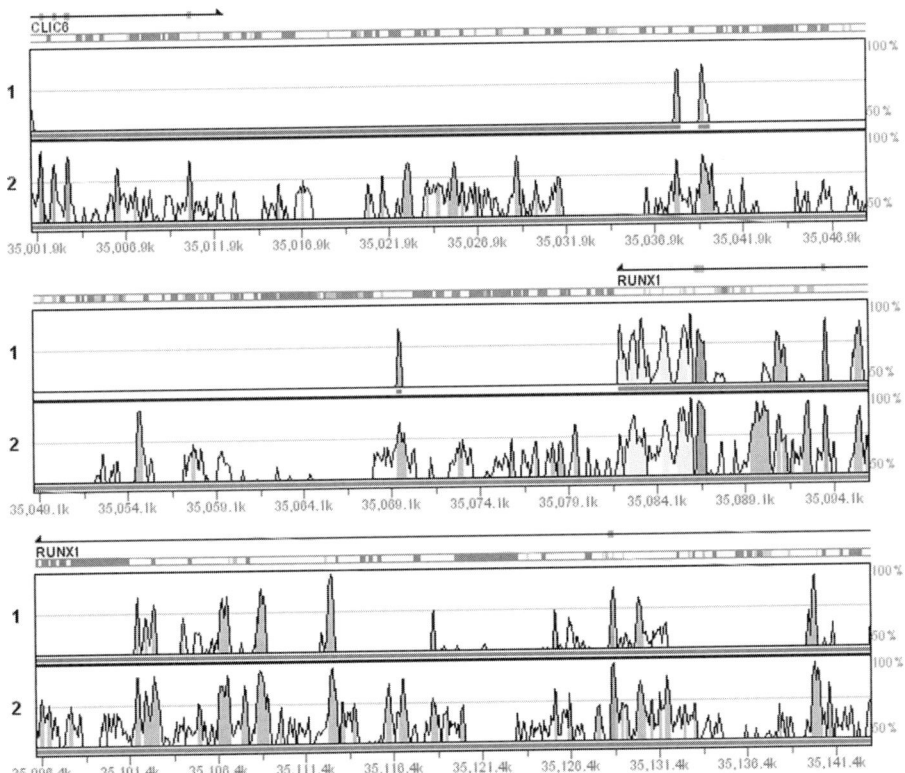

Fig. 3. Five opossum contigs submitted to the GenomeVISTA server are aligned to the human gene *RUNX1*, a disease-related gene encoding for the α-subunit of core binding factor. GenomeVISTA alignment is marked 1 on the figure. The areas matched by opossum sequences are marked by the gray bars underneath the conservation curve. Conserved regions are colored according to the annotation, the darker areas denoting exons. Human-mouse whole genome alignment, marked 2 on the figure, appears underneath the human-opossum curve to provide additional information for the researcher. Other whole genome alignment windows can be added if needed for comparison.

Compared with the human-mouse alignment, the peaks on the human-opossum curve are fairly sparse; this can mean that opossum is a good organism to use in analyzing this section of the human genome—it is phylogenetically far enough to reduce the noise factor but close enough to clearly identify regions of interest (*see* **Note 6**). It is clear that all the annotated exons are conserved in the opossum sequence; interestingly, there are also other highly conserved intervals in the intronic region of the gene. These peaks allow the researcher to narrow down the search for regulatory elements or other regions of functional importance. To extract the coordinates of these conserved regions, one only needs to click the "I" button and then click on the CNS link on the Text Browser.

4. Using mVISTA

mVISTA is a set of programs for comparing DNA sequences up to megabases long from two or more species and visualizing these alignments along with annotation information. It consists of two components—an alignment program (several are available) and a visualization module. The AVID program produces global pairwise alignments and can work with finished sequences, as well as with sequences in a draft format. LAGAN produces global pairwise as well as multiple alignments of finished sequences. Shuffle-LAGAN generates global pairwise alignments of finished sequences (*see* description of the method in **ref. *30***), taking into account rearrangements.

The stand-alone versions of the programs can be downloaded from the web site (*see* **Subheading 6**); here we describe how to use the web server interface.

4.1. Submission Instructions

1. Prepare sequence files. mVISTA accepts one file per sequence, plain text FASTA files (*see* **Note 5**), or GenBank identifiers. At the time of this writing, the maximum sequence size is 2 megabases, and the total maximum size of all the sequences is 10 megabases.
2. Go to the VISTA web site at http://genome.lbl.gov/vista and click on the mVISTA button.
3. Enter the number of sequences you wish to align (the server accepts up to 100 sequences), and click "submit." A form will now be generated that allows you to enter all the sequences you need.
4. Enter the e-mail address you wish the results to be sent to, and the files or GenBank identifiers of the sequences to be aligned.
5. (Optional) Select the alignment program.
6. (Optional) Name sequences, provide annotations in the format described at http://genome.lbl.gov/vista/mvista/instructions.shtml, and specify the RepeatMasker options. A check box is provided for sequences to be reverse-complemented.
7. (Optional) At the very bottom of the form is a check box for submitting the sequences to rVISTA transcription factor binding site analysis. This service is described in **Subheading 5**.
8. (Optional) You can choose to use translated anchoring in LAGAN/Shuffle-LAGAN, which can improve the alignment of distant homologs.
9. (Optional) Input a pairwise phylogenetic tree for the sequences (used by LAGAN), for example **(((human baboon) (mouse rat)) chicken)**. If you do not enter a tree, it will be calculated by our program. The names that you use in the phylogenetic tree should exactly match the names given to the submitted sequences.

4.2. Results

Once the submitted sequences are aligned, you will receive an e-mail with a link to the results of the computation. The results page has a table, each row of which corresponds to one of the submitted sequences. If the sequence aligned to any of the other sequences, its row will have links to the alignments in Text Browser (the detailed information window described in **Subheading 2.3.**), VISTA Browser, and a simple PDF graph of the alignments. If you wish, you can adjust the visualization and conservation parameters for any pair of sequences involved in the alignments by clicking the link directly underneath the table. The meaning of all of the adjustable parameters is described in **Note 2**.

4.3. Sample Analysis

For this example, we show how to align opossum, chicken, and human sequences for the *RUNX1* gene. First, prepare the sequences. Go to GenBank and save sequences with the accession numbers AC146773 and AC146486 as plaintext FASTA files. The first one should be saved as "opossum.fasta" (it is the *D. virginiana* clone used earlier to demonstrate GenomeVISTA) and the second as "chicken.fasta."

To obtain the human sequence, go to the NCBI web site, click on "Map Viewer" in the right-hand column, select "Homo Sapiens" from the drop-down menu, and enter "RUNX1" in the search field. Click on the "Genes Seq" link next to the RUNX1 map element, then on "dl" link in the highlighted line, and then "Display." To save the sequence, click the "Send" button.

Submit the sequences for mVISTA analysis using the procedure described above. Make sure to name the sequences (opossum, chicken, and human), and provide the human annotation. Once you receive an e-mail with a link to the results, view the alignments in VISTA Browser using the human sequence as base. You can now analyze the aligned sequences—look for conserved areas, consider the implications of the differences between the human-chicken and human-opossum alignments, and so on. It is frequently useful to adjust the conservation analysis parameters (*see* **Note 6**). **Figure 4** shows the results of this mVISTA analysis. The top two rows show what the human-chicken and human-opossum alignments look like when one is using the default parameters. The bottom two rows show the same alignments visualized using a 50-bp calculation window, a 50-bp minimum conserved region width, and an 80% minimum conservation identity.

5. Using rVISTA

The rVISTA (regulatory VISTA) is designed to predict TFBS in the submitted sequences by combining a search of the major transcription binding site database TRANSFAC Professional from Biobase (http://www.gene-regulation.com/) with comparative sequence analysis. It can be used directly or through links in mVISTA, Genome VISTA, or VISTA Browser.

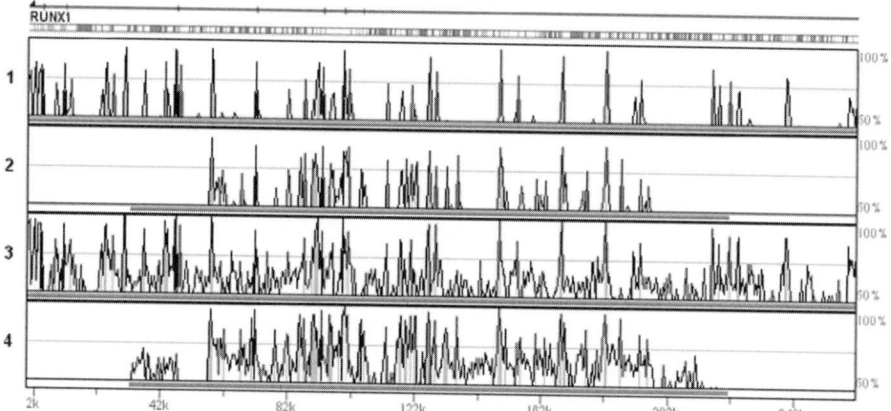

Fig. 4. Human, opossum, and chicken sequences submitted to the mVISTA server are aligned with each other, and the resulting alignment is visualized using the human sequence as base. Annotation for the human sequence, provided during the submission process, is shown above the curves. The first two curves that appear show human-opossum and human-chicken alignments graphed using the default parameters (70%/100 bp conservation, 100-bp calculation window). The third and fourth curves show the same alignments graphed using a 50-bp calculation window, a 50-bp minimum conserved region width, and an 80% minimum conservation identity. Reducing the size of the calculation window can be useful when annotating phylogenetically distant species.

5.1. Submission Instructions

The beginning of the rVISTA submission is the same as for mVISTA (*see* **Subheading 4.2.**). It is important to provide annotation files, because rVISTA is searching for TFBS in noncoding intervals of submitted sequences. The rVISTA box should be checked at the bottom of the mVISTA submission form to start the analysis. rVISTA will perform a separate analysis for each pairwise alignment produced.

With the rVISTA box checked, clicking "Submit" will take you to a new, rVISTA-specific form. This form is also accessed through the rVISTA link in the Text Browser. Here you can select from three different options for finding candidate TFBS. These options are: TRANSFAC matrices, user-defined consensus sequences, or user-defined matrices.

If you choose to use the TRANSFAC matrices, you can select from several options to refine your search. The line of check boxes allows the user to select the type of organism being analyzed (vertebrates, plants, nematodes, insects, fungi, or bacteria). You can also select one of the following options for cutoff Transfac matrix parameters for predicting TFBS: (1) recommended threshold (appropriate in most cases); (2) cutoff to minimize false-positive matches (minFP); (3)

Table 1
IUPAC One-Letter Codes

Symbol	Name	Symbol	Name
A	Adenine	N	A or G or C or T
B	G or T or C	R	G or A
C	Cytosine	S	G or C
D	G or A or T	T	Thymine
G	Guanine	V	G or C or A
H	A or C or T	W	A or T
K	G or T	Y	T or C
M	A or C		

cutoff to minimize false-negative matches (minFN); or (4) cutoffs for core and matrix similarity. Details on these parameters are provided at the TRANSFAC web site http://www.gene-regulation.com/.

After cutoff selection, a form with available TRANSFAC TFBS matrices will be presented. Choosing the factors of interest is the last step in the submission process (*see* **Note 7**).

Instead of using the TRANSFAC matrices, you may decide to supply your own DNA motif. A motif search can be specified in the form of IUPAC one-letter codes. The IUPAC recommendations (**Table 1**) include letters to represent all possible ambiguities at a single position in the sequence except a gap.

The third option is to supply your own position weight matrix. To search for motifs defined by position weight matrices, specify the name of a matrix and enter your matrix in the format shown below:

```
A| 1 0 0 3 1 2 1 1
C| 0 0 0 1 0 0 0 1
T| 1 4 1 0 0 0 0 0
G| 2 0 3 0 3 2 3 2
```

If you want to search for several motifs in one request, you can press "Another matrix," enter more motifs, and press "Done" after you have entered all of them.

5.2. Results

Once the analysis is complete, you will receive an e-mail with a link to a table very similar to that described in **Subheading 4.2**. It will have one new column —"rVISTA." Clicking on the rVISTA link will take you directly to the rVISTA results (if there was only one alignment for a given sequence) or a list of alignments (if there were several) from which you can go to rVISTA analysis of each alignment by clicking a link in the rightmost column.

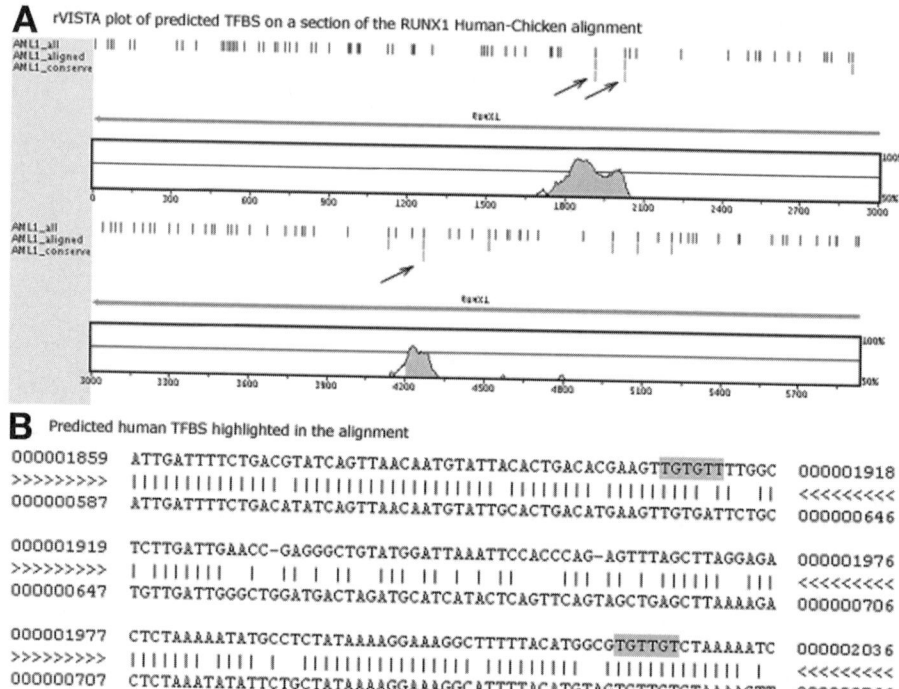

A rVISTA plot of predicted TFBS on a section of the RUNX1 Human-Chicken alignment

B Predicted human TFBS highlighted in the alignment

```
000001859   ATTGATTTTCTGACGTATCAGTTAACAATGTATTACACTGACACGAAGTTGTGTTTTGGC   000001918
>>>>>>>>>   |||||||||||||| ||||||||||||||||||| ||||||||| ||||||| || ||   <<<<<<<<<
000000587   ATTGATTTTCTGACATATCAGTTAACAATGTATTGCACTGACATGAAGTTGTGATTCTGC   000000646

000001919   TCTTGATTGAACC-GAGGGCTGTATGGATTAAATTCCACCCAG-AGTTTAGCTTAGGAGA   000001976
>>>>>>>>>   | |||||||  |   ||  |  || ||| || | | ||    ||| || | |||||||  |||   <<<<<<<<<
000000647   TGTTGATTGGGCTGGATGACTAGATGCATCATACTCAGTTCAGTAGCTGAGCTTAAAAGA   000000706

000001977   CTCTAAAAATATGCCTCTATAAAAGGAAAGGCTTTTTACATGGCGTGTTGTCTAAAAATC   000002036
>>>>>>>>>   |||||||| |||| |  ||||||||||||||||||| ||||||||||  |||||||||||| |   <<<<<<<<<
000000707   CTCTAAATATATTCTGCTATAAAAGGAAAGGCATTTTACATGTAGTGTTGTCTAAAACTT   000000766
```

Fig. 5. A subregion of the human-chicken alignment is analyzed by rVISTA, a transcription factor binding site (TFBS) prediction program. The predicted TFBS are marked above a VISTA graph with tick marks. The top row of tick marks shows all the sites predicted by the MATCH program from TRANSFAC. The second row shows only those sites that aligned to predicted sites on the second sequence. The third row shows only those sites that, in addition to being aligned, also fall within a region that shows at least 80% conservation over a 24-bp window. The user can download the underlying pairwise alignment with conserved predicted TFBS highlighted for ease of identification.

A form for selecting TFBS for visualization will appear. Clicking "Go" leads to the rVISTA form with a number of options: the length of sequence that will be depicted on a single row of the VISTA graph, the width of the picture produced, clustering (*see* **Note 8**), and which binding sites will be visualized. Next to the name of each of the TFBS is a link called "view in alignment." By clicking on this link, the user can retrieve the original alignment, with each occurrence of a predicted conserved TFBS highlighted on the base sequence (**Fig. 5B**).

rVISTA graphic display can show its predictions of conserved, aligned, or all the binding sites. "All binding sites" shows all sites predicted by the TRANSFAC Match program. "Aligned binding sites" are those in which core positions of the predicted binding sites on the sequences corresponded to each other in the alignment. "Conserved binding sites" are defined to be aligned binding sites, located in the sequence fragments conserved between two species at the level of over 80% over a local 24-bp window. Depending on your selection of the type of TFBS to visualize, from one to three tracks for each factor will appear on the top of the VISTA plot with associated gene annotation (**Fig. 5A**). This image allows you to see different TFBS in the context of the conservation of the analyzed genomic region.

5.3. Sample Analysis

For this example, we continue using the alignment created in **Subheading 4**. Given the human-chicken and human-opossum alignments, we need to select a region for rVISTA analysis, using the conserved regions shown on the graph as a rough guideline. The double peaks at 127.7 Kbp look like an interesting area to explore, as they are conserved all the way through chicken and opossum. Highlight that area to zoom into it, being careful to select less than 20 Kbp. Once you are zoomed in, click the "I" button to see the alignment details. The link to submit to rVISTA is on the right. Select the human-chicken alignment, and proceed with the submission as described above. Use the recommended threshold matrix cutoff, and select the following factors, which we can reasonably expect to show up given the description of the gene *RUNX1* at NCBI and Ensembl: AML, AML1, PEBP, PEA3 (*see* **Note 7**). The analysis shows three predicted conserved transcription factor binding sites for the AML1 factor (**Fig. 5**).

6. Using Stand-Alone VISTA

In addition to using VISTA through the web server portal, one can also download the visualization module as a stand-alone Java program to create alignment visualization and perform conservation analysis locally. The program is available free for academic and nonprofit research and for a licensing fee to commercial entities. The alignment programs, such as AVID or LAGAN, can be downloaded separately under similar conditions. A detailed Readme file accompanies the package.

The input to stand-alone VISTA consists of one or more alignment files, a "plotfile" that contains parameters for running VISTA, and several optional annotation files (gene annotation, single-nucleotide polymorphisms, repeats, and so on). The stand-alone VISTA is highly customizable, and provides the user with a variety of options for visualization and output. Although the sheer number of options described in the documentation for the stand-alone version of the program can intimidate the user, it is important to note that almost all these parameters are optional. It is possible to get acceptable output with just the following bare-bones plotfile:

```
COORDINATE human
ALIGN alignment.out
   SEQUENCES human mouse
END
```

The stand-alone version of the VISTA program provides certain features that are not available through the online tools described earlier, such as a "DIFFS" mode, in which the plot represents percent difference, not percent identity—useful for analyzing closely related species. On the other hand, it also only provides static output and does not show any of the precomputed whole-genome alignments available through the VISTA Browser. Researchers are encouraged to take these differences into account when choosing the appropriate tool for a particular task.

7. New Capabilities and Tools

Our group is constantly working on the development of new comparative genomics algorithms and associated tools using the successful VISTA concept as the basis. A few new directions are described here—an advanced visualization technique for multiple alignments, novel analysis of cross-species conservation, and examination of rVISTA TFBS predictions on a whole-genome scale.

The Phylo-VISTA program (short for Phylogenetic VISTA) allows a user to visualize submitted multiple sequence alignment data while using a phylogenetic tree as a framework to guide the display and analysis of conservation levels across tree nodes *(31)*.

In particular, Phylo-VISTA supports interactive visual analysis of prealigned multispecies sequences by performing the following functions: display of a multiple alignment sequences with the associated phylogenetic tree; computation of a similarity measure over a user-specified window for any node of the tree; visualization of the degree of sequence conservation by a line plot; and presentation of comparative data together with annotations. Phylo-VISTA can be used through the VISTA Web site for visualizing results of multiple alignments.

RankVISTA conservation plots (**Fig. 6**) depict evolutionarily conserved segments in pairwise or multiple alignments as a bar graph, where the heights scale with statistical significance $[-\log_{10}(p \text{ value})]$. For example, a height of 4 indicates that the probability of seeing that level of conservation by chance in a neutrally evolving 10-kb segment of the base sequence is less than 10^{-4}. RankVISTA graphs are based on the Gumby algorithm (S. Prabhakar, in preparation), which estimates neutral evolutionary rates from nonexonic regions in the multiple sequence alignment and then identifies local segments of any length in the alignment that evolve more slowly than the background. Gumby has no window-size parameter and no fixed percent-identity threshold. Since the algo-

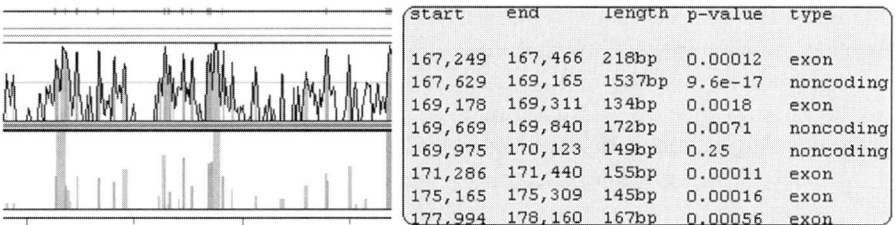

start	end	length	p-value	type
167,249	167,466	218bp	0.00012	exon
167,629	169,165	1537bp	9.6e-17	noncoding
169,178	169,311	134bp	0.0018	exon
169,669	169,840	172bp	0.0071	noncoding
169,975	170,123	149bp	0.25	noncoding
171,286	171,440	155bp	0.00011	exon
175,165	175,309	145bp	0.00016	exon
177,994	178,160	167bp	0.00056	exon

Fig. 6. A sample of a RankVISTA graph and text output.

rithm uses a more-conserved-than background paradigm, it can perform phylogenetic shadowing for close species and footprinting for distant species with equal facility (*see* the examples in **ref. 32**). RankVISTA plots are precalculated for several whole-genome alignments, and a user can access it through the "select/add" option in VISTA Browser. RankVISTA plots are also generated for user-submitted sequences through mVISTA and GenomeVISTA.

Whole Genome rVISTA provides access to the computational tool that allows for evaluation of which TFBS are overrepresented in upstream regions in a group of genes, for example, those obtained by the analysis of microarray experiments data. This evaluation is based on a database of all TFBS conserved in the alignment between the mouse and human genomes using the rVISTA method (*see* **Subheading 5.**). A user can input a set of genes using locus link IDs or RefSeq names. The programs will calculate which TFBS located in 5-Kbp upstream regions of these genes are overrepresented in the group (at the *p*-value cutoff 0.006) in comparison with all RefSeq genes as an outgroup.

8. Conclusions

VISTA tools have been widely used by the biological community. A number of biological studies have utilized these programs to answer various questions, such as comparing genes from the same gene families *(33,34)*, discovering functional noncoding elements *(35,36)*, finding patterns of conservation on a whole-genome scale *(5,37)*, and comparing genome assemblies of different depths *(38)*. VISTA is an ongoing project, so all your positive and negative comments are welcome as they help us to improve and expand our suite of programs. Please contact us at vista@lbl.gov.

9. Notes

1. Since there are many ways to evaluate conservation in the alignment, we provide here a short description of the method used by VISTA. The VISTA curve is calculated as a windowed-average identity score for the alignment. A variable-sized window (Calc Window) is moved along the alignment, and a score is calculated at

each base in the coordinate (base) sequence. That is, if the Calc Window is 100 bp, then the score for every point is the percentage of exact matches between the two sequences in a 100-bp-wide window centered on that point. Owing to resolution constraints when one is visualizing large alignments, it is often necessary to condense information about a hundred or more base pairs into one display pixel. This is done by only graphing the maximal score of all the base pairs covered by that pixel.

2. Regions are classified as "conserved" by analyzing scores for each base pair (*see* **Note 1** for how the scores are calculated). Two parameters control this analysis, "Min Cons Width" (default value 100 bp) and "Cons Identity" (default value 70%). A region is considered conserved if the conservation over this region is greater than or equal to the Conservation Identity and it has the minimum length of "Minimum Conserved Width." After all the regions that satisfy these conditions are calculated, they are modified using the annotation information—UTRs and exons are considered conserved based on Cons Identity without taking into consideration their length. Thus we are not missing short highly conserved exons and UTRs. CNS (Conserved Noncoding Sequences) are trimmed when the alignments at the edge of the region contribute little to the overall conservation score. This can sometimes lead to CNS that are a bit shorter than the assigned "Min Cons Width."

3. Sometimes the researcher will encounter "overlaps"—areas in which multiple regions from one sequence were aligned to the same area on another sequence. This can happen because of a variety of reasons—repeats, duplications, and so on. When this happens, VISTA Browser will draw a "best conservation" curve—for every point in the overlapped region, it will evaluate every participating alignment and plot the highest score. This creates an optimistic view of the conservation. To view each alignment separately, click on the "alignment details" button. You will be able to download PDF plots of each separate alignment.

4. If you have pop-up blocking software (external, such as the Google toolbar, or built-in, in Internet Explorer 6 for example), you might need to temporarily disable it—this is usually done by holding down the CTRL key while clicking the button.

5. The GenomeVista and mVISTA servers only accept plain-text sequence files in the FASTA format. The FASTA format consists of a header that starts with the greater than symbol (>), followed by the sequence name (one word) and the sequence itself on the following lines. If you have the sequence in a Word document, make sure that when you save it, the "Save as type" field says "Plain text (*.txt)."

6. The parameters used for conserved region analysis can have a significant impact on the number and quality of conserved regions identified by VISTA. When comparing distant or very closely related species, one may want to change default parameters by varying percent identity and window size when finding conserved sequences. **Figure 4** illustrates how different the same alignment can look when the visualization parameters are altered.

7. We recommend that the user carry out preliminary studies to determine transcription binding factors that are most likely to occur in a given sequence or are most interesting for your research. This information can be found through reading rele-

vant literature or looking up gene entries in public databases such as GenBank and Ensemble. A surprising amount of relevant information can often be found by performing a simple Google search on the gene name (try running a search on *RUNX1*). The new Google Scholar website at http://scholar.google.com can also be extremely useful for finding articles pertaining to your sequence of interest.

8. "Clustering" allows the users to identify TFBS that are present in groups or clusters. For an individual cluster to occur, K number of these binding sites must occur within N base pairs. K and N can be varied for different sites. In order for a group cluster to occur, K number of *any* TFBS need to occur within N base pairs.

Acknowledgments

The VISTA project is an ongoing collaborative effort of a large group of scientists and engineers. It has been developed and maintained in the Genomics Division of Lawrence Berkeley National Laboratory. You can find the names of all contributors at the VISTA web site http://genome.lbl.gov/vista.

The project was partially supported by the Programs for Genomic Applications (PGA) funded by the National Heart, Lung, and Blood Institute (NHLBI/NIH) and by the Office of Biological and Environmental Research, Office of Science, US Department of Energy.

References

1. Batzoglou, S., Pachter, L., Mesirov, J. P., Berger, B., and Lander, E. S. (2000) Human and mouse gene structure: comparative analysis and application to exon prediction. *Genome Res.* **10,** 950–958.
2. Brent, M. R. and Guigo, R. (2004) Recent advances in gene structure prediction. *Curr. Opin. Struct. Biol.* **14,** 264–272.
3. Guigo, R., Dermitzakis, E. T., Agarwal, P., et al. (2003) Comparison of mouse and human genomes followed by experimental verification yields an estimated 1,019 additional genes. *Proc. Natl. Acad. Sci. USA* **100,** 1140–1145.
4. Pennacchio, L. A., Olivier, M., Hubacek, J. A., et al. (2001) An apolipoprotein influencing triglycerides in humans and mice revealed by comparative sequencing. *Science* **294,** 169–173.
5. Woolfe, A., Goodson, M., Goode, D. K., et al. (2004) Highly conserved noncoding sequences are associated with vertebrate development. *PLoS Biol.* **3,** e7.
6. Gottgens, B., Barton, L. M., Chapman, M. A., et al. (2002) Transcriptional regulation of the stem cell leukemia gene (SCL)—comparative analysis of five vertebrate SCL loci. *Genome Res.* **12,** 749–759.
7. Waterston, R. H., Lindblad-Toh, K., Birney, E., et al. (2002) Initial sequencing and comparative analysis of the mouse genome. *Nature* **420,** 520–562.
8. Gibbs, R. A., Weinstock, G. M., Metzker, M. L., et al. (2004) Genome sequence of the Brown Norway rat yields insights into mammalian evolution. *Nature* **428,** 493–521.

9. Kirkness, E. F., Bafna, V., Halpern, A. L., et al. (2003) The dog genome: survey sequencing and comparative analysis. *Science* **301**, 1898–1903.

10. Hillier, L. W., Miller, W., Birney, E., et al. (2004) Sequence and comparative analysis of the chicken genome provide unique perspectives on vertebrate evolution. *Nature* **432**, 695–716.

11. Hardison, R. C. (2003) Comparative genomics. *PLoS Biol.* **1**, 156–160.

12. Miller, W., Makova, K. D., Nekrutenko, A., and Hardison, R. C. (2004) Comparative genomics. *Annu. Rev. Genom. Hum. Genet.* **5**, 15–56.

13. Dubchak, I. and Frazer, K. (2003) Multi-species sequence comparison: the next frontier in genome annotation. *Genome Biol.* **4**, 122–128.

14. Frazer, K. A, Elnitski, L., Church, D. M., Dubchak, I., and Hardison, R. C. (2003) Cross-species sequence comparisons: a review of methods and available resources. *Genome Res.* **13**, 1–12.

15. Wei, L., Liu, I., Dubchak, I., Shon, J., and Park, J. (2002) Comparative genomics approaches to study organism similarities and differences. *J. Biomed. Inform.* **35**, 142–150.

16. Frazer, K. A., Pachter, L., Poliakov, A., Rubin, E. M., and Dubchak, I. (2004) VISTA—computational tools for comparative genomics. *Nucleic Acids. Res.* **32**(Web Server issue):W273.

17. Dubchak, I., Brudno, M., Pachter, L. S., et al. (2000) Active conservation of noncoding sequences revealed by 3-way species comparisons. *Genome Res.* **10**, 1304–1306.

18. Mayor, C., Brudno, M., Schwartz, J. R., et al. (2000) VISTA: visualizing global DNA sequence alignments of arbitrary length. *Bioinformatics* **16**, 1046–1047.

19. Loots, G., Ovcharenko, I., Pachter, L., Dubchak, I., and Rubin, E. (2002) rVISTA for comparative sequence-based discovery of functional transcription factor binding sites. *Genome Res.* **12**, 832–839.

20. Bray, N., Dubchak, I., and Pachter, L. (2003) AVID: a global alignment program. *Genome Res.* **13**, 97–102.

21. Brudno, M., Do, C. B., Cooper, G. M., et al., and NISC Comparative Sequencing Program (2003) LAGAN and Multi-LAGAN: efficient tools for large-scale multiple alignment of genomic DNA. *Genome Res.* **13**, 721–731.

22. Couronne, O., Poliakov, A., Bray, N., et al. (2003) Strategies and tools for whole genome alignments. *Genome Res.* **13**, 73–80.

23. Brudno, M., Poliakov, A., Salamov, A., et al. (2004) Automated whole-genome multiple alignment of rat, mouse, and human. *Genome Res.* **14**, 685–692.

24. Schwartz, S., Zhang, Z., Frazer, K. A., et al. (2000) PipMaker—a web server for aligning two genomic DNA sequences. *Genome Res.* **10**, 577–586.

25. Schwartz, S., Elnitski, L., Li, M., et al., and NISC Comparative Sequencing Program (2003) MultiPipMaker and supporting tools: alignments and analysis of multiple genomic DNA sequences. *Nucleic Acids Res.* **31**, 3518–3524.

26. Schwartz, S., Kent, W. J., Smit, A., et al. (2003) Human-mouse alignments with BLASTZ. *Genome Res.* **13**, 103–107.

27. Pollard, D. A., Bergman, C. M, Stoye, J., Celniker, S. E., and Eisen, M. B. (2004) Benchmarking tools for the alignment of functional noncoding DNA. *BMC Bioinformatics* **5,** 6–22.

28. van der Helm-van Mil, A. H., Wesoly, J. Z., and Huizinga, T. W. (2005) Understanding the genetic contribution to rheumatoid arthritis. *Curr. Opin. Rheumatol.* **17,** 299–304.

29. Yamada, R. and Ymamoto, K. (2005) Recent findings on genes associated with inflammatory disease. *Mutat. Res.* **573,** 136–151.

30. Brudno, M., Malde, S., Poliakov, A., et al. (2003) Glocal alignment: finding rearrangements during alignment. *Bioinformatics* **Suppl 1,** I54–I62.

31. Shah, N., Couronne, O., Pennacchio, L. A., et al. (2004) Phylo-VISTA: interactive visualization of multiple DNA sequence alignments. *Bioinformatics* **20,** 636–643.

32. Martin, J., Han, C., Gordon, L. A., et al. (2004) The sequence and analysis of duplication-rich human chromosome 16. *Nature* **432,** 988–994.

33. Parent, S. A., Zhang, T., Chrebet, G., et al. (2002) Molecular characterization of the murine SIGNR1 gene encoding a C-type lectin homologous to human DC-SIGN and DC-SIGNR. *Gene* **293,** 33–46.

34. Chen, J., Kitchen, C. M., Streb, J. W., and Miano, J. (2002) Myocardin: a component of a molecular switch for smooth muscle differentiation. *J. Mol. Cell. Cardiol.* **34,** 1345–1356.

35. Premzl, M., Delbridge, M., Gready, J. E., et al. (2005) The prion protein gene: identifying regulatory signals using marsupial sequence. *Gene* **349C,** 121–134.

36. Anguita, E., Sharpe, J. A., Sloane-Stanley, J. A., Tufarelli, C., Higgs, D. R., and Wood, W. G. (2002) Deletion of the mouse α-globin regulatory element (HS 26) has an unexpectedly mild phenotype. *Blood* **100,** 3450–3456.

37. Cooper, G. M., Brudno, M., Stone, E. A, Dubchak, I., Batzoglou, S., and Sidow, A. (2004) Characterization of evolutionary rates and constraints in three mammalian genomes. *Genome Res.* **14,** 539–548.

38. Margulies, E. H., Vinson, J. P., Miller, W., et al. (2005) An initial strategy for the systematic identification of functional elements in the human genome by low-redundancy comparative sequencing. *Proc. Natl. Acad. Sci. USA* **102,** 4795–4800.

7

Computational Prediction of *cis*-Regulatory Modules from Multispecies Alignments Using Galaxy, Table Browser, and GALA

Laura Elnitski, David King, and Ross C. Hardison

Summary

One major goal of genomics is to identify all the functional sequences in genomes, including sequences that regulate the expression of genes. Sequence conservation is a good, albeit imperfect, guide to these functional elements. We describe how to use publicly available servers (Galaxy, the UCSC Table Browser, and GALA) to find genomic sequences whose alignments (from blastZ and multiZ) show properties associated with *cis*-regulatory modules, such as high conservation score, high regulatory potential score, and conserved transcription factor binding sites. Links to these servers can be accessed at http://www.bx.psu.edu/ and http://genome.ucsc.edu/.

Key Words: Enhancers; promoters; gene regulation; multispecies sequence alignments; blastZ; multiZ; UCSC Genome Browser; GALA; Galaxy; human genome.

1. Introduction

With complete, or almost complete, genome sequences from a large number of species becoming available, the issue of assigning a function, if any, to each string of nucleotides has now moved to the forefront of activity in the human genome project *(1)*. A string of nucleotides involved in a physiological process, such as encoding part of a protein (an exon) or specifying the spatiotemporal pattern of gene expression (e.g., a binding site for a transcription factor), is referred to here as a functional element in the genome. Much progress has been made in identifying genes using either *ab initio* predictions or evidence-based predictions, but a complete set of genes for most organisms cannot be unambiguously assigned *(2)*. Computational detection of noncoding functional elements is even less well developed, mainly because of the limited understanding of the

From: *Methods in Molecular Biology, vol. 338: Gene Mapping, Discovery, and Expression: Methods and Protocols*
Edited by: M. Bina © Humana Press Inc., Totowa, NJ

role of DNA sequences in the molecular mechanisms of gene regulation or other noncoding functions *(3–5)*. However, methods of comparative genomics succeed at a sufficiently high rate that they are commonly used to predict candidate *cis*-regulatory elements for experimental validation (e.g., *6,7–10*).

cis-regulatory modules (CRMs) are sets of functional elements that are clustered to form a regulatory unit (such as a promoter or enhancer) that acts in *cis* to a gene to control its expression level, timing, or tissue specificity. A large number of bioinformatic approaches have been developed to help investigators predict CRMs. This chapter describes how to use publicly available, web-based bioinformatic servers developed in our research group and those of our collaborators to predict CRMs based on properties of vertebrate genomic sequence alignments. Additional excellent servers are described in other chapters in this book; some are listed in **Table 1**.

The Methods section (**Subheading 3.**) refers to several functions computed from genomic sequence alignments to bring out different features associated with regulatory functions. For instance, a fundamental observation is whether a sequence falls within an alignment. The methods discussed in this chapter utilize precomputed, whole-genome alignments of sequences from several species, generated with the programs blastZ *(11)* and/or multiZ *(12)*. Several other alignment algorithms and servers have been developed, as described in a recent review *(13)*. More recently servers with improved features have been developed, which provide enhanced abilities to align and analyze sequences provided by the user (**Table 1**).

Purifying (or negative) selection is one of the most general genomic features that indicate function. The precomputed, whole-genome alignments have been analyzed for evidence of purifying selection following their divergence from a common ancestor. This type of selection can be inferred using the phastCons program *(14)*, which computes the likelihood that a given nucleotide in a sequence (represented as a column in the alignment) is in the 10% most slowly changing sequences in the genome. Scores associated with phastCons analyses are visualized in the "conservation" track on display at the UCSC Genome Browser *(15)*. Presented as highly resolved scores with wide dynamic range, the scores increase with stronger evolutionary constraint. Higher scores are implicated in function, but they provide no insight into the nature of the function.

The precomputed, whole-genome alignments have also been analyzed for the likelihood of involvement as a CRM, computed as a regulatory potential (RP) score *(16,17)*. Considered as short runs of columns (containing from two to five aligned positions), regions are analyzed for their frequency of appearance in a training set of known regulatory elements vs a training set of ancestrally derived neutral DNA. This function is influenced by the degree of evolutionary constraint, as is phastCons, but it also incorporates information about patterns in

Table 1
URLs for Servers Used to Predict CRMs

Property	Server	URL
Genome sequences, alignments, and annotations	UCSC Genome Browser and Table Browser	http://genome.ucsc.edu/
	GALA	http://www.bx.psu.edu/
	Galaxy	http://www.bx.psu.edu/
	ECR Browser	http://www.dcode.org/
Aligners	zPicture, Mulan, eShadow	http://www.dcode.org/
	PipMaker, MultiPipMaker	http://www.bx.psu.edu/
	VISTA	http://genome.lbl.gov/vista/index.shtml
	MAVID	http://baboon.math.berkeley.edu/mavid/
	LAGAN	http://lagan.stanford.edu/lagan_web/index.shtml
Phylogenetic footprints	FootPrinter2.0	http://wingless.cs.washington.edu/htbin-post/unrestricted/FootPrinterWeb/FootPrinterInput2.pl
	ConSite	http://mordor.cgb.ki.se/cgi-bin/CONSITE/consite
	rVista 2.0, multiTF	http://www.dcode.org/
Gene expression data	Gene Expression Omnibus	http://www.ncbi.nlm.nih.gov/geo/
	ArrayExpress	http://www.ebi.ac.uk/arrayexpress/
Motif discovery	Meme	http://meme.sdsc.edu/meme/website/intro.html
	MotifSampler	http://www.esat.kuleuven.ac.be/~thijs/Work/MotifSampler.html
	Weeder	http://159.149.109.16:8080/weederWeb/
	AlignAce	http://atlas.med.harvard.edu/
	Crème 2.0	http://www.dcode.org/

the alignments *(16)*. Empirical evaluations of the effectiveness of this approach for finding regulatory regions of proven function show that both RP and PhastCons work well with some highly conserved datasets, such as enhancers of developmental genes *(18)*. RP performs better than phastCons on a very difficult reference set containing all the CRMs in the human *HBB* gene complex.

None of the alignment-derived scores, including phastCons and RP scores, are sufficiently specific for highly reliable predictions of CRMs *(19)*. Therefore, it is prudent to combine these with other features commonly found in CRMs, such as binding sites for transcription factors. Many binding site motifs have been discovered and are recorded in resources such as TRANSFAC *(20)* and JASPAR *(21)*. Tools to identify motifs, based on overrepresentation of sequence strings in a given set of sequences, are also widely used *(5,22)*. In general, any approach to find motifs in one single sequence returns an excess of false positives. Requiring strict conservation in alignments of human, mouse, and rat sequences reduces the number of hits to binding sites for transcription factors by a factor of about 40 *(23)*. This chapter describes how to access matches to conserved transcription factor binding sites (cTFBS) computed by the program *tffind* *(24)*.

Precomputed binding sites allow a user to look for sites of interest that fall within a neighborhood of a genomic locus, without setting strict limitations on the amount of sequence being submitted in the search. In contrast, someone using a server to find matches to TFBS in a sequence will typically extract a few kilobases upstream and downstream of a gene to submit. The limitation of the analysis to a certain distance around a gene may inadvertently exclude important regions. The use of precomputed binding sites allows a user to select a larger region and subsequently reduce it through queries of a more refined region.

The data discussed in this chapter are stored in databases at the University of California at Santa Cruz (UCSC) Genome Browser *(15)* and GALA, a database of genome sequence alignments and annotations *(25,26)* (**Table 1**). A recently released metaserver, Galaxy, provides a platform for integrative analyses of genomic sequences and annotations *(27)*. The metaserver uses the query engines from remote databases such as the UCSC Table Browser *(28)* and other resources to retrieve primary data, and it provides operations and tools to filter, combine, and analyze the data. The Galaxy metaserver project is new and should grow to connect to many data repositories and provide a large suite of operations and tools. GALA is a more mature database project that also provides access to alignment and annotation results. GALA follows the traditional approach of recording all the data in a database on one large machine, whereas Galaxy accesses data from remote sites. Instructions for acquiring and analyzing data to predict CRMs using both Galaxy (in conjunction with the UCSC Table Browser) and GALA are presented in the Methods section (**Subheading 3.**).

The basic method described in this chapter is to retrieve candidate CRMs in erythroid cells as noncoding DNA segments with a high phastCons score or high RP score and a conserved match to a GATA-1 binding site. GATA-1 is a transcription factor that is essential for proper gene expression during late erythroid maturation *(29)*. A description is given of how to obtain noncoding genomic

DNA segments with the desired phastCons or RP scores, how to obtain conserved GATA-1 binding sites, and how to identify all the conserved or high-RP intervals with a conserved GATA-1 binding site in close proximity. A similar approach could be followed for any binding site of interest, when some information is known regarding preferential tissue specificity of the factor.

Although the approach described using premapped matches to binding sites for the entire genome is useful, other computational tools are being developed to discover motifs (short nucleotide strings). These extensions of basic pattern matching require a given motif to be enriched in, for example, sequences immediately upstream from a set of coexpressed genes. Thus, they are frequently used to find candidates for common regulatory elements controlling similarly expressed genes. Clusters of coexpressed genes are commonly deduced from transcriptional profiles based on microarray or other experiments measuring expression. Two large public databases of gene expression data are located at the Gene Expression Omnibus and ArrayExpress (**Table 1**). A sample of motif-finding servers is listed in **Table 1**. These simply require that users submit a list of sequences, such as the promoters (known or predicted) for a set of coexpressed genes. The servers use different methods *(5)* for motif discovery. An evaluation of the performance of these methods was recently published and provides further information on the subject *(22)*.

2. Materials

The only material required is a computer connected to an Internet service provider and running an Internet browser (such as Internet Explorer, Safari, Mozzilla, or Netscape).

3. Methods

3.1. Retrieving Strongly Conserved, Noncoding Genomic Intervals With Galaxy/UCSC Table Browser

1. Enter the Galaxy portal by pointing your Internet browser to the URL http://www. bx.psu.edu/ (**Table 1**) and clicking on "Galaxy."
2. At the Galaxy portal, you are presented with a few options. The first is to go to the UCSC Table Browser to retrieve any of the rich variety of data recorded there and automatically upload it to Galaxy. However, for phastCons and RP scores, it is more efficient to choose "Galaxy featured datasets" (*see* **Note 1**). On the new page, select the genome of the species of interest (e.g., Human) and the desired sequence assembly (e.g., hg17: May 2004) (*see* **Note 2**). The available options are specific to the genome assembly; for example, hg17 currently offers:
   a. Known regulatory regions [93 regions].
   b. phastCons (stringent, top approx 5%) [1,313,584 regions].

    c. phastCons (sensitive, ≥0.2) [26,277,600 regions].

    d. Regulatory potential (3way, human-mouse-dog, >0) [5,800,931 regions].

3. Regarding phastCons scores, select option b for regions under intense constraint or option c for increased sensitivity (*see* **Note 3**). Then click on the button labeled "Go." The results are added to your history page, which is displayed on your computer.

4. The next step is to retrieve the locations of all exons so that they can be removed from the high phastCons intervals. Users should return to the Galaxy Portal by clicking on "Portal" on the top row of the window in your Internet browser. At the Portal, click on the link to the UCSC Table Browser.

5. To retrieve exons, use the Table Browser pull-down menus to select "Genes and Gene Prediction Tracks" under the category of "group" and "Known Genes," found under "track" (*see* **Note 4**). If desired, the query can be limited to a particular genomic interval using the window labeled "position" (*see* **Note 5**). Because you entered the Table Browser via Galaxy, the default for "output format" is "send data to Galaxy." Now click on "get output."

6. A window appears that gives you the option to select whole genes, exons, coding exons, and so on. Select "Exons," and click on "Send query to Galaxy" (*see* **Note 6**).

7. This returns the user automatically to the Galaxy History Page (*see* **Note 7**), where each query appears as a short description (*see* **Note 8**) followed by the number of results retrieved.

8. In preparation for performing an operation, you need to select the desired datasets. Select the boxes for the queries of high phastCons scores and exons. Now select "Perform operations like intersection, etc." and click on "Go."

9. On the Query Operations page, the two queries now appear, and you should click on the box next to the operation "Subtraction." The screen automatically refreshes. Use the pull-down menus to determine the order and type of subtraction. In this case, it should be the query for phastCons intervals minus the query for the Known-Genes exons, removing "only overlapping segments." Click on "Go" (*see* **Note 9**).

10. The user is returned automatically to the History Page, which will show the number of results when the operation has completed. If the operation is listed as "running," the user should click "Refresh" periodically until the operation is finished. The resulting genomic intervals are noncoding, highly conserved DNA segments, which is one class of candidates for CRMs.

11. Galaxy provides several forms of output, which are accessed by clicking "Get output" followed by "Go." At the Display Options page, select "Genome Browser" to view each of the returned intervals in the UCSC Genome Browser (*see* **Note 10**), or "Raw result file" to obtain a file with the desired genomic intervals. Other options include viewing the results in the Ensembl browser (*see* **Note 11**).

3.2. Retrieving High-RP, Noncoding Genomic Intervals With Galaxy/UCSC Table Browser

The procedure for finding high-RP intervals via Galaxy is the same as outlined in **Subheading 3.1.**, except that when using the "Galaxy featured datasets" (accessed through the Galaxy portal), the user should choose option D "Regulatory potential (3way, human-mouse-dog, >0) [5,800,931 regions]" (*see* **Note 12**).

3.3. Retrieving Conserved Matches to Transcription Factor Binding Sites (cTFBS) Using Galaxy/UCSC Table Browser

Conserved matches to binding sites for transcription factors with weight matrices in TRANSFAC *(20)* can be obtained via the UCSC Table Browser and retrieved into Galaxy. The software used is an update of the *tffind* program *(24)*.

1. From the Galaxy Portal page, take the link to the UCSC Table Browser, where the user should select the group "Expression and Regulation" and the track "TFBS Conserved." Under region, select position and enter the chromosome name and coordinates of interest. Otherwise, select "genome."
2. To restrict the search to a single binding site, you should filter by name. Next to "filter," click on "create." This brings up a new page, on which you should select "does" match, next to "name." The name of the binding site matrix should be entered after "match." For instance, using the term "V$GATA*" will return conserved matches to a set of weight matrices for GATA-1 and GATA-3 binding sites (*see* **Note 13**).
3. Press "submit" to upload the filter to the Table Browser query page, and then click on "get output." Results will appear on the Galaxy history page.

3.4. Integrating the Conservation or RP Data With cTFBS at Galaxy

1. At the Galaxy history page, the user now has the noncoding intervals with high phastCons scores, the noncoding intervals with high RP scores, and intervals with conserved GATA-1 binding sites. Select two of the results to combine, e.g., noncoding high-RP intervals and conserved GATA-1 binding sites, by clicking on the buttons next to each query.
2. Under "Action to Perform," click on the button for "Perform operations like intersection, etc." and click "Go." This takes the user to the Query Operations page. Only the queries selected from the history page are transferred to the operations page. For a given number of queries, only a certain set of operations is allowed. Those that are not allowed are dimmed.
3. To find all the noncoding, high-RP intervals that have a conserved GATA-1 binding site in proximity to them, under "Operation," click on the button next to "Proximity" (*see* **Note 14**). After the screen refreshes, use the pull-down options to return regions from the noncoding, high-RP query results that lie less than 50 bp in either direction from a region in the query for conserved GATA-1 binding sites. Click on "Go," which returns you to the history page. The page initially returned frequently shows the new query as "running." Again, periodically click "Refresh" to obtain the results.
4. The results are the predicted CRMs, based on three criteria—they have a high RP score, they are not exons, and they are close to or encompass a conserved match to

a GATA-1 binding site. To retrieve the results of a selected query, select "Get output" from the list of "Actions to Perform" and hit "Go." For viewing the results, select "UCSC Browser custom track" or "Ensembl Genome Browser custom track." For a plain text file, select "Raw result file (bed)." The desired action is taken when you click "Go." Other features can be combined, such as high phastCons scores, and other operations can be performed on the data, using the utilities at Galaxy.

3.5. Retrieving High-RP or High-phastCons Intervals in Noncoding Sequences Using GALA

1. The GALA database is accessed at http://www.bx.psu.edu/ (**Table 1**) by selecting the link for "GALA." On the home page, the user finds links to GALA databases built for genomes of five different species (human, mouse, rat, chimp, and chicken), with up to three assemblies for each (*see* **Note 15**). Click on "Query page" under the appropriate species and assembly (e.g., Human July 2003 data release).

2. The query page is presented as an expandable selection of choices for categories, i.e., genes and gene predictions, expressed sequence tags and mRNA, comparative genomics, variation and repeats, expression and regulation, and mapping and sequencing, which are compatible with groups on the UCSC Genome Browser (*see* **Note 16**). Halfway down the page, you will find the query boxes for "Regulatory potential scores based on multiple alignments," with options for filtering the results by a minimum and maximum score. A good score for the minimal threshold is 0.001; leave the "less than or equal to" box blank. Alternatively, you may wish to query on the next item, PhyloHMM Cons (an earlier name for phastCons). A good score for the minimal threshold is 0.4 *(18)* (*see* **Note 17**).

3. Users wishing to investigate only a small genomic locus can choose the button to "Restrict search to interval" (near the bottom of the form). Otherwise, proceed to select the choice of output. "Text list" is the preferred choice when preparing datasets for use with subsequent operations.

4. Click "run query in background," so the server will save the results for 48 h. The results are returned on the GALA history page, where they can be combined with other queries (*see* **step 6**).

5. To collect exons, return to the GALA query page, and for the category "Genes and gene models," click on "Show the fields for this category" and then "Refresh" (toward the bottom of the page). The new page has many options for obtaining genes or parts of genes. Under "Protein Coding Genes, GALA's default set of genes," go to "Other gene fields" and click the box for "exons." Scroll to the bottom of the page, restrict the query to a chromosomal interval if desired, choose "text file" under "Output," and click on "run query in background."

6. Use the GALA history page to remove the exons from the high-RP intervals. Click the box next to each query on which you want to perform an operation (such as subtraction). Under "Compound queries," choose "SUBTRACTION." If you follow the steps in the order covered in here, choose the option to subtract "earlier minus later query" to subtract exonic intervals from the high-RP intervals. Using the pull-down menu, specify that "only overlapping segments" should be removed.

Click on "Run compound query in the background," located almost at the bottom of the page. The results are noncoding, high-RP intervals.

3.6. Retrieving Conserved Matches to TFBS Using the GALA Server

1. On the GALA query page, under the category of "Expression and Regulation," go to "Transcription factor binding sites" and choose, e.g., "only binding sites conserved in hg16Mm3Rn3, cutoff used was 0.85" (*see* **Note 18**).
2. Click on the button after "To select/add factor names," which opens a new page with all the choices. Select those of interest, and press the button "add selections to main form," which is at the bottom of the selection page (*see* **Note 19**).
3. The user is returned to the GALA query page. As before, users can limit the query by entering a restricted genomic interval, or they can query the entire genome. After selecting the desired output (e.g., "text list"), the user should click on "Run query in the background." A results page appears, after which the user can go to the history page.

3.7. Integrating the Conservation or RP Data With cTFBS Data at GALA

1. The GALA history page lists the queries that have been run, such as noncoding high-RP intervals and conserved GATA-1 binding sites, along with the number of results obtained for each. To find features that are in proximity to others, scroll down the page under "Compound queries" to "Proximity."
2. Enter the appropriate query numbers in the boxes under "Proximity," specifying that the noncoding, high-RP intervals "lie within 50bp" of regions in the conserved GATA-1 binding site query (*see* **Note 20**).
3. Select the type of output (such as "text list"), and then click "Run compound query in the background."
4. The results returned are the CRMs predicted by having a high-RP score, not being exons, and being close to a GATA-1 binding site that is conserved among human, mouse, and rat. Other criteria can be applied, and other operations (such as intersections or clustering) can be used for alternative predictions.

4. Notes

1. Instead of using the "Galaxy featured datasets," the user can follow the link to the Table Browser and retrieve genomic intervals whose phastCons scores exceed a desired threshold. However, this step takes a rather long time for the entire genome (searching through about 800 million records), and it is likely to time-out. Thus a user should limit this search to a specific interval (megabases should be no problem), or one can use the preselected intervals deposited in the "featured datasets." A similar logic holds for the RP scores.
2. It is often the case that most recent assembly is more complete and better annotated. However, it takes some time for annotations to be "lifted" onto new assemblies, and thus for some time after a new assembly is released, more information will be available on the previous assembly. As of this writing, the very extensive

data on the ENCODE regions are available only for hg16, the July 2003 assembly of human.

3. Selecting the more sensitive threshold for phastCons score (≥0.2) returns a large set of intervals that does the best job of finding known CRMs in the *HBB* gene complex *(18)*. However, it almost certainly returns many false positives, and for some purposes, the more stringent threshold may be more appropriate.

4. The choice of the collection of genes used is, of course, up to the user. The Known Genes track is very extensive and quite reliable, but it misses some genes. Users may prefer RefSeq, Ensembl, or other sets. Users should be aware that despite the considerable overlap in these gene sets, there are many differences, and these will affect the results of subtracting them from a set of intervals to find noncoding conserved sequences.

5. In this step, and in all steps in which the user has an option to limit a query to a particular interval, it is important to realize that the larger the interval examined, the more time it takes for the database to complete the query. Thus, searching the entire genome (approx 3000 Mb) takes considerably longer than searching the ENCODE regions (approx 30 Mb), which will take longer than a given locus (perhaps 0.3 Mb). Likewise, the number of features in the intervals searched is a major determinant of time to complete the query. phastCons and RP scores are given for every aligning nucleotide, and thus there are almost 800 million of these records to search. In contrast, the number of exons in the KnownGenes set is about 400,000, and thus a query to retrieve them takes less time. For full data on dense features like phastCons or RP, downloading files is much more efficient.

6. Users may instead wish to choose exons with an additional short interval, e.g., 10 bases, at each end. By doing so, the user will include regions that may be indirectly under selection because of their proximity to exons.

7. The Galaxy history page will load immediately, even if the query has not finished running at the Table Browser. In this case, at the end of the query, the notation "running" appears. The user should periodically click on "refresh" to see when the query has been completed and the results sent to Galaxy.

8. In this step, or any time the user is on the history page, one of the options is to edit the descriptions. Select a query, and click on "More." The screen refreshes, and now the option to "Edit query descriptions" is displayed. This editing is particularly helpful for the results of operations, for which Galaxy simply refers to the queries by number, not by content. A similar feature is implemented in GALA.

9. The time it takes for an operation to complete is determined primarily by the number of intervals that are in each query.

10. All the returned intervals can easily be viewed in the UCSC Genome Browser. On the left of the Genome Browser display is a list of all the returned intervals, which are hyperlinks to new views that show each region. The text file that can be returned is in BED format, in which the first three columns are chromosome, start position, and stop position for each interval.

11. After seeing the results, if the user decides that the genomic regions selected for the queries requires optimization, e.g., it was too small or too large, return to **step 5** and enlarge or reduce the coordinate distance.

12. Selecting "Regulatory potential (3way, human-mouse-dog, >0)" returns a set of 5.8 million intervals that does the best job of finding known CRMs in the *HBB* gene complex *(18)*. It probably also returns some false positives. To increase the stringency of the search, users can go the UCSC Table Browser, and select "Expression and Regulation" as the group and "3x Reg Potential" as the track (currently only available on the human July 2003 assembly). By clicking on the "create" button for "filter," the user gets to a page at which the threshold can be set higher, e.g., dataValue is ≥0.001. By clicking on "submit," this filter will be applied to the query when it is run.

13. The first set of filters is for the table of conserved binding sites, and the "name" refers to the name (or ID) of the weight matrix for a binding site. Thus one could enter a TRANSFAC ID for a particular weight matrix, such as "V$GATA1_02." Of course, this requires that the user know these IDs, which can be obtained from TRANSFAC. In the example given here, a wild card character ("*") was used to filter on "V$GATA*," which will include multiple binding sites for GATA-1 and GATA-3 (which have very similar binding sites). In order to filter based on the name of the transcription factor (not the binding site), users can take advantage of the ability of the Table Browser to filter on fields in related tables. On the filter page, choose the option to allow filtering on hg17.tfbsConsFactors, and choose "factor does match GATA-1" (or the name of the desired factor).

14. Users can find features in proximity to other features, such as described here, and the distance between them is set by the user. Alternatively, users may elect to perform a simple intersection. Note that the screen refreshes for each newly selected operation, because the parameters and choices relevant to each operation differ. In our research, we have found that using proximity has predicted some active CRMs that were missed by the intersection operation, but this is not frequent.

15. On the GALA home page users may want to access "Annotation statistics" to see the all the different types of data recorded, the number of records in each, and a partial list of fields in each table. Users can also go directly to their history page.

16. The default GALA query page lists only minimal or no choices for categories such as genes and gene models. Users who want to query on information within these should click on "Show the fields for this category" and then "Refresh" (toward the bottom of the page).

17. Users may wish to choose alignments computed between different species or filtered in various ways. These options are all under the comparative genomics section of the query page.

18. The options available for binding sites in GALA differ by the species and genome assembly. Here we selected binding sites conserved in human-mouse-rat alignments (hg16Mm3Rn3), but users can select other alignments, such as a pairwise human-chicken (hg16Gg2) or five-way human-chimp-mouse-rat-chicken (hg16Pt1Mm3 Rn3Gg2). The threshold scores ("cutoff") for the matches to the weight matrices are adjusted in each case.

19. Users can select by ID for weight matrices for factors instead of by name of the factor. Queries of all binding sites (not just the conserved ones) must be a limited

to a chromosomal interval because of the very large number of sites in the entire genome (about 212 million for the human genome).

20. Users can elect to do intersections or other operations. Clustering is also supported, e.g., requiring that each high-RP interval have at least two conserved factor-binding sites within it. This set of operations is supported in both Galaxy and GALA.

Acknowledgments

This work was supported by NIH grants DK65806 (to R.H.) and HG02325 (to L.E.).

Note Added in Proof

A new version of Galaxy, accessed at the same URL, has a different interface but provides access to the functions described here and many more. The version described here is still maintained.

References

1. Collins, F. S., Green, E. D., Guttmacher, A. E., and Guyer, M. S. (2003) A vision for the future of genomics research. *Nature* **422,** 835–847.
2. Parra, G., Agarwal, P., Abril, J. F., Wiehe, T., Fickett, J. W., and Guigo, R. (2003) Comparative gene prediction in human and mouse. *Genome Res.* **13,** 108–117.
3. Pennacchio, L. A. and Rubin, E. M. (2001) Genomic strategies to identify mammalian regulatory sequences. *Nat. Rev. Genet.* **2,** 100–109.
4. Hardison, R. C. (2003) Primer on comparative genomics. *Public Library of Science, Biology* **1,** 156–160.
5. Wasserman, W. W. and Sandelin, A. (2004) Applied bioinformatics for the identification of regulatory elements. *Nat. Rev. Genet.* **5,** 276–287.
6. Gumucio, D., Shelton, D., Zhu, W., et al. (1996) Evolutionary strategies for the elucidation of *cis* and *trans* factors that regulate the developmental switching programs of the beta-like globin genes. *Mol. Phylog. Evol.* **5,** 18–32.
7. Loots, G. G., Locksley, R. M., Blankespoor, C. M., et al. (2000) Identification of a coordinate regulator of interleukins 4, 13, and 5 by cross-species sequence comparisons. *Science* **288,** 136–140.
8. Hardison, R. C. (2000) Conserved noncoding sequences are reliable guides to regulatory elements. *Trends Genet.* **16,** 369–372.
9. Nobrega, M. A., Ovcharenko, I., Afzal, V., and Rubin, E. M. (2003) Scanning human gene deserts for long-range enhancers. *Science* **302,** 413.
10. Miller, W., Makova, K. D., Nekrutenko, A., and Hardison, R. C. (2004) Comparative genomics. *Annu. Rev. Genomics Hum. Genet.* **5,** 15–56.
11. Schwartz, S., Kent, W. J., Smit, A., et al. (2003) Human-mouse alignments with *Blastz*. *Genome Res.* **13,** 103–105.
12. Blanchette, M., Kent, W. J., Riemer, C., et al. (2004) Aligning multiple genomic sequences with the threaded blockset aligner. *Genome Res.* **14,** 708–715.

13. Frazer, K. A., Elnitski, L., Church, D., Dubchak, I., and Hardison, R. C. (2003) Cross-species sequence comparisons: a review of methods and available resources. *Genome Res.* **13**, 1–12.

14. Siepel, A., Bejerano, G., Pedersen, J. S., et al. (2005) Evolutionarily conserved elements in vertebrate, fly, worm and yeast genomes. *Genome Res.* **15**, 1034–1050.

15. Kent, W. J., Sugnet, C. W., Furey, T. S., et al. (2002) The human genome browser at UCSC. *Genome Res.* **12**, 996–1006.

16. Elnitski, L., Hardison, R. C., Li, J., et al. (2003) Distinguishing regulatory DNA from neutral sites. *Genome Res.* **13**, 64–72.

17. Kolbe, D., Taylor, J., Elnitski, L., et al. (2004) Regulatory potential scores from genome-wide three-way alignments of human, mouse and rat. *Genome Res.* **14**, 700–707.

18. King, D. C., Taylor, J., Elnitski, L., Chiaromonte, F., Miller, W., and Hardison, R. C. (2005) Evaluation of regulatory potential and conservation scores for detecting cis-regulatory modules in aligned mammalian genome sequences. *Genome Res.* **15**, 1051–1060.

19. Berman, B. P., Pfeiffer, B. D., Laverty, T. R., et al. (2004) Computational identification of developmental enhancers: conservation and function of transcription factor binding-site clusters in *Drosophila melanogaster* and *Drosophila pseudoobscura*. *Genome Biol.* **5**, R61.

20. Wingender, E., Chen, X., Fricke, E., et al. (2001) The TRANSFAC system on gene expression regulation. *Nucleic Acids Res.* **29**, 281–283.

21. Sandelin, A., Alkema, W., Engstrom, P., Wasserman, W. W., and Lenhard, B. (2004) JASPAR: an open-access database for eukaryotic transcription factor binding profiles. *Nucleic Acids Res.* **32**, (Database issue) D91–D94.

22. Tompa, M., Li, N., Bailey, T. L., et al. (2005) Assessing computational tools for the discovery of transcription factor binding sites. *Nat. Biotechnol.* **23**, 137–144.

23. Gibbs, R. A., Weinstock, G. M., Metzker, M. L., et al. (2004) Genome sequence of the Brown Norway rat yields insights into mammalian evolution. *Nature* **428**, 493–521.

24. Schwartz, S., Elnitski, L., Li, M., et al. (2003) MultiPipMaker and supporting tools: alignments and analysis of multiple genomic DNA sequences. *Nucleic Acids Res.* **31**, 3518–3524.

25. Giardine, B. M., Elnitski, L., Riemer, C., et al. (2003) GALA, a database for genomic sequence alignments and annotations. *Genome Res.* **13**, 732–741.

26. Elnitski, L., Giardine, B., Shah, P., et al. (2005) Improvements to GALA and dbERGEII: Databases featuring genomic sequence alignment, annotation and experimental results. *Nucleic Acids Res.* **32**, (Database issue) D466–D447.

27. Giardine, B., Riemer, C., Hardison, R. C., et al. (2005) Galaxy: a platform for interactive large-scale genome analysis. *Genome Res.* **15**, 1451–1455.

28. Karolchik, D., Hinrichs, A. S., Furey, T. S., et al. (2004) The UCSC Table Browser data retrieval tool. *Nucleic Acids Res.* **32**, D493–D496.

29. Weiss, M. J. and Orkin, S. H. (1995) GATA transcription factors: key regulators of hematopoiesis. *Exp. Hematol.* **23**, 99–107.

8

Comparative Promoter Analysis
in Vertebrate Genomes With the CORG Workbench

Christoph Dieterich and Martin Vingron

Summary

CORG is a versatile web-based workbench for comparative promoter analysis in verte-brate model organisms. Two kinds of information are explicitly considered in the automated annotation process. First, local conservation patterns in upstream regions of homologous genes: These phylogenetic footprints are likely to stem from sequence elements that are under selective pressure. The CORG pipeline detects and exploits patterns of local simi-larity to annotate promoter regions. Second, experimental data on transcription start sites: exon positions and DNA binding site descriptions complete the promoter annotation. These data are made available via an interactive web portal. Individual promoter studies are supported by a JAVA applet that supplies all data down to the nucleotide level.

Key Words: Promoter; phylogenetic footprinting; vertebrates; comparative sequence analysis; regulatory elements; DNA binding site; conserved promoter elements; sequence alignment.

1. Introduction

Cells have to react and adapt to various external constraints like temperature, oxygen supply, or mechanical stress. Evidently, external conditions require com-plex responses involving the coordinate expression of many genes. Unlike in bacteria, coordinately expressed genes are not spatially linked. Nevertheless, it should be possible to activate genes from disjoint parts of the genome, simulta-neously. Britten and Davidson (*1*) proposed a model for coordinate expression in unlinked genes. Genes regulated in parallel with one another would contain common control elements. As specific signals have to be met by individual re-sponses, cells require a set of freely combinable control elements. The product

From: *Methods in Molecular Biology, vol. 338: Gene Mapping, Discovery, and Expression:*
Methods and Protocols
Edited by: M. Bina © Humana Press Inc., Totowa, NJ

of an integrator gene would recognize a specific control element. This product would then activate all genes containing one particular control element.

Today, we know that DNA sequence elements about 6 to 30 bp in length serve as regulatory elements *(2)*. Such elements can be divided into two classes: general and specific ones. General ones like the TATA box (consensus sequence TATAWAW) are constituents of many promoters, whereas combinations of specific binding site make up the identity of a promoter region.

Regulatory elements close to the start site of transcription constitute the promoter region. Incoming signals from many different sources can be integrated at this level. Activating factors either ease complex assembly of the transcription machinery or stimulate the activity of the already assembled complex. Activators have a multitude of targets to exert their function, and multiple interactions are the reason for strong synergistic activation.

Speaking of gene regulation, it has been known for a long time that there is considerable sequence conservation between species in noncoding regions in general and promoter regions in particular. Sequence conservation within promoter regions often stems from transcription factor binding sites that are under selective pressure *(3)*.

The CORG workbench *(4)* provides access to precompiled annotation of promoter regions. Information of two kinds is explicitly considered. First, biological meaningful cross-species conservation is detected within upstream regions of homologous genes. An upstream region is defined by a sequence window of 15 kb upstream of the translation start site. If other data (validated transcription start sites or exon annotation) suggest a different extension of an upstream region, this is taken into account, and the corresponding region is adjusted. Pairwise as well as multiple sequence alignments are computed employing motifs as alignment anchors. Second, binding site descriptions (position-weight matrices) are used to predict conserved regulatory elements with a novel approach. Binding site description stems from the TRANSFAC database *(5)*. Exon annotations and verified transcription start sites are incorporated to distinguish exonic from nontranscribed sequence. CORG is built on top of the EnsEMBL database *(6)* and utilizes protein homology and gene structure information from this resource.

CORG is fitted with an intuitive interface that leads the user to the information of her choice. CORG contents are accessible as graphical and textual information. Various export functions exist to obtain and process any CORG data locally.

The subsequent sections will guide you step by step through the CORG protocol for upstream region analysis. An entire example session dealing with the analysis of an example gene, *E2F2*, is shown there. The protein encoded by this gene is a member of the E2F family of transcription factors. The E2F family plays a crucial role in the control of the cell cycle and the action of tumor suppressor proteins and is also a target of the transforming proteins of small DNA tumor viruses *(7)*.

2. Materials

CORG as a web-based tool has only modest requirements. A common web browser (e.g., Firefox, Mozilla, or Internet Explorer) is required to visit and access most of the CORG website (http://corg.molgen.mpg.de). To make sure that the interactive graphical display of sequence alignments works, a JAVA-compliant web browser is mandatory (JDK 1.1). No JavaScript or StyleSheet support is needed.

3. Methods

A personal CORG session is simply started by pointing a browser to http://corg.molgen.mpg.de. There, the latest news on recent CORG developments is presented on the welcome screen. Choose an appropriate CORG version, which will usually be the latest one (version 5, at the time of writing).

3.1. Querying CORG

A search form similar to **Fig. 1** should appear now. Please enter your gene or transcript identifier in the designated form field. Valid identifiers include HUGO symbols, Marker symbols, SWISSPROT identifier, EntrezGene identifier, RefSeq identifiers, and ENSEMBL gene or transcript identifiers.

Example session:

1) Type in E2F2 in the blank form field, select "Homo sapiens" as species if it is not highlighted yet, and hit the submit button.

Makes sure that you have selected the right species before you hit the submit button. Allow some time to process your request. The result page (**Fig. 2**) is structured into the following sections:

1. The search request that was sent to the CORG server is restated in the top line.
2. All entries matching the search request are listed below in tabular form, which contains a gene description field.

Check whether your query led to the right result by reading the gene's description and continue by clicking on the gene identifier.

Example session:

2) Click on the single text link "ENSG00000007968" that appears in column "gene identifier."

3.2. Query Gene Overview

Subsequently, a gene information page shows up that summarizes all available data for that particular gene in CORG. The image (**Fig. 3**) in the upper part

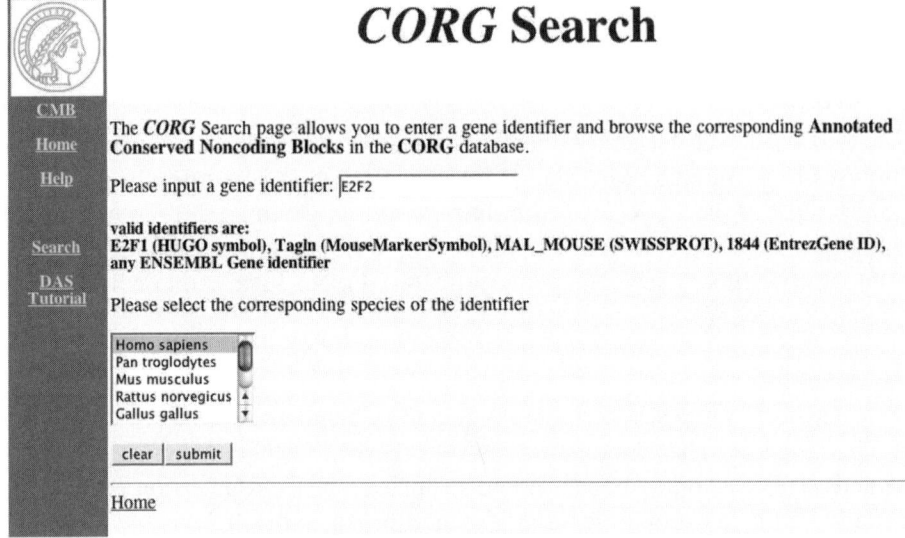

Fig. 1. CORG search screen. Identifiers of any gene of interest may be entered into the text field. Valid identifiers include HUGO symbols, Marker symbols, SWISSPROT accessions, and EntrezGene identifiers. The corresponding species must be selected from the list below.

CORG Search results

The following hits have been found for your request:
Identifier: *E2F2*   Database: **Homo sapiens**

Species	Gene identifier	External name	External database	Gene description
Homo sapiens	ENSG00000007968	E2F2	HUGO	Transcription factor E2F2 (E2F-2). [Source:SWISSPROT;Acc:Q14209]

Home

Fig. 2. Result screen. A query for the HUGO symbol "E2F2" returns the result screen above. At this step, the user should consult the gene description column to make sure that she has selected the right gene. To continue, one has to click on the link in the gene identifier column.

displays the gene structure and the upstream region in the selected reference genome. Alternative transcripts, validated transcription start sites, local sequence conservation to other species, and the repeat structure are shown simultaneously. One may quickly check now whether the gene structure corresponds to the anticipated one. It occasionally happens that the suggested EnsEMBL structure differs from what one would expect. Right below the image is the first export facility of CORG. By clicking on the bold link, a Generic Feature Format 3 (GFF3) dump of the overview image can be downloaded to the local computer.

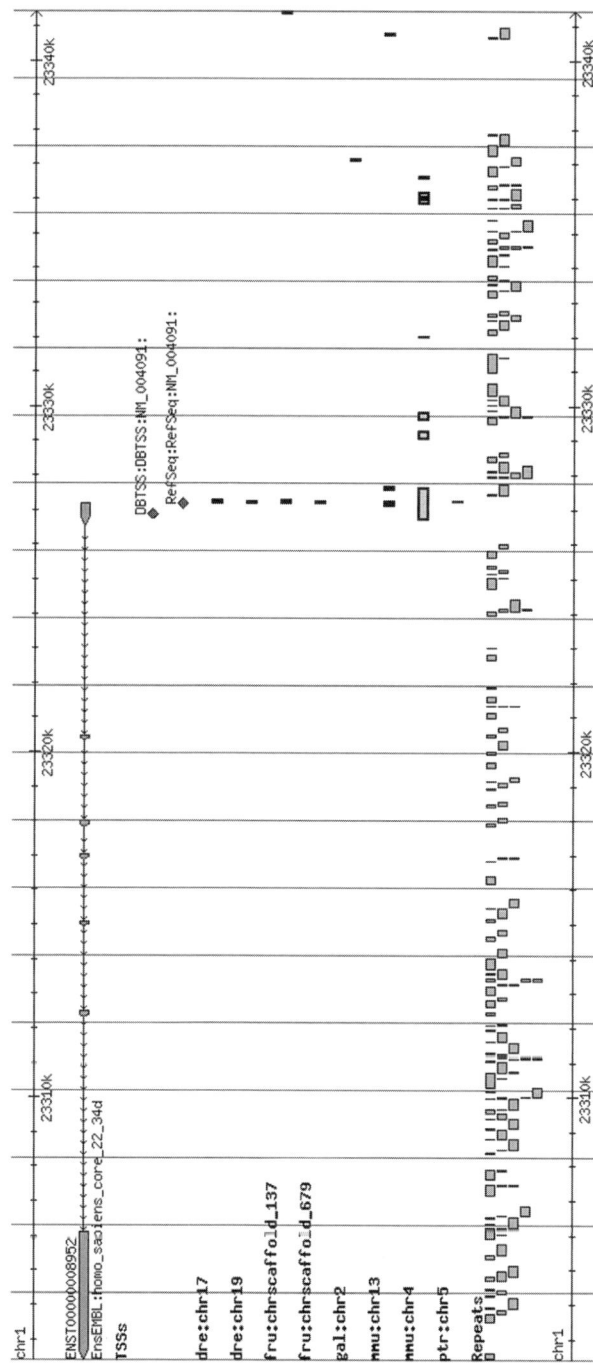

Fig. 3. Genomic context information page. The genomic context of E2F2 is displayed above. The structure of a single transcript is shown in the first row. Little arrows pointing left indicate the gene orientation. Validated transcription start sites are represented by little diamonds below. Local similarities to various species are shown as well as annotated repeats.

The following alignments have been found for your request:

Pairwise alignments of upstream regions:

Query Transcript	Options	Target Transcript	Chromosome	Upstream Start	Upstream Stop	Gene identifier
ENST00000008952	Chromosome: 1 Upstream Start: 23326760 Upstream Stop: 23341761					
	View Applet Get Alignments	ENSDART00000002343	17	19335530	19341427	ENSDARG00000011314
	View Applet Get Alignments	ENSDART00000011217	19	9312550	9317551	ENSDARG00000004035
	View Applet Get Alignments	ENSGALT00000020689	2	59238976	59253977	ENSGALG00000012680
	View Applet Get Alignments	ENSMUST00000016621	13	29873160	29888084	ENSMUSG00000016477
	View Applet Get Alignments	ENSMUST00000061721	4	134942921	134957922	ENSMUSG00000018983
	View Applet Get Alignments	ENSMUST00000066995	13	29808032	29888084	ENSMUSG00000016477
	View Applet Get Alignments	ENSPTRT00000032831	5	20890670	20905668	ENSPTRG00000017762
	View Applet Get Alignments	ENSPTRT00000032832	5	20890670	20905668	ENSPTRG00000017762
	View Applet Get Alignments	SINFRUT00000160939	scaffold_679	76468	81469	SINFRUG00000151346
	View Applet Get Alignments	SINFRUT00000161895	scaffold_137	280008	295006	SINFRUG00000152207

Fig. 4. Springboard page with overview of pairwise alignments. Here, all pairwise similarities in upstream regions of a user's query gene are listed in tabular form (example is shown for E2F2). Two alternative options are offered at this step. One may either visualize annotated phylogenetic footprints or export them in text format.

GFF is a well-recognized format for sequence annotation and is therefore suited to be imported into other software packages.

Example session:

3) The query gene is reported to lie on chromosome 1 and has seven exons. The gene structure seems to be correct. E2F2 has two annotated transcription start sites in CORG: one from the DBTSS project and another from RefSeq. Most species show some sequence similarity close to the RefSeq transcription start site. Confusingly, little sequence similarity is shown to chimp and none to rat. This may occur for many reasons, and the reader is referred to the Notes section.

3.3. Pairwise Sequence Conservation

All pairwise sequence similarities (local alignments) between the query sequence and assigned homologous sequences are listed in **Fig. 4**. The genomic position and EnsEMBL identifiers are given for each target sequence. From the option column, one may retrieve a graphical summary given in the JAVA-based Alignment block viewer (choose "View Applet") or obtain the alignment in textual form (choose "Get Alignments"). Additionally, one could check the identity of target homologs in EnsEMBL by following the link in the "Gene Identifier" column. Further down the page, multiple sequence similarities are listed.

Example session:

4) **Figure 3** clearly shows that the upstream region of E2F2 is most similar to a mouse region on chromosome 4, which is assigned to identifier "ENSMUSG 00000016477." Less similarity is observed for another mouse upstream region on chromosome 11, which is assigned to identifier "ENSMUSG00000018983." In terms of similarity, the mouse gene on chromosome 4 is most likely the orthologous gene and is therefore selected for further investigation.

5) Click on the "View Applet" link in the row of transcript "ENSMUST 00000061721." A new browser window should now appear within which the applet is initialized. Please be patient, as this may take a few seconds.

3.3.1. Instructions for Use of JAVA Applet

The left bar of the applet window is the legend or explanation guide to what is displayed. Each individual track can be activated or deactivated by clicking on a legend symbol. Inactive tracks are marked by the crossed-out legend symbol. Each legend symbol has a lever assigned to it. This lever can be used to gradually fade out the sequence features of the corresponding track according to their rank or score.

The central display area is to be read from left to right ("ATG" stands for translation start site) and holds the core information: connected colored boxes depict local similarities. Note the conserved synteny of most sequence similarity blocks in man-mouse alignments. Sequence annotation is only given for the query species. Annotated features include binding site predictions with position-weight matrices and consensus sequences, transcription start sites, and exons. The levers in the legend may be used to reduce the high number of some annotated features (e.g., PWM predicted binding sites). Considerably fewer motif hits are found for IUPAC consensus motifs, as this is a more stringent description of a DNA binding site. The magnification of the central display is adjustable down to the nucleotide level. Zooming in and out is attained by clicking on the corresponding buttons in the lower left region of the applet panel. To zoom in, the mouse must be dragged over the desired area.

Example session:

6) **Figure 5** depicts the applet start-up screen that one gets for the selected man-mouse comparison of E2F2.

7) Zoom into the region around the transcription start site. **Figure 6** shows that CORG identified the previously described autoregulatory feedback loop among other putative binding sites close to the transcription start (*6*).

Further information is available by clicking on sequence feature with the right mouse button (Ctrl + mouse button for Mac users). The user (in case of alignments) either sees a nucleotide-level display of the alignment or links out to some external web resource related to the sequence feature (**Fig. 7**).

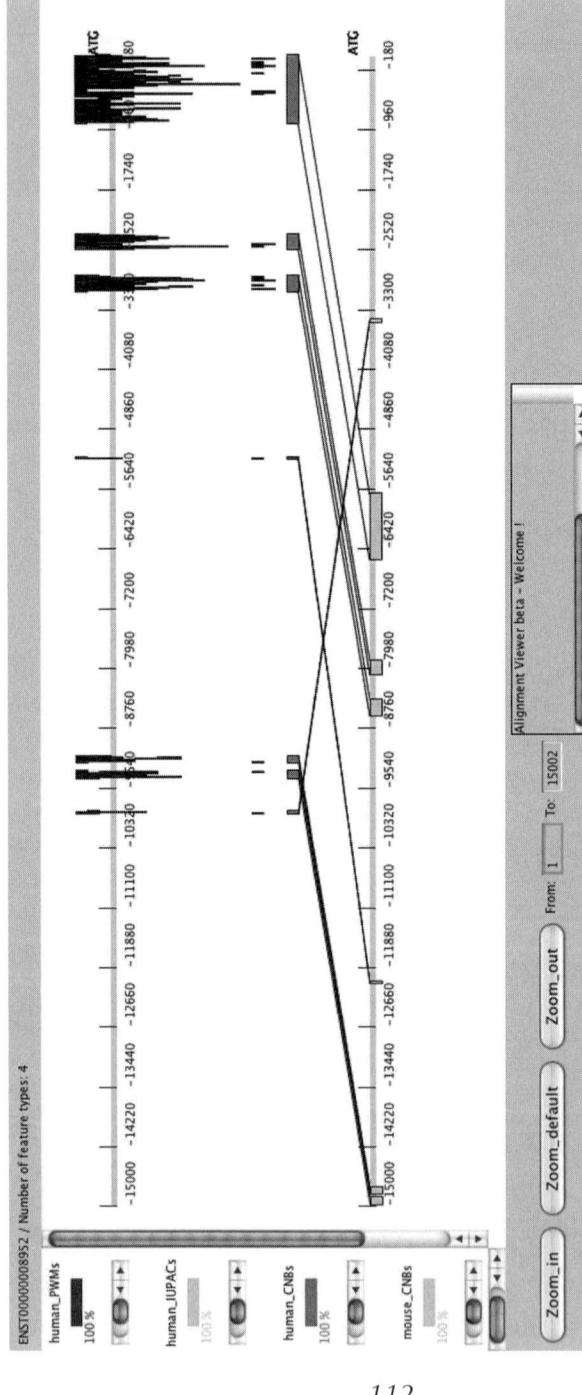

Fig. 5. Local pairwise alignments of the upstream region of man-mouse E2F2 orthologs. Linked colored boxes visualize the arrangement of phylogenetic footprints for man-mouse sequence comparison. The annotation of the query (reference) sequence is stacked on top of the footprints. A legend in the left panel describes the different features shown. The gene start in mouse seems to be shifted upstream relative to the human gene. This either indicates a wrong annotation or the shift of the translation start site to the second exon for mouse.

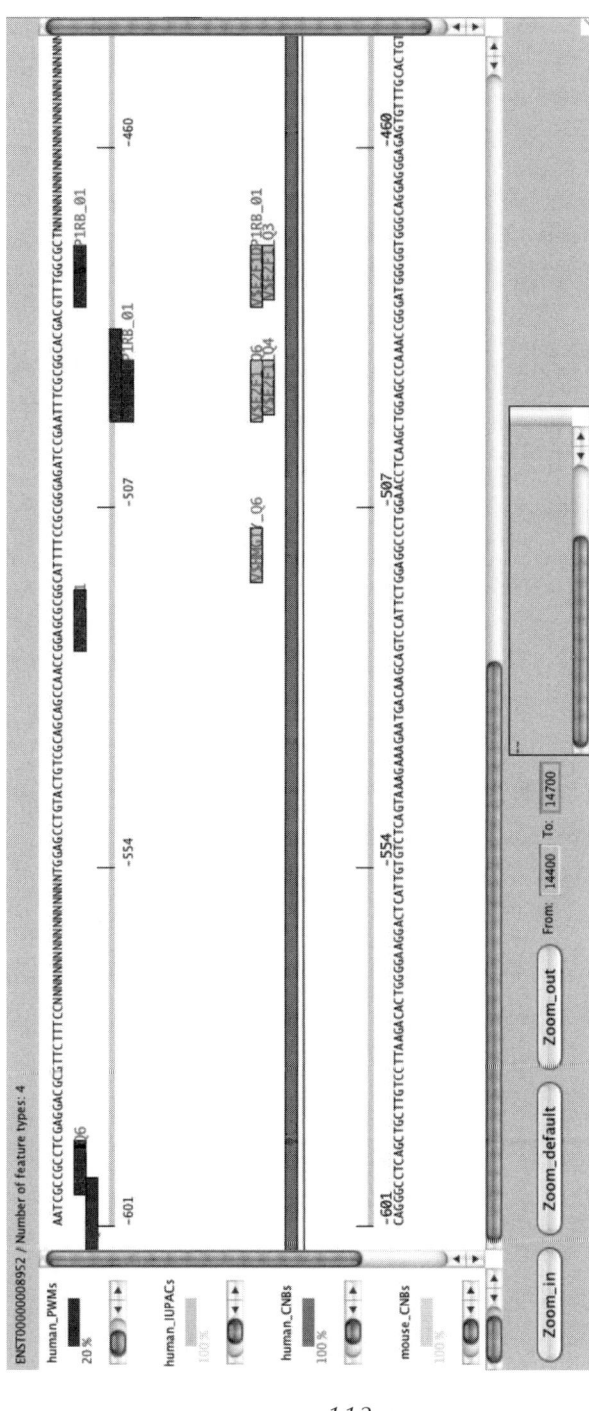

Fig. 6. Enlarged view of promoter region with two E2F sites in tandem. If one zooms into the rightmost footprint seen in **Fig. 5**, one gets to see the constituents of an experimentally validated feedback loop for E2F2. Two conserved E2F2 sites appear in close vicinity.

Fig. 7. Textual view of pairwise alignment. Click on the corresponding footprint with the right mouse button (Ctrl + mouse button for Mac users) and select the menu item "Alignment View" to activate this nucleotide-level view.

8) A compilation of all man-mouse conserved blocks upstream of E2F2 is available for download by clicking on the "Get Alignments" link on the springboard page.

3.4. Sequence Conservation in Multiple Species

Multiple pairwise comparisons of a reference sequence to various target sequences can be shown on the same screen by clicking on the multiple target sequence identifiers and pressing the "View Alignment (Star Topology)" button. This is particularly useful if one wants to contrast patterns of local similarity of a reference sequence to its homologs in other species. The option "View Alignments (Max clique topology)" lets one see all consistent multiple alignments. Consistent multiple alignments are built from pairwise alignments in such a way that each member sequence is sufficiently similar to all other members. If one is less interested in a graphical representation of the positioning of local similarities, all consistent alignments can be dumped via the text link "Dump multiple alignments."

Example session:

9) Return to the lower part ("Multiple alignments of upstream regions") of the overview page (**Fig. 8**).

Multiple alignments of upstream regions:

Query Transcript	Target Transcript
ENST00000008952	Dump multiple alignments
	☐ ENSDART00000002343
	☐ ENSDART00000011217
	☐ ENSGALT00000020689
	☐ ENSMUST00000016621
	☐ ENSMUST00000061721
	☐ ENSMUST00000066995
	☐ ENSPTRT00000032831
	☐ ENSPTRT00000032832
	☐ SINFRUT00000160939
	☐ SINFRUT00000161895
	clear | View Alignment (Star topology) | View Alignment (Max clique topology)

Fig. 8. Springboard page for multiple alignments. Two ways of presenting local multiple cross-species similarities are offered at the bottom table of the springboard page. One may either pick a set of upstream regions by their transcript identifier from the list and display all pairwise similarities to the reference sequence simultaneously or view consistent multiple alignments (*see* **Subheading 3.4.**).

10) Select the four fish entries ("ENSDART00000002343," "ENSDART 00000011217," "SINFRUT00000160939," and "SINFRUT00000161895") from the multiple alignment table.

11) Activate the simultaneous display of the corresponding pairwise alignments by clicking on the "View Alignment (Star Topology)" button.

12) By employing the previously mentioned applet functions like zooming and textual alignment display, it becomes apparent that all four fish promoters share the autoregulatory feedback loop (**Fig. 9**).

13) Going back to the multiple alignment table, one gets another confirmation by browsing through the textual representation of multiple alignments ("Dump multiple alignments" link).

4. Notes

This section deals with challenges and problems that may arise while one is working with CORG. Evidently, there are errors that are caused by browser configuration problems. The user can fix those in most cases. Other errors are rooted in external resource that are essential to CORG (e.g., EnsEMBL and TRANSFAC).

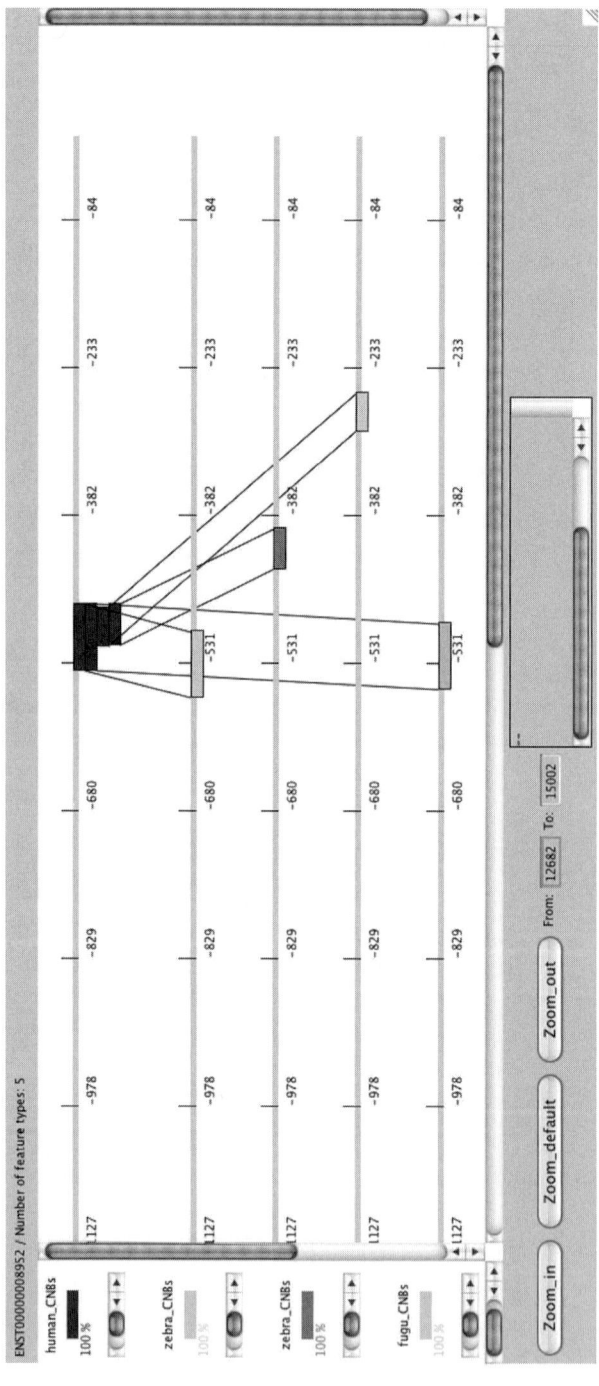

Fig. 9. Multiple pairwise similarities for the E2F binding site region. After selecting the four fish homologs ("ENSDART 0000002343," "ENSDART00000011217," "SINFRUT00000160939" and "SINFRUT00000161895") and choosing the star topology option for multiple alignments, one should get a display similar to the one above. The E2F binding sites in tandem are only conserved for the E2F2 orthologs in fish. Single binding sites were detected for E2F3 homologs close to the translation start site in fish.

Lastly, CORG itself it still under development and is by no means error free. CORG developers are keen on feedback from users. Please send your comments to corg@molgen.mpg.de.

4.1. General Remarks

The most problematic part of CORG is the JAVA-driven graphical display. If problems are encountered here, we advise the user to try a different web browser or make sure that JAVA is supported on your client computer.

4.2. Troubleshooting

4.2.1. Orthologous/Homologous Gene is Absent From Gene Group

The CORG database takes a gene-centered view of phylogeny. CORG employs single linkage clustering on the graph of EnsEMBL orthologous gene pairs to define the CORG gene groups. This approach is error prone in the sense that the grouping of genes is occasionally ambiguous, and homology relationships are not clearly resolved.

4.2.2. Gene Structure Looks Bizarre

First exon prediction is a difficult business. Therefore, it happens that the 5' end of a gene does not look like what one would expect. If the sequence region is still contained in the CORG upstream region, one may circumvent this problem by directly targeting this region.

4.2.3. Unexpected Degree of Similarity in Upstream Region

This could be either too little or too much conservation. For the latter case, consider using a more distantly related species to attain more discriminatory power for the footprinting approach (e.g., for mouse-rat comparisons, try man-mouse instead). In case of too little conservation, make sure that a true ortholog was considered. Alternatively, select a more closely related species. Since evolutionary parameters vary from locus to locus, the possibility cannot be excluded that some comparisons are uninformative.

References

1. Britten, R. J. and Davidson, E. H. (1969) Gene regulation for higher cells: a theory. *Science* **165**, 349–357.
2. Michael, L. and Robert, T. (2003) Transcription regulation and animal diversity. *Nature* **424**, 147–151.
3. Zhang, Z. and Gerstein, M. (2003) Of mice and men: phylogenetic footprinting aids the discovery of regulatory elements. *J. Biol.* **2**, 11.
4. Dieterich, C., Grossmann, S., Tanzer, A., et al. (2005) Comparative promoter region analysis powered by CORG. BMC *Genomics* **6**, 24.

5. Matys, V., Fricke, E., Geffers, R., et al. (2003) TRANSFAC: transcriptional regulation, from patterns to profiles. *Nucleic Acids Res.* **31,** 374–378.

6. Hubbard, T., Andrews, D., Caccamo, M., et al. (2005) Ensembl 2005. *Nucleic Acids Res.* **32,** (Database issue) D447–453.

7. Sears, R., Ohtani, K., and Nevins, J. R. (1997) Identification of positively and negatively acting elements regulating expression of the E2F2 gene in response to cell growth signals. *Mol. Cell. Biol.* **17,** 5227–5235.

9

cis-Regulatory Region Analysis Using BEARR

Vinsensius Berlian Vega

Summary

Genome-wide studies are fast becoming the norm, partly fueled by the availability of genome sequences and the feasibility of high-throughput experimental platforms, e.g., microarrays. An important aspect in any genome-wide studies is determination of regulatory relationships, believed to be primarily transacted through transcription factor binding to DNA. Identification of specific transcription factor binding sites in the *cis*-regulatory regions of genes makes it possible to list direct targets of transcription factors, model transcriptional regulatory networks, and mine other associated datasets for relevant targets for experimental and clinical manipulation. We have developed a web-based tool to assist biologists in efficiently carrying out the analysis of genes from studies of specific transcription factors or otherwise. The batch extraction and analysis of *cis*-regulatory regions (BEARR) facilitates identification, extraction, and analysis of regulatory regions from the large amount of data that is typically generated in genome-wide studies. This chapter highlights features and serves as a tutorial for using this publicly available software. The URL is http://giscompute.gis.a-star.edu.sg/~vega/BEARR1.0/.

Key Words: Regulatory region analysis; promoter analysis; motif search; microarray; transcription factor.

1. Introduction

Batch extraction and analysis of *cis*-regulatory regions (BEARR) *(1)* is a web-based tool devised with biologists as end users for assisting them in the regulatory sequence analysis surrounding their genes of interest. It was designed for high-throughput motif searches around regulatory regions of genes. BEARR can identify putative transcription factor binding sites of a large set of genes, such as those coming from microarray studies or other genome-wide investigations. The system contains two main parts: the regulatory region sequence extraction module and the sequence analysis module.

From: *Methods in Molecular Biology, vol. 338: Gene Mapping, Discovery, and Expression: Methods and Protocols*
Edited by: M. Bina © Humana Press Inc., Totowa, NJ

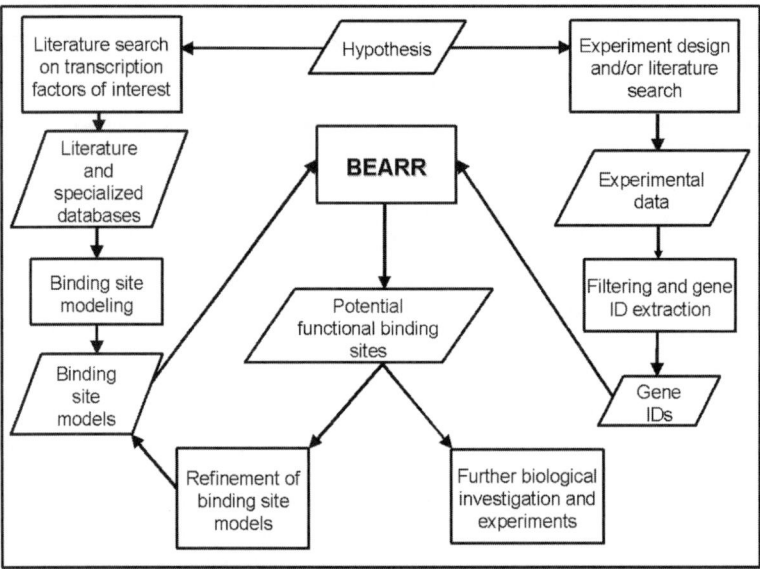

Fig. 1. A typical flow analysis employing BEARR. Fundamental to this analysis is the conception of the main hypothesis. Note that binding site modeling and prediction in fact forms a cycle, depicting the iterative nature of the process.

The following sections discuss problem abstraction and question formulation prior to running analysis with BEARR, describe how to perform an analysis, expound what can be expected from the analysis, explore features of BEARR, and suggest potential additional analyses.

2. Preanalysis Strategy

The value of BEARR analysis outcome greatly depends on the appropriateness of the questions being asked and the working hypothesis of the overall investigation (**Fig. 1**). Outlined in this section are a number of research questions that would benefit from BEARR, as well as other key components of the analysis.

2.1. Questions and Hypothesis

Two research frameworks could generally utilize BEARR. One deals with the task of predicting direct target genes of certain transcription factors. The goal is to identify genes that are potentially regulated by the transcription factors of interest. Typically, a genome-wide scan is required to exhaustively identify potential targets of the transcription factor, although in some cases the gene set might have already been restricted based on certain additional knowledge coming from experimental results or otherwise.

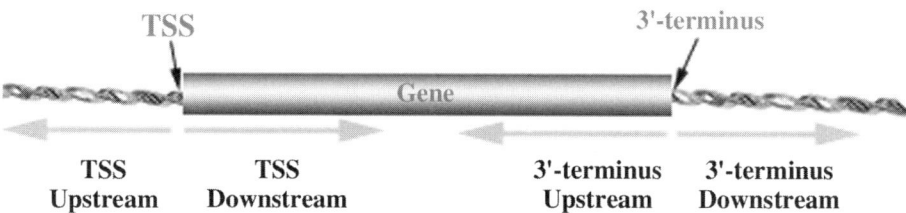

Fig. 2. Pictorial representation of the *cis* region definitions used in BEARR. Traditionally, only upstream regions of the transcription start sites (TSS) are considered *cis*-regulatory regions, but discovery of functional binding sites within introns or around the 3' terminus have dispelled that restrictive definition. BEARR supports automatic extraction of up/downstream regions around TSS and the 3' end.

The second type of investigation concerns more the question of what might regulate a given set of genes. Here, a much smaller set of genes than that of the above framework is usually involved, and usually users possess a list of potential transcription factors of the gene set and their corresponding binding site motifs.

2.2. Gene Sets, cis-Regions, and Motifs

Depending on the goal of the study, appropriate gene sets need to be generated. One needs to understand that such an in silico analysis is actually an experiment. A hypothesis must be formulated. Suitable control sets also need to be constructed. For example, if one is interested in determining whether a specific transcription factor governs the observed differential expression of a gene set, the nondifferentially expressed genes could be employed as the negative set or control set. This would also help in downstream statistical assessment of the findings.

Also important is the definition of *cis* regions. Transcription factors are generally expected to bind region upstream of the transcription start sites (TSS), although in many instances they bind downstream of the TSS up to the first intron, as well around 3' end untranslated regions (3'-UTRs) (**Fig. 2**). Thus, depending on the assumptions made and the purpose of the analysis, one might opt to examine large regions (approx 10 kbp) around the two ends of the genes (which is normally done at the beginning of an analysis exercise) or conservatively probe a few thousand nucleotides around the TSS (e.g., 3 kbp upstream and 500 bp downstream of TS). All of these are conveniently supported by BEARR.

The final key piece is the binding site model. Binding site consensus sequences are generally available for the known transcription factor, as it is the most basic and easily understood model. For well-studied factors, more complex models, such as the position-weight matrix (PWM) (*2*), might also exist. As much as possible, information on the sites should be gathered, for example: how much deviations from the consensus sequence can still be tolerated for binding, whether there are key positions that have to be conserved, and where binding is usually located. These might be found in public databases, e.g., TRANSFAC *(3)*.

3. Working With BEARR

This section details the key components of BEARR and highlights additional features. The BEARR website (http://giscompute.gis.a-star.edu.sg/~vega/BEARR 1.0) is publicly available for nonprofit research usage. Heavy and frequent users are advised to request the source code for a local installation within their own computing resources.

Figure 3 is a screen capture of the BEARR home page; numbered regions are key input areas. Frequent references will be made to this figure throughout the rest of the chapter. Readers are also encouraged to read and follow the tutorial provided in the web site.

3.1. Understanding the Pipeline

Upon the submission of an analysis request, BEARR will:

1. Collect all the information filled in the form.
2. Extract sequences surrounding the given genes to the extent defined by the user.
3. If consensus sequence(s) has been defined, possibly with the amount of acceptable nucleotide mismatches, search in the sequences for sites that matches.
4. If a PWM is also given, use it to score sites in the sequences.
5. Produce a summary report, detailing the sites found.

Thus, for an accurate analysis result, users need to pay additional attention while providing information for the following areas:

1. *Sequence extraction.*
   a. The two key parts in sequence extraction are the gene list (**Fig. 3**, input area 1) and the desired *cis* region for analysis (**Fig. 3**, input area 2).
   b. The **GeneIDs** input box accepts gene identifiers, delimited with white spaces (spacebar, tab, and newline characters). Strip the identifiers off extra characters (e.g., semicolons ";," commas ",") that are not part of recognizable identifiers. BEARR will not check or fix such errors. Although BEARR currently supports three different IDs, we only recommend the use of NCBI RefSeq ID *(4)* for stability reasons.
   c. In defining the *cis* regions to be investigated, make sure that upstream and downstream are correctly defined. The most common mistake is to confuse the meaning of upstream and downstream around the 3' terminus and for reverse-strand genes.
2. *Consensus search.*
   a. Users can specify the consensus patterns as single patterns (**Fig. 3**, input area 3) and/or two half-sites (**Fig. 3**, input area 4). The second option is for the numerous tandem binding sites. In both, users might also specify the amount of acceptable mismatches (or nucleotide deviations) from the inputted consen-

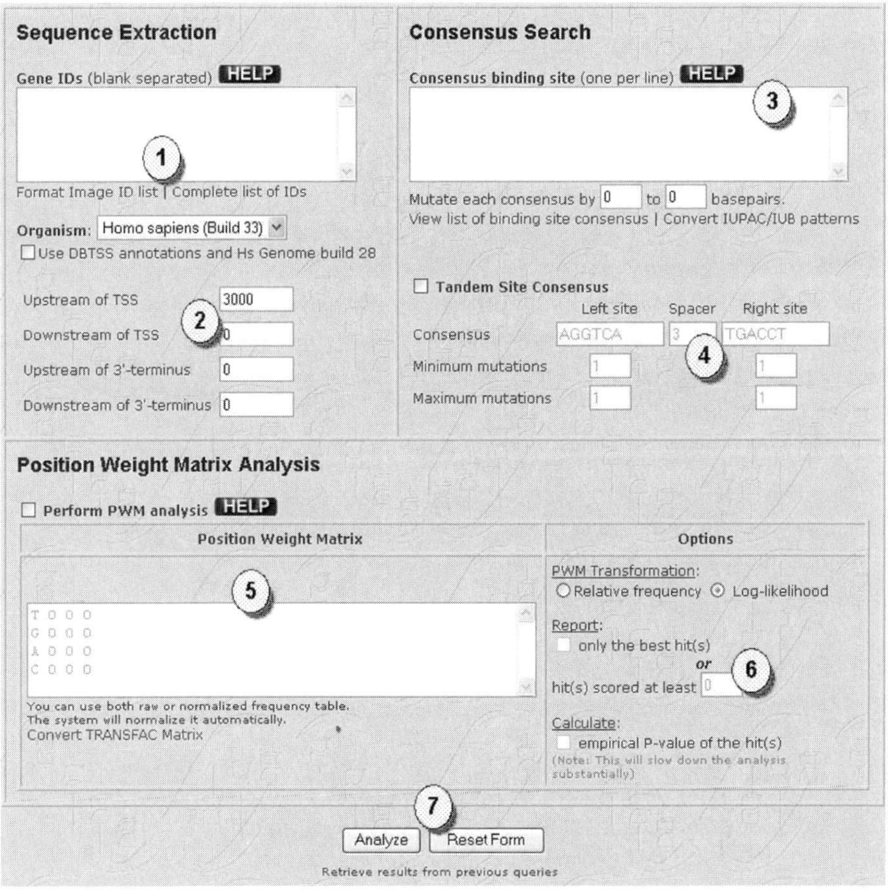

Fig. 3. BEARR main web site and interface. It consists of three clearly demarcated main parts: Sequence Extraction, Consensus Search, and Position-Weight Matrix Analysis. The input areas, which will be referred to frequently in this chapter, are numbered.

sus sequences. The two-half-site specification permits explicit control of the mutations on each half-site.

b. For pattern searching, BEARR adopts the use of Regular Expression *(5)*, as it is highly flexible and easy to understand, but slightly adjusts it to meet the need of nucleotide motif searches. Three main, yet less widely known, operators are the dots ".," square brackets "[]," curly brackets "{}," and brackets "()."

c. Dots represent the wild-card characters, which is the same as the "N" under IUB/IUPAC nucleic acid nomenclature, for example, **.TA** matches ATA, CTA, GTA, and TTA.

d. A set of nucleotides enclosed in a pair of square brackets signifies only a single position and that the listed nucleotides are allowed to appear in that position, for example, **TA[TA]A** will find TATA and TAAA.

e. In many instances, repeats of same nucleotides (or set of nucleotides) are prevalent. Curly brackets indicate that the preceding character (or set of character) is to be repeated *n* time (using "{n}") or at least *n* times and at most *m* times (using "{n,m}"), for example, **TA{2}** searches for TAA, **TA{2,3}** finds TA and TAA, and **[GA][GA][GA]C[AT][AT]G** can be written as **[GA]{3}C[AT]{2}G**.

f. Brackets are similar to square brackets, in that they define possible nucleotides at the given position. To further illustrate, **TATA** will only match TATA, **TA[TA] A** will find TATA and TAAA, and **TA{1,2}T** will search for TAT and TAAT.

3. *Position-weight matrix analysis.*

a. The PWM *(2)* models the strength of protein-DNA binding at each position for each nucleotide. Such a matrix can thus be used to assess the binding likelihood of a given site. BEARR asks users to input the raw nucleotide counts or relative nucleotide frequencies at each position of the desired binding sites (**Fig. 3**, input area 5). This can be derived from samples of known functional sites.

b. Users have the option to use the relative frequency for scoring or to transform it into log-likelihood scores. It is advisable that log-likelihood transformation is used. Should it be desired, one might ask BEARR to keep only the best site for each sequence or, alternatively, a cutoff threshold could also be specified (**Fig. 3**, input area 6).

3.2. Exploring Additional Features

A number of additional useful features are found in BEARR:

1. For users who rely on Incyte mouse IMAGE, the web site provides a tool to ensure that the identifiers being inputted are correctly formatted (**Fig. 3**, under input area 1).

2. We have incorporated DBTSS' experimentally determined transcription start sites *(6)*. The use of this information is optional (**Fig. 3**, under input area 1).

3. A list of some common transcription factors binding sites is also available, for easy reference (**Fig. 3**, under input area 3).

4. The IUPAC/IUB pattern to Regular Expression conversion tool is useful for users with IUPAC/IUB-defined binding site consensus sequences (**Fig. 3**, under input area 3).

5. TRANSFAC PWMS could readily be used in a BEARR analysis by first converting it into the BEARR PWM format using the online tool provided (**Fig. 3**, under input area 5).

6. As PWM raw output score might be too cryptic to be interpreted directly, users can optionally ask BEARR to estimate the *p* values of the putative sites (based on the PWM scores assigned to the sites) through Monte Carlo simulation *(7)* (**Fig. 3**, input area 6).

7. BEARR also allows users to retrieve previous analysis results, based on the unique **queryID** assigned to each analysis. Do note that owing to storage constraints, the BEARR server keeps the results for only a short period. Users are advised to download the results into their local computers (**Fig. 3**, under input area 7).

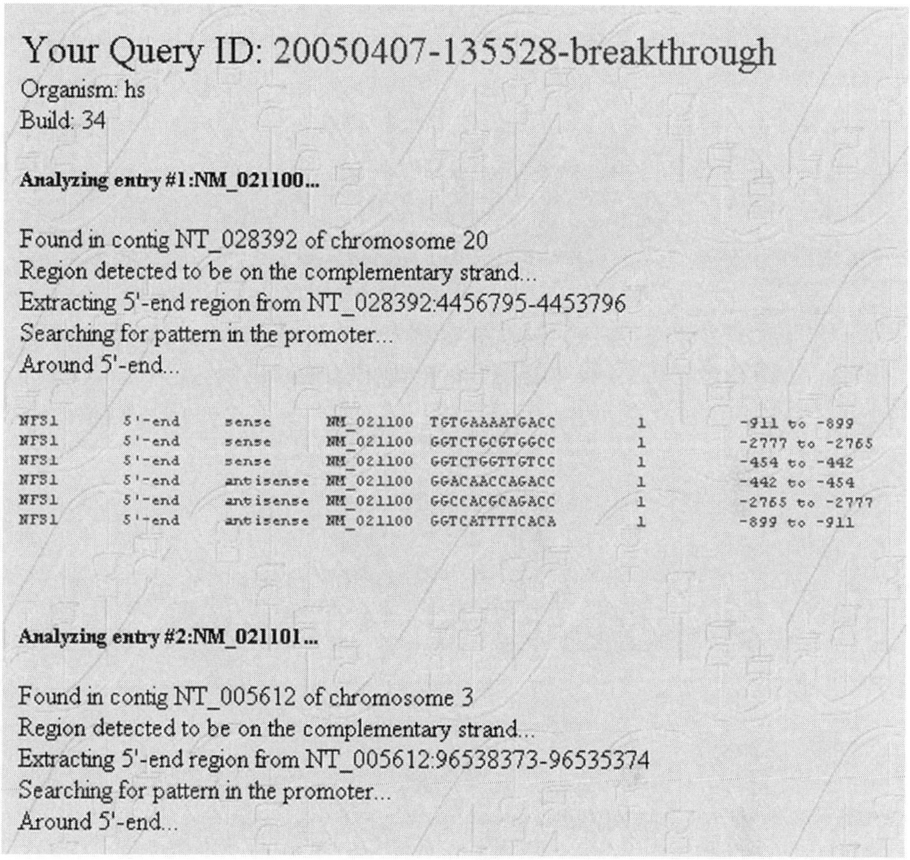

Your Query ID: 20050407-135528-breakthrough

Organism: hs
Build: 34

Analyzing entry #1:NM_021100...

Found in contig NT_028392 of chromosome 20
Region detected to be on the complementary strand...
Extracting 5'-end region from NT_028392:4456795-4453796
Searching for pattern in the promoter...
Around 5'-end...

NFS1	5'-end	sense	NM_021100	TGTGAAAATGACC	1	-911 to -899
NFS1	5'-end	sense	NM_021100	GGTCTGCGTGGCC	1	-2777 to -2765
NFS1	5'-end	sense	NM_021100	GGTCTGGTTGTCC	1	-454 to -442
NFS1	5'-end	antisense	NM_021100	GGACAACCAGACC	1	-442 to -454
NFS1	5'-end	antisense	NM_021100	GGCCACGCAGACC	1	-2765 to -2777
NFS1	5'-end	antisense	NM_021100	GGTCATTTTCACA	1	-899 to -911

Analyzing entry #2:NM_021101...

Found in contig NT_005612 of chromosome 3
Region detected to be on the complementary strand...
Extracting 5'-end region from NT_005612:96538373-96535374
Searching for pattern in the promoter...
Around 5'-end...

Fig. 4. A snapshot of the progress page. Shown here are the head of the progress page, which bears information regarding the QueryID, and the genome used. Intermediate results are output, allowing users to browse the results while waiting for the analysis to be completed.

3.3. Running the Analysis and Interpreting the Results

Once the necessary information has been supplied through the input fields and the form has been submitted by clicking the **Analyze** button (**Fig. 3**, input area 7), the analysis progress screen will appear, such as that shown in **Fig. 4**.

The progress page contains much useful information. The **QueryID** assigned for the ongoing analysis is prominently displayed at the top of the page. This ID is useful for future retrieval of the results. Information about the genome is also shown. This page is in fact an active page that keeps receiving data from the BEARR server until the analysis is complete. Closing the browser window of a running analysis might terminate the analysis. The page shows the real-

	A	B	C	D	E	F	G
1	Gene name	Region	Strand	Accession number	Pattern	Number of matches	Position of matches
2	ADCY1	5'-end	antisense	NM_021116	GGTCGCGGAGACC	1	-677 to -689
3	ADCY1	5'-end	sense	NM_021116	GGTCTCCGCGACC	1	-689 to -677
4	BAT5	5'-end	antisense	NM_021160	GGTCAGGCTGGTC	1	-1929 to -1941
5	BAT5	5'-end	sense	NM_021160	GACCAGCCTGACC	1	-1941 to -1929
6	C19orf15	5'-end	antisense	NM_021185	GGTCACATGGAGC	1	-1848 to -1860
7	C19orf15	5'-end	sense	NM_021185	GCTCCATGTGACC	1	-1860 to -1848
8	C20orf97	5'-end	antisense	NM_021158	GGTCAGGCTGGTC	1	-1182 to -1194
9	C20orf97	5'-end	sense	NM_021158	GACCAGCCTGACC	1	-1194 to -1182
10	C6orf47	5'-end	antisense	NM_021184	AGTCATAATGACT	1	-2779 to -2791
11	C6orf47	5'-end	antisense	NM_021184	GCTCAGTGTGGCC	1	-1619 to -1631
12	C6orf47	5'-end	sense	NM_021184	AGTCATTATGACT	1	-2791 to -2779
13	C6orf47	5'-end	sense	NM_021184	GGCCACACTGAGC	1	-1631 to -1619
14	CD209	5'-end	antisense	NM_021155	GACCAGTCTGACC	1	-2429 to -2441

Fig. 5. A sample output from consensus sequence search.

time progress of the analysis. After analysis of each gene ID, its relevant information and the analysis outcome are output, which allows the user to scrutinize the results even while the rest of the analysis is still in progress. Once the analysis has been completed, users are directed to the result page. The result page contains links to the various files, including the report file(s) and the extracted sequences in FASTA format. **Figure 5** shows an example of consensus sequence search results.

BEARR motif search outputs typically include the gene names, the *cis* region under study, the gene's accession ID, the patterns found, and their respective count, as well as the positions of the patterns. Note that negative positions denote upstream of the reference point (i.e., TSS or 3' terminus), positives mean downstream, whereas position 0 is undefined. For PWM analysis, the results might include PWM scores and (optionally) the estimated *p* value of the site. The length of DBTSS extension is also reported, for analysis that employs DBTSS-defined TSS.

4. Further Analysis

BEARR was not intended to be an all-in-all one-stop sequence analysis portal. A number of more specialized tools are available for more advanced analysis. Following a BEARR analysis, users should statistically assess the significance of the findings, to see whether they agree with the prior hypotheses. The right statistical tests need to be chosen, or an ad hoc one should be properly devised. More than that, at the sequence level, users can easily download the extracted sequences and analyze those using different tools. Discovery of overrepresented motifs, using algorithms like MEME *(8)*, GLAM *(9)*, or MITRA *(10)*, could potentially uncover important novel binding sites. On top of additional in silico analyses, relevant biological assays should be accurately designed to both validate the findings and discover new biology.

5. Summary

BEARR offers ease and convenience for experimental scientists who wish to perform high-throughput large-scale basic *cis*-regulatory region analysis. Developed as a web-based tool, it is extremely accessible and consumes virtually no processing power at the user end. Its two main modules, sequence extraction and sequence analysis, support automatic rapid extraction of regulatory regions based on gene identifiers and provide the first step of binding site analysis. Results of BEARR are easily construed. The raw output could also be effortlessly subjected to further in silico investigations.

Acknowledgments

Development of this tool and the computational resource described here are supported by funding from the Biomedical Research Council (BMRC) of the Agency for Science, Technology, and Research (A*STAR) in Singapore.

References

1. Vega, V. B., Bangarusamy, D. K., Miller, L. D., Liu, E. T., and Lin, C.-Y. (2004) BEARR: batch extraction and analysis of *cis*-regulatory regions. *Nucleic Acids Res.* **32,** W257–260.
2. Stormo, G. D. (1990) Consensus patterns in DNA. *Methods Enzymol.* **183,** 211–221.
3. Wingender, E., Dietze, P., Karas, H., and Knuppel, R. (1996) TRANSFAC: a database on transcription factors and their DNA binding sites. *Nucleic Acids Res.* **28,** 316–319.
4. Pruitt, K. D. and Maglott, D. R. (2001) RefSeq and LocusLink: NCBI gene-centered resources. *Nucleic Acid Res.* **29,** 137–140.
5. Kleene, S. (1956) Representation of events in nerve nets and finite automata, in *Automata Studies* (Shannon, C. and McCarthy, J., eds.), Princeton University Press, Princeton, NJ, pp. 3–42.
6. Suzuki, Y., Yamashita, R., Nakai, K., and Sugano, S. (2002) DBTSS: DataBase of human transcriptional start sites and full-length cDNAs. *Nucleic Acids Res.* **30,** 328–331.
7. Metropolis, N. and Ulam, S. (1949). The Monte Carlo method. *J. Am. Stat. Assoc.* **44,** 335–341.
8. Bailey, T. L. and Elkan, C. (1994) Fitting a mixture model by expectation maximization to discover motifs in biopolymers, in *Proceedings of the Second International Conference on Intelligent Systems for Molecular Biology*, Stanford University, AAAI Press, Menlo Park, CA, pp. 28–36.
9. Frith, M. C., Hansen, U., Spouge, J. L., and Weng, Z. (2004) Finding functional sequence elements by multiple local alignment. *Nucleic Acids Res.* **32,** 189–200.
10. Eskin, E. and Pevzner, P. A. (2002) Finding composite regulatory patterns in DNA sequences, in *Special Issue Proceedings of the Tenth International Conference on Intelligent Systems for Molecular Biology* (ISMB-2002). *Bioinformatics* **Suppl. 1,** S354–363.

10

A Database of 9-Mers from Promoter Regions of Human Protein-Coding Genes

Minou Bina, Phillip Wyss, and Syed Rehan Shah

Summary

Discovery of lexical characteristics of specific sequence motifs in human genomic DNA can help with predicting and classifying regulatory *cis* elements according to the genes they control. In lexical models, some "words" may serve as downstream targets of signaling systems, whereas other "words" may specify sequences that selectively control the expression of a subset of genes to produce the various cell types and tissues. To discover lexical features of potential regulatory "words," we have created a database of 9-mers derived from the promoter regions of a subset of human protein-coding genes. This report describes the procedure for extracting information from that database through the web.

Key Words: Human genome; regulatory motifs; gene regulation; *cis*-element prediction; regulatory words.

1. Introduction

The information in the human genomic DNA has been referred to as the book of life *(1)*. Linear arrays of three-letter words (codons) specify the coding regions of genes and thus the amino acid sequence of proteins. The language metaphor can be extended to state that genomic DNA should include regulatory words (*cis* elements) that specify the signals controlling the expression of genes *(2)*. *cis*-regulatory elements often specify functional transcription factor binding sites but may also include signals with currently unknown functions *(2–5)*.

Control signals may correspond to sequence elements that occur frequently in the regulatory regions of genes. Based on this hypothesis, researchers have compiled several collections of sequence motifs derived from the promoter regions of protein-coding genes *(2,6–8)*. It is thought that collections of this type could help with discovery of the sequence context of "words" that exert

From: *Methods in Molecular Biology, vol. 338: Gene Mapping, Discovery, and Expression: Methods and Protocols*
Edited by: M. Bina © Humana Press Inc., Totowa, NJ

control over gene expression *(2)*. Furthermore, discovery of lexical characteristics of specific sequence motifs in genomic DNA can help with classification of *cis* elements according to the genes that they control.

To discover the lexical features of regulatory words, previously we created a database of 9-mers derived from the promoter regions of a subset of the human protein-coding genes *(2)*. In this report we describe how information can be extracted from that database through the web: http://bina-grid.chem.purdue.edu/genome/.

2. Materials

To create a reference, we have computationally generated all possible 9-mers, producing 4^9 sequences *(2)*. To facilitate data utilization, each 9-mer was assigned an identifier (ID) *(2)*. To make the 9-mers independent from their orientation in DNA, the complementary 9-mers are considered as pairs. Therefore, the set contains 131,072 pairs of 9-mers. This set was named BINA_RF_sorted_master9 and can be downloaded from the web (*see* **Note 1**).

Each pair in that set defines the forward (f) and the reverse complement (r) of a 9-mer. For example, RF107075f corresponds to TCGCGAGCG, and RF107075r defines its reverse complement (CGCTCGCGA). The sequences are written in the 5' to 3' direction, by convention (*see* **Note 2**).

The human promoters that we have analyzed for data collection *(2)* correspond to a subset of sequences described in a previous report *(9)*. The sequences were derived with respect to the 5' end of a relatively large set of human expressed sequence tags (ESTs) *(9)*.

To facilitate data collection and management, we created a database. For the database engine, we chose MySQL (*see* **Note 3**). The data were collected by comparing each 9-mer in human promoter sequences (between positions −500 to +50) with the reference set consisting of all possible 9-mers (*see* **Note 1**).

A summary table was created to include the 9-mers that occurred between positions −500 to +50; between positions −50 to +50 (considered as basal promoter), and between positions −500 to −50 (considered as proximal promoter). The 9-mers were ranked according to their relative abundance in promoters and in total human genomic DNA *(2)*. The 9-mers that occur equally in the specified promoter regions and in the total genomic DNA are expected to have a ranking in the vicinity of 1. Statistically significant 9-mers would have a ranking of 5 or higher *(2)*. The data for proximal promoters can be downloaded from http://www.chem.purdue.edu/bina/Data.htm.

The database also includes the accession numbers of the ESTs used for localization of the predicted transcription initiation sites of genes. Perl scripts were developed for correlating the accession numbers to the definitions described in GenBank files (*see* **Note 3**).

3. Methods

The page at http://bina-grid.chem.purdue.edu/genome/ provides a link to an interface that would allow you to query the database. Several queries were tailored to facilitate retrieval of information about the rank of each 9-mer, in a given segment in the promoter region of human genes, and to obtain a listing of the genes with which each 9-mer was associated. Gene association was deduced from the definitions listed in GenBank files, on February 2004 (*see* **Note 4**).

Figure 1 shows the control keys through which queries can be made. Below, we explain the type of information that can currently be retrieved.

1. Use "Search 9_mers by ID" to obtain information about a 9-mer according to its ID (**Fig. 1**). The format is RF, followed by a number, followed by r or f (providing the orientation of the 9-mer). For example, type RF129572f and click on submit Query next to the query box. The output will provide:
    a. The ID of the sequence (in that example, RF_ID = RF129572f).
    b. The sequence of the 9-mer (TTTCGCGCC) associated with that ID.
    c. The number of times the sequence was found between positions +50 to –50 [Hits (+50/–50)], between positions –50 to –500 [Hits (–50/–500)], and between positions +50 to –500 [Hits (+50/–500)], of the analyzed genes.
    d. The rank of each 9-mer, reflecting its statistical significance in the specified region.
2. Use "Search 9_mers by sequence" to obtain information about a 9-mer according to its nucleotide sequence (**Fig. 1**). For example, in the query box, type TTTCGCGCC. The query typed in that box should contain exactly 9 bases (*see* **Note 5**). You will get the same information as that obtained from the query "Search 9_mers by ID," described above. You can use the "save as" option in your web browser to save the results as a text file.
3. Use "List genes corresponding to a 9_mer" to obtain a listing of genes whose promoter region includes a specific 9-mer (*see* **Notes 5** and **6**). For example, TTTCGC GCC includes the consensus sequence (TTTSSCGC) for interactions with the E2F family of transcription factors. Therefore, you might be interested in determining whether the occurrences of that 9-mer (in promoter regions of human genes) could be correlated with genes associated with the control of the cell-cycle. This information can be obtained by using the control key named "List genes corresponding to a 9_mer" (**Fig. 1**). The listing is for the occurrences between positions –500 to +50 of each gene. This region of some genes may contain more than one occurrence of a given 9-mer. In that case, the gene will be listed twice.

   As an example, **Table 1** shows a subset of the gene list obtained for TTTCGCG CC. The list includes the ID associated with that 9-mer and the GenBank accession number for the cDNA sequence with respect to which the promoter region of the gene has been defined. The list also includes the definition (Annotation) in the GenBank files (*see* **Note 7**). From the inspection of the listing in **Table 1**, we can identify several genes that are candidates for regulation during the cell cycle. Examples include dihydrofolate reductase; MCM5 minichromosome maintenance

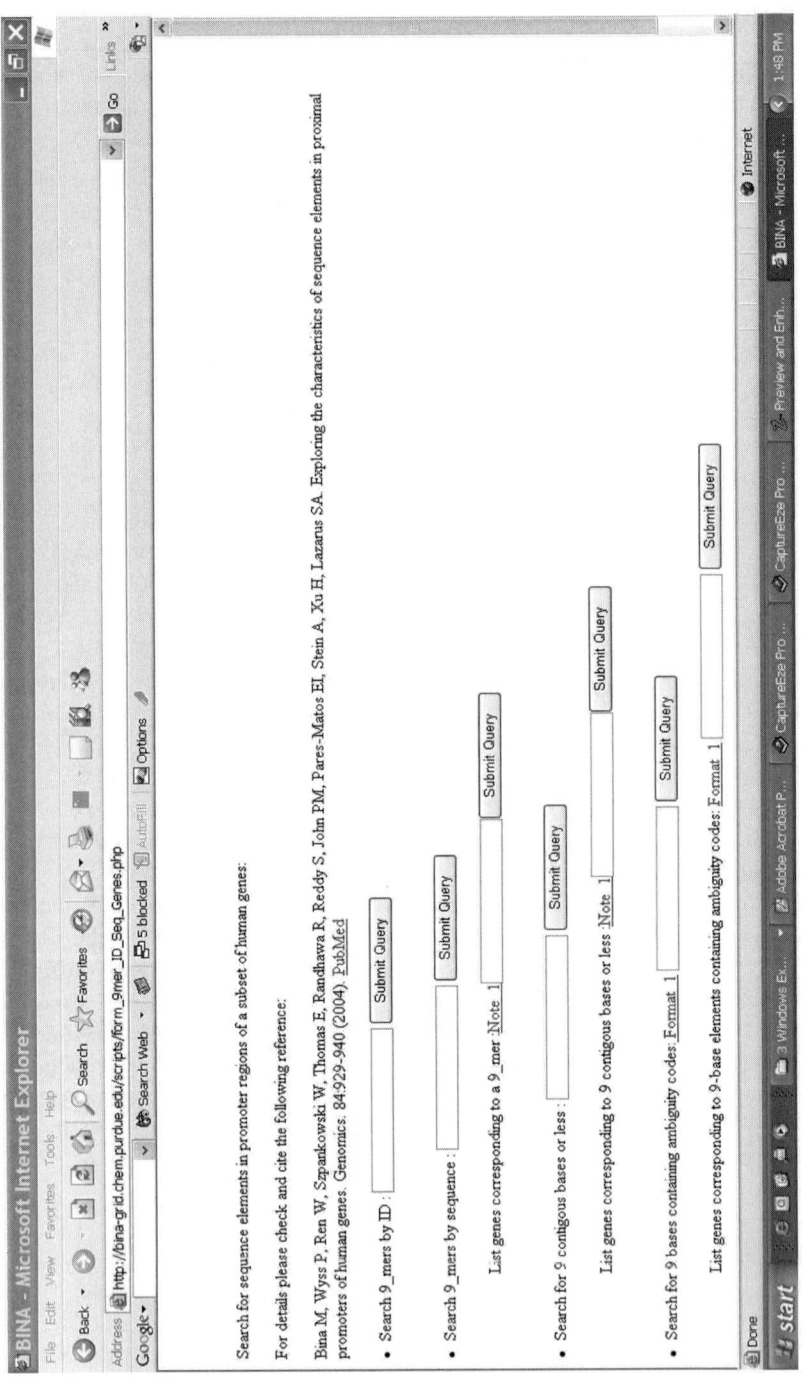

Fig. 1. The web-interface to a database of 9-mers from promoter regions of human genes.

Table 1
Example of Output of Gene List

RF_ID	9-mer	Accession no.	Annotation
RF129572f	TTTCGCGCC	BC000123	Pyridoxal (pyridoxine, vitamin B_6) kinase
RF129572f	TTTCGCGCC	BC142	MCM5 minichromosome maintenance deficient 5
RF129572f	TTTCGCGCC	BC000192	Dihydrofolate reductase
RF129572f	TTTCGCGCC	BC000893	Histone 1, H2bk
RF129572f	TTTCGCGCC	BC011685	Transcription factor DP-1

deficient 5, cell division cycle 46; and transcription factor DP-1 (**Table 1**). From the listing we can infer that TTTCGCGCC may function as a *cis* element in the regulation of genes associated with the cell cycle. Therefore, the results offer hypotheses that could be tested in experimental studies, for example, DNA binding assays and analyses of data obtained from microarrays.

4. Use the control key "Search for 9 contiguous bases or less" to obtain information about sequences that are shorter than 9 bases (*see* **Note 8**). For example, you may wish to identify sequence elements that include a potential E2F site. The results will be a listing similar to that obtained in **step 2**, but the list will include more hits (*see* **Note 9**).

5. Use "List genes corresponding to 9 contiguous bases or less" to type a sequence that is shorter than 9 to obtain a listing of genes whose promoters contain the sequence used as query (*see* **Note 9**).

6. Use "Search for 9 bases containing ambiguity codes" to type a sequence that includes ambiguity codes. To include ambiguity codes, use brackets. For example, the query GGGG[CT]GGGG will search for GGGGCGGGG and GGGGTGGGG. TT[ATGC]AA will search for TTAAA, TTTAA, TTGAA, and TTCAA. You can include ambiguity at several positions. The total length of the sequence in the query should add up to 9 bases or less (*see* **Note 8**). The bases in the brackets are counted as one base. For example, for the consensus E2F site (TTTSSCGC), the query would be TTT[GC][GC]CGC. The result of that query will list all 9-mers that contain TTTSSCGC, as well as the ranking of each 9-mer in the specified regions. From the ranks, you can identify the statistically significant sequences.

7. Use "List genes corresponding to 9-base elements containing ambiguity codes" to type a sequence that includes ambiguity codes to obtain a listing of genes that contain that sequence in their promoter region. The format of the sequence used as query is the same as that described in **step 6**. Clearly, short sequence motifs for TF sites would produce a long list. This would increase the number of the candidate genes. The tradeoff is that the list could include false predictions. Nonetheless, the query can narrow down the list of genes that could be tested for experimental validation (*see* **Note 9**).

4. Notes

1. The complete set (BINA_RF_sorted_master9) can be downloaded from http://www.chem.purdue.edu/bina/Data.htm.
2. Referring to 9-mers as pairs would allow their identification, irrespective of their orientation in DNA. This scheme also eliminates problems arising from redundancy, since it considers the complementary pairs to represent the same sequence element in the genomic DNA.
3. For more details, see Chapter 11 in this volume.
4. The definition in a GenBank file may change when it is updated. Therefore, for genes of interest check GenBank for the updated definitions. To do so, go to the nucleotide database at NCBI. For batch retrieval, go to Batch Entrez and upload a text-file that includes several accession numbers.
5. Other query boxes can be used to type a sequence containing fewer bases or a sequence that includes ambiguity codes. However, it will take a longer time to obtain the results.
6. The current database includes the promoter regions of nearly 4500 human protein-coding genes.
7. If the gene no longer exists in GenBank, you may get accessionnumber.promoter, with no associated definition.
8. Queries that are shorter than 9 bases and those that include ambiguity codes may take several minutes to produce an output.
9. To save the output in a text file, use the "Save As" option in the File menu in your browser.

References

1. Collins, F. S., Green, E. D., Guttmacher, A. E., and Guyer, M. S. (2003) A vision for the future of genomics research. *Nature* **422,** 835–847.
2. Bina, M., Wyss, P., Ren, W., et al. (2004) Exploring the characteristics of sequence elements in proximal promoters of human genes. *Genomics* **84,** 929–940.
3. Baldi, P., Brunak, S., Chauvin, Y., and Pedersen, A. G. (1999) The biology of eukaryotic promoter prediction—a review. *Comput. Chem.* **23,** 191–207.
4. Bina, M. and Crowley, E. (2001) Sequence patterns defining the 5' boundary of human genes. *Biopolymers* **59,** 347–355.
5. Lemon, B. and Tjian, R. (2000) Orchestrated response: a symphony of transcription factors for gene control. *Genes Dev.* **14,** 2551–2569.
6. Hutchinson, G. B. (1996) The prediction of vertebrate promoter regions using differential hexamer frequency analysis. *Comput. Appl. Biosci.* **12,** 391–398.
7. Marino-Ramirez, L., Spouge, J. L., Kanga, G. C., and Landsman, D. (2004) Statistical analysis of over-represented words in human promoter sequences. *Nucleic Acids Res.* **32,** 949–958.
8. FitzGerald, P. C., Shlyakhtenko, A., Mir, A. A., and Vinson, C. (2004) Clustering of DNA sequences in human promoters. *Genome Res.* **14,** 1562–1574.
9. Trinklein, N. D., Aldred, S. J., Saldanha, A. J., and Myers, R. M. (2003) Identification and functional analysis of human transcriptional promoters. *Genome Res.* **13,** 308–312.

11

A Program Toolkit for the Analysis
of Regulatory Regions of Genes

Phillip Wyss, Sheryl A. Lazarus, and Minou Bina

Summary

A major challenge in systems biology is to discover and reconstruct the *cis*-regulatory networks through which the expression of genes is controlled. Even though a variety of sequences have been shown to interact with the transcription factors that bind DNA, extensive work is needed to discover and classify regulatory "codes" and to elucidate the role played by the sequence context of genomic DNA in the regulation of genes. Databases of sequence elements extracted from regulatory regions may facilitate this process. This report provides a Toolkit and instructions for creating a database for collecting and analyzing 9-base elements (9-mers) from a large collection of DNA sequences. A reference set consisting of all possible 9-mers is included for extracting potential control elements, irrespective of their orientation and order in DNA.

Key Words: Regulatory code; human genome; gene regulation; transcription factor binding sites.

1. Introduction

Novel strategies could help with discovery and mapping of the "codes" in DNA that regulate the expression of genes. The completed sequence of the human genome *(1–3)* and the genomic sequences of model organisms offer a rich source of data for addressing this problem. Not surprisingly, most efforts have focused on discovery of codes that control the expression of protein-coding genes since these codes are the key components of the complex networks and pathways through which various cell types are produced.

We have aimed at discovering regulatory codes, irrespective of their orientation and order in genomic DNA *(4)*. The underlying hypothesis that drives our strategy is that short-sequence motifs occurring frequently in promoter regions

From: *Methods in Molecular Biology, vol. 338: Gene Mapping, Discovery, and Expression: Methods and Protocols*
Edited by: M. Bina © Humana Press Inc., Totowa, NJ

of genes may correspond to codes that act in *cis* to regulate the expression of linked genes *(4)*. This research problem has also been addressed by others (*see*, for example, **ref. 5–7**).

Our strategy relies on collecting 9-base elements that occur in the promoter regions of human genes *(4)*. The hypothesis is that these elements may encompass, include, or overlap with "words" through which the regulatory codes could be described *(4)*. To examine this hypothesis, we have created a "dictionary" consisting of all possible 9-mers, irrespective of their orientation in DNA *(4)*. Previously we used the dictionary as a reference for collecting 9-mers from proximal promoter regions that were experimentally defined and from promoter regions that were deduced with respect to the 5' end of a nonredundant set of human expressed sequence tags (ESTs) *(4)*. This latter dataset was obtained from a previous publication *(8)*.

In this chapter, we describe the schema of the database and the associated programs developed for data collection. The philosophy of the design was to create a set of relatively small programs that could be used along with system utilities and ad hoc scripts to create databases. The database was constructed to provide multifaceted views of the data as well as a mechanism to relate ancillary data to the originally processed data without having to reprocess the original sequences. Although this approach requires the researcher to be more comfortable in a shell/command-line environment, it allows flexibility in research direction. Furthermore, multiple contributors to the Toolkit need only know the general design philosophies instead of intricate knowledge of one or two monolithic programs.

The Toolkit for creating the database can be downloaded from the following web site: http://www.chem.purdue.edu/bina/Data.htm. Our studies have focused on analyzing the promoter regions of human protein-coding genes *(4)*. The strategy is general and can be applied to collect data from the regulatory segments of other species. Since the sequences in our dictionary are associated with identifiers (IDs), we propose that collections from various species could help with creating a framework for examining the evolution of regulatory codes in DNA. The reference table and the strategy may also help with a general format for organizing and studying the single-nucleotide polymorphisms (SNPs) associated with regulatory codes in DNA.

2. Materials

Data processing was done on a Red Hat Linux x86, version 6.2, operating system (OS). For the database engine, we used MySQL, version 3.24.40 (*see* **Note 1**). The programs for processing the sequences were written in Perl and executed under version 5.6.0 of the Perl interpreter *(4)*. Perl is well known for its text-processing capabilities as well as its wealth of add-on "modules" freely contributed to the Perl community (*see* **Notes 2** and **3**). Our Perl programs (*see* **Subheading 3.5.**) utilize several of the BioPerl 1.0 modules (*see* **Note 4**).

3. Methods

The database schema includes six tables (*see* **Tables 1–6** for a description of the fields). The database was designed to collect 9-mers, irrespective of their orientation in DNA (**Table 1**). In our studies, the 9-mers were collected using nonredundant datasets (*4*), that is, one gene per promoter DNA. The promoter regions were defined to include positions −500 to +50 with respect to the transcription initiation site (TI, +1 by convention). In data analyses, we examined three DNA segments: region A (positions −50 to +50); region B (positions −500 to −50); and region C (positions −500 to +50) (**Table 3**). In data collection, the programs also determine the location and direction of a 9-mer with respect to the transcription initiation site (**Table 6**). The database includes a table for recording the number of times a given 9-mer is found in total genomic DNA, the number of times a 9-mer is found in repetitive DNA, and the number of times a 9-mer is found in the coding regions (CDS) regions (**Table 2**). These data can be useful for statistical evaluations.

The command lines described in the following section begin with the characters *shell>*. This generically indicates the prompt that a command line interpreter (also know as a command shell) outputs to indicate it is ready for a command. The prompt is dependent of individual system configurations.

3.1. Setups for the General Procedures

1. To create a database, download the Toolkit from http://www.chem.purdue.edu/ bina/Data.htm. The programs in the Toolkit were written to populate and access the database tables via the DBI module, which is database independent (*see* **Note 3**). Other database engines (e.g., Oracle) may be used by installing the correct Perl DBD, for database-dependent modules (*see* **Note 3**), and defining the resulting database name in the environment variable DBI_DSN. **Subheading 3.5.** provides a description of the programs included in the Toolkit. In addition to the programs, the Toolkit includes two Perl modules that contain special-function packages used by the scripts.
2. Place the two Perl modules (Util9mer and mySequence::Experiment) in Perl's site library directory (*see* **Note 5**). Put Util9mer.pm directly in the site/lib directory and put Experiment.pm in a subdirectory named mySequence.
3. Download and install the Getopt::Long module used to parse the command line parameters of the Toolkit's programs (*see* **Note 3**). The general method for installing Perl modules is demonstrated by the installation of Getopt::Long module. Assuming that the tape archive file (tar) file is in your current working directory, the following procedure is needed for testing and installing the module:
   shell> tar −xvf Getopt-Long-2.28.tar

```
shell> cd Getopt-Long-2.28
shell> perl Makefile.PL
shell> make
shell> make test
shell> make install
```

The last of these commands require administrative (root) access to place the files properly in the Perl directory tree.

4. Download and Install the BioPerl modules in the lib/site subdirectories (*see* **Note 4**). BioPerl modules come with an installer, quite similar to the one described in **Subheading 3.1., step 3**, to place the packages correctly into the Perl directory tree (*see* **Notes 2** and **5**). The Toolkit uses several of these modules in the package. BioPerl requires several other modules that you will need to install (*see* **Note 3**). The installation for each is similar to the Getopt::Long and BioPerl installations.

5. Set the environment variables required for the Toolkit programs: SEQUENCE_ARCHIVE (root directory of the promoter sequences to be registered in the database); SEQUENCE_SOURCE (default directory of genomic DNA sequences that the user may want to examine (i.e., for research or publication); STORED_PROCEDURES (default directory for prewritten queries used by the scripts), i.e., DBI_USER (database user name) and DBI_DSN (database interface and database name), e.g., mysql: RFgenomeDB (*see* **Note 6**).

6. Download a copy of the genome of interest (e.g., human genome) in FASTA format. The sequence of the chromosomes can be retrieved from the ftp site at the Genome Browser at the University of California at Santa Cruz (UCSC; *see* **Note 7**). The page includes links to the listed genomes (*see* **Note 8**). Follow the instructions at the UCSC website to download the data that you want. In a previous publication, we analyzed an older version of the sequences of the human chromosomes *(4)*. For human sequences you can find all FTP downloads at ftp://hgdownload.cse. ucsc.edu/goldenPath/. The May 2004 sequence set is at ftp://hgdownload.cse.ucsc. edu/goldenPath/hg17/bigZips; the dataset split up into one file per chromosome is at ftp://hgdownload.cse.ucsc.edu/goldenPath/hg17/chromosomes/.

7. Download the sequences of promoter regions of interest in a single FASTA-formatted file (*see* **Note 9**). These can be downloaded from the Genome Browser at UCSC *(9–11)*.

3.2. Initializing the Database

In MySQL, as in any other SQL database engine, you use the "CREATE DATABASE" statement to make an empty database (e.g., RFgenomeDB) that can hold the database tables to be populated with records (*see* **Note 10**). Subsequently, you grant privileges for access, to yourself (e.g., joe) and other users. The "GRANT" command provides access to the database for all or specified users (*see* **Note 11**).

```
shell> mysql –u root
mysql> CREATE DATABASE "RFgenomeDB"
```

mysql> GRANT ALL PRIVILEGES ON RFgenomeDB.* TO "joe@localhost"
mysql> quit

3.3. Creating Database Tables

After creating an empty database, create the tables within that database. Use a script (named maketables.sql) that is included in the Toolkit. This script contains a series of CREATE TABLE commands to construct six tables: Sequence_ name, Expt, RF_code_locate, RF_code_hits, Genome_count, and Ranking (**Tables 1–6**).

shell> mysql –u joe
mysql> USE RFgenomeDB
mysql> source maketables.sql
mysql> quit

3.4. Populating the Tables

The order with which the tables are populated is based on convenience. It is possible to populate the tables in batches or manually, i.e., one sequence at a time. In this section, the emphasis is on procedures for analyzing large batches of sequences.

1. Use the following commands to populate three of the tables (RF_code_hits, Genome_ count, and Ranking; **Tables 1, 2,** and **3**) with the information stored in master9. This file contains all possible 9-mers (*see* **Note 12**). In the following command line, myquery corresponds to a general utility in the Toolkit (*see* **Subheading 3.5., step 8**). The procedure executed by the command line also sets the initial counts of each 9-mer to zero. The Ranking table uses only the RF_id, so the cut utility is used to create a separate temporary file.

   shell> myquery --parameters=master9 --query="INSERT INTO RF_code_hits (RF_id,RF_9mers) VALUES (?,?)"
   shell> myquery --parameters=master9 --query="INSERT INTO Genome_ count (RF_id,RF_9mer) VALUES (?,?)"
   shell> cut --f1 master9 >RFid
   shell> myquery --parameters=RFid --query="INSERT INTO Ranking (RF_id) VALUES (?)"
   shell> rm RFid

   Genome_count will hold information about the occurrences of each 9-mer in total genomic DNA and in sequences classified as repetitive DNA. RF_code_hits will summarize the data collection counts of the 9-mers in the promoter analyzed. Rankings summarize a calculated ranking for each 9-mer (**Tables 1, 2,** and **3**).

2. Update the counts of 9-mers in the Genome_count table using the downloaded genome files (in our case, human genome). This is done using the study_chromo program (*see* **Subheading 3.5., step 11**). In the total genomic sequences that are downloaded from the genome center at the UCSC (*see* **Note 7**), the regions that

correspond to repetitive DNA are shown in lower case letters. The study_chromo program counts the total occurrence of 9-mers in a specified chromosome file as well as the counts of all 9-mers that overlap a region of repetitive DNA. This program requires the files be a random access format. A file of this format can be generated from FASTA format using the fasta2random_access Toolkit program (*see* **Subheading 3.5., step 10**). Shown below is an example for creating a random access file and collecting the 9-mers that occur in human chromosome 1.

> *shell>* fasta2random_access chr1.fa
> *shell>* study_chromo chr1.ra >chr1.cnt
> *shell>* myquery --parameters=chr1.cnt --query="UPDATE Genome_count
>   SET Mer_count = Mer_count+?, Repeat_count=Repeat_count+? WHERE
>   RF_9mer = ?"
> *shell>* rm chr1.cnt chr1.ra

Repeat the procedure for each chromosome file (chr2.fa, chr3.fa, …) to build the total genomic count. The output of study_chromo is a file to be used by myquery to update the counts in Genome_count table. The example assumes the commands are executed with the current working directory set to the directory in which the chromosome files reside. Relative and absolute path names can be used to specify input and output if that is not the case (*see* **Note 13**). The chr1.cnt and chr1.ra file may be deleted afterward.

3. Use the utility split_seq for creating a separate file for each promoter sequence from the downloaded multisequence file. This provides for more efficient processing of the promoter sequences in subsequent steps (*see* **Subheading 3.5., step 9**). When splitting the multipromoter-sequence file, the program names the resulting promoter files according to the accession number of cDNAs from which the promoter regions of genes were derived. Files are placed in the current working directory. In the example, the downloaded FASTA file is called promoter_sequences, located in the home (default) directory, although it may be located via any path name (*see* **Note 13**). The noprompt option instructs the program to extract all files without any additional user intervention. The example places promoter files in a subdirectory called promoters of the SEQUENCE_ARCHIVE directory (*see* **Subheading 3.5., step 1**).

> *shell>* cd $SEQUENCE_ARCHIVE
> *shell>* mkdir promoters
> *shell>* cd promoters
> *shell>* split_seq --noprompt $HOME/promoter_sequences
> *shell>* cd $HOME

4. From GenBank, obtain the files for the cDNA sequences used for localization of the promoter regions. Each promoter has been named according to the accession number of the cDNA to which it corresponds *(8)*. The actual cDNA sequence files are useful for analyzing the number of times a 9-mer occurs in the coding regions (CDS) and for obtaining information about the encoded proteins. This information (DEFINITION in GenBank files) is added to the database. To do so, the first step is to collect a listing that contains the cDNA accession numbers associated with the

promoter sequences. This file can then be uploaded in Batch Entrez (*see* **Note 14**). Following the example in **Subheading 3.4., step 3**, to get a listing of cDNA accessions, use this command:

    shell> ls $SEQUENCE_ARCHIVE/promoters/* >cDNA_accession

The file cDNA_accession should be uploaded to Batch Entrez (*see* **Note 14**). You will obtain a multisequence GenBank file. Place this file in the $SEQUENCE_ ARCHIVE directory. Batch Entrez will also provide a file of any sequences that were withdrawn from GenBank. Remove them from the $SEQUENCE_ARCHIVE/promoters directory before continued processing.

5. Use the Toolkit program gb_titles to extract the accession number to gene definition (title) mapping. This mapping is used for adding to the database the definitions in GenBank files. In the continuing example, assume that the downloaded multisequence file is named cDNA_sequences_march22_05.gb. The following command extracts the information to a file called "titles.map":

    shell> gb_titles cDNA_sequences_march22_05.bg >titles.map

The output, titles.map, will be used below, in the next step of data processing.

6. For data processing, analyze the promoter sequences you have generated in **step 3**. The analysis uses several programs for populating specific tables. The description shown below is for collecting the data in batch.

In **step 3**, a multisequence promoter file was split, creating a file for each promoter sequence. For data collection, each promoter sequence must be registered. Registration populates the Sequence_name table (*see* **Table 4**) with information extracted from each sequence file and augmented by information provided in the command line (*see* **Subheading 3.5., step 2**). In the example below, the information extracted in **Subheading 3.4., step 5** is inserted as part of the registration process via the command line. This registration is associated with a "labname" to which it is referred by study_seq (*see* **Subheading 3.5., step 3**). The study_seq program collects data about the location of 9-mers occurring in a registered sequence. The program populates the Expt table (**Table 5**) with its input process parameters. The program also populates RF_code_locate (**Table 6**) with location and orientation of 9-mers. Study_seq can flag each 9-mer location as being in a repetitive versus nonrepetitive DNA region. In promoter sequences downloaded from the genome browser at UCSC, regions that correspond to repetitive DNA are shown in lower case letters.

In the course of registering and analyzing thousands of promoter regions, it is wise to automate the process. It is simple to do this in the example case as all promoter sequences have the same transcription initiation site (+1) and are all of the same length. The following *ad hoc* shell script does this for the promoter sequence files created in the example of **Subheading 3.4., step 3** incorporating the data from the GenBank files obtained in **Subheading 3.4., step 4**. The labnames created are Promoter.1, Promoter.2, and so on (*see* **Note 15**).

```
#!/bin/csh
#
@ x=1
```

```
foreach file ($SEQUENCE_ARCHIVE/promoters/*)
    set filename='basename $file'
    echo $filename
    set title='grep "^$filename" titles.map | sed –e "s/^[^\t]\t]//"'
    register_seq --noprompt --format=fasta --idtag=gb promoters/$filename --
title="$title" "Promoter.$x"
        study_seq --noprompt --start=-500 --TI_site=551 --flag Promoter.$x
        @ x += 1
end
```

Assuming the script file is called process_promoters, an easy way to execute the script is

    shell> csh process_promoters

7. Use the information in RF_code_locate table (**Table 6**) to obtain the number of times a 9-mer was encountered in all analyzed sequences to update the RF_code_ hits table. The update_hits program will do this (*see* **Subheading 3.5., step 4**).

    shell> update_hits

8. Populate Genome_count (*see* **Table 2**) to include the number of times a 9-mer has been found in the coding regions of genes. This information is obtained as aggregate 9-mer counts from the coding region of the cDNA sequences downloaded from GenBank (*see* **Subheading 3.4.**).

    Use the study_cds (*see* **Subheading 3.5., step 12**) to extract the information into file cds.par. The input to study_cds is the GenBank file obtained in **Subheading 3.4.** There is no repetitive DNA indication in the CDS regions of the GenBank file, so that field is cut out. Use myquery with the cds.par file as parameter input to update the Genome_count table's CDS_count field with the extracted data.

    shell> study_cds cDNA_sequences_march22_05.gb | cut –d" " –f1,3 >cds.par
    shell> myquery --parameters=cds.par --query="UPDATE Genome_count
        SET CDS_count = ? WHERE RF_9mer = ?"

9. Compute rankings of 9-mers based on their over/underrepresentation in the promoter region. To do this we need: the total number of 9-mers in the human genome (G), the total number of 9-mers in the promoter regions of interest (E), the number of times the ith 9-mer occurs in the human genome (G_i), and the number of times the ith 9-mer occurs in the proximal promoter region of interest (*4*). The ranking is

$$\frac{\dfrac{E_i}{E}}{\dfrac{G_i}{G}} = \frac{GE_i}{EG_i}$$

To get the total number of 9-mers in the human genome (G), simply sum up the count of all the 9-mers:

    shell> myquery –query="SELECT SUM(Mer_count) FROM Genome_count
    SUM(Mer_count)
       5731480962

1 row(s) selected

In this example, the promoter region of interest is from −500 to −50 with respect to the transcription initiation site. Since each record in the RF_code_locate represents one record, all records with location from −500 to −50 will be counted to get E.

> shell> myquery –query="SELECT COUNT(Location) FROM
> RF_code_locate WHERE Location >= -500 AND Location <= -50"
> COUNT(Location)
> 19181030
> 1 row(s) selected

This gives

$$\frac{G}{E} = 298.81$$

This is used to generate the ranking of each 9-mer using the rank.sql file in the Toolkit. In this file, the phrase FACTOR is replaced with the value, and the resulting file is placed in the STORED_PROCEDURE directory (*see* **Subheading 3.1., step 5**). The following will generate a ranking for the −500 to −50 promoter region for all 9-mers in the file rankings.

> shell> myquery -noheader rank.sql -- . −500 −50 >rankings

The -- in the command prevents the −500 and −50 from being processed as command options. The dot is a regular expression indicating that all flags (i.e., both R and N, repetitive DNA and nonrepetitive DNA regions) are to be counted. See more about regular expressions and the REGEXP operator in MySQL documentation. For convenience, the rankings are placed in the Rankings table (**Table 3**). The region −500 to −50 is associated with the Rank_B and Rank_B_Ei fields. The same procedure can be used to associate other regions with other rank fields.

> shell> myquery –parameters=rankings --query="UPDATE Rankings SET
> Rank_B_Ei=?, Rank_B=? WHERE RF_id=?"

3.5. Programs in the Toolkit

1. Use the explain program in the Toolkit to list a description of command line option and environment variables required for each perl script. The explain program extracts the first segment of POD (plain old documentation) from the programs. Explain searches the PATH environment for the program specified in the command line. Example:

   > shell> explain study_seq

2. Use register_seq to register a sequence selected for data collection. The result of this program is to populate the Sequence_name table (*see* **Table 4** for a listing and descriptions of the fields in Sequence_name). The Sequence_name table is the main table to which almost all other tables refer, either directly or indirectly. Other data analysis programs that analyze the sequence place the sequence identifier (Sequence _id in the Sequence_name table) in other tables along with the results of the analysis. This allows the results to be linked back to all the information in the Sequence _name table.

The information stored in the sequence name table is extracted from the sequence file registered. Command line parameters may give values to fields not extractable from the sequence file itself. For example, FASTA files do not always have a description of gene function, so the command line option named title is used to specify it. Command line options also specify the format of the file. In the previous section, the files were analyzed in batch. In the example shown below, a single-sequence file (U24128) is analyzed to specify the format, and to include a title (or definition) and a labname (PC1). From a GenBank-formatted file, the definition is automatically extracted.

In the example shown below, the file should be located in the SEQUENCE_ ARCHIVE directory or subdirectory thereof (*see* **Subheading 3.1.5.**).

shell> register_seq --format=fasta --title="prohormone convertase" U24128 PC1

Otherwise, the subdirectory must be specified along with the sequence file name. For example, if the sequence file were in the subdirectory hs (for homo sapiens) in the SEQUENCE_ARCHIVE directory, the command line would be

shell> register_seq --format=fasta –title="prohormone convertase" hs/ U24128 PC1

3. Use study_seq for extracting the 9-mers from a registered promoter sequence. The output places the 9-mers according to their name, their sequence, and their location within the promoter sequence into the RF_code_locate table (**Table 6**). The operation identifies the 9-mers in both forward and reverse direction (*see* **Note 12**). In the RF_code_locate table, the field named location captures the position of the 9-mer, and its complement, with respect to the transcription start site. Another field (named Direction in the RF_code_locate Table) provides the orientation (+ or −) with respect to the chromosome strand.

The RF_code_locate table also includes a field to flag a 9-mer as special. The study_seq program will record R in the flag field if that instance of the 9-mer is found in a repetitive DNA sequence in the promoter region and flagged as N if there is no overlap with DNA marked as repetitive.

The processing parameters of study_seq and the sequence_id of the analyzed promoter sequence are stored as records in the Expt table (*see* **Table 5** for a listing and descriptions of the fields in Expt). When you invoke the program, include the required criteria. In the example:

shell> study_seq --flag --start=-300 --end=+55 --TI_site=4375 PC1

The program is instructed to use the sequence with labname PC1, to start from position −300, and to end at position +55 with respect to the transcription initiation site (TI). The TI_site, in the command line, informs the program about the position of the transcription initiation site in the promoter sequence that is being analyzed for data collection. This feature is particularly useful for collecting data from GenBank files since these files are numbered with respect to the beginning of a submitted sequence. For historic reasons, the study_seq program also handles RF_ code_locate tables without the flag field. The flag option must be used with study_seq if the flag field is included; a database error will be generated if it is not included.

4. Use update_hits to update the collective hit counts (number of occurrences) for each 9-mer in RF_code_locate into RF_code_hits table. The program counts the number of times each 9-mer occurs in the RF_code_locate table and updates the Num_of_hits field in the record with the matching 9-mer. A query file with parameters can be specified to return a subset. Suppose that a file in the STORED_PRO-CEDURES (*see* **Subheading 3.1., step 5**) directory named hit_filter contained the following query:

> SELECT COUNT(RF_id), RF_id, FROM RF_code_location WHERE Flag = ? GROUP BY RF_id

The command line line

> *shell>* update_hits --query=hit_filter:N

Would count only the 9-mers flagged with N, that is, ones not overlapping with a region of repetitive DNA.

5. Use hit_summary to produce a GCG-formatted report (*see* **Note 16**) on the 9-mers using both the RF_code_hits and RF_code_locate database tables. The report lists each 9-mer, RF-id of the 9-mer, the number of registered sequence in which it appears (by labname), and the location of the 9-mer in each of the sequences in which they appear. Since study_seq populates the RF_code_locate table, only registered sequences that have been studied will appear in the report. Command line options for restricting the report to a specific hit count or 9-mers with hit counts greater than some threshold are available.

6. Use myquery to execute predefined or on-the-fly SQL statements. Query parameters can be specified on the command line or in a parameter file. Predefined queries are stored in the directory defined by the environment variable STORED_PROCEDURES. When the statements are executed, parameter place-holders, indicated using question marks in the SQL statement, are replaced by the parameters in the command line or parameter file in the order in which they occur. The parameters on the command line are the last items on the command line. The query is executed once for command line parameters. The query is executed as many times as there are lines in a parameter file. The program allows not only for selecting some dataset from the database, but also for operators to populate and update arbitrary tables in the database.

7. Use split_seq to extract one or more sequences in a multisequence file to separate sequence files (*see* **Note 17**). By default the files are named by the accession number of the sequence at each position within the combo-file. The individual files are named by their accession number or other identifier in file. In FASTA files, there are sometimes several identification tags listed (gi and gb), and any can be used to name the files using the idtag option. In interactive mode, the user is asked whether the file should be extracted. The --noprompt command option skips the user prompt. The format of the multisequence file can be any that the BioPerl package can interpret, FASTA and GenBank being the two obvious formats.

8. Use fasta2random_access to convert a FASTA-formatted file to a random access file. This format makes it easier to convert base coordinates to file coordinates as

they are equal. Programmatic access to locations within the file are done with *seek* or other similar file system calls (*see* **Note 18**). Random access files are created for large sequence files, typically representing an entire chromosome. Programs to extract subsequences from the chromosome given the sequence coordinates are extremely simple to program.

9. Use study_chromo to obtain a count of all 9-mers in a random access sequence file. The program name implies that the file represents a large sequence, representing an entire chromosome. The results are outputted in column format that is structured to be used as input as a parameter file to myquery to update tables. As with other programs that produce only aggregate count values, study_chromo does not require the chromosome sequence to be registered in the database. Instead, the file name of the file containing the sequence is specified in the command line. The program provides a count of each 9-mer in a chromosome. The program also produces a count of the 9-mers in segments with lower case bases, denoting regions of repetitive DNA. The program does have a command line option to count oligomers of different sizes.

10. Use study_cds to count 9-mers in the CDS region of a GenBank-formatted sequence file. The format must be GenBank as the program uses the features documented in the file to determine the CDS region. If the file is a multisequence file it will count each 9-mer in all the CDS regions in total (not per sequence). As with other programs that produce only aggregate count values, study_cds does not require the chromosome sequence to be registered in the database. Instead, the file name of the file containing the sequence is specified in the command line. The program does have a command line option to count oligomers of different sizes.

11. The gb_titles script is more of an ad hoc script than the other more flexible programs provided. It has no command line options. It also removes the string "Homo Sapiens CDS" and similar strings from the description as the project involved human genome analysis. It is, however, extremely simple as it uses the BioPerl functions to do the extraction of the description line. Only a simple knowledge of Perl is needed to modify the script to handle another specific species.

12. The retrieve_seq and download_seq programs are command-line programs that will download a sequence file from GenBank given an accession number. The output option allows the file to be named and placed to non-default locations. The default name is the accession number. The default place is SEQUENCE_SOURCE (retrieve_seq) and SEQUENCE_ARCHIVE (download_seq). Specification of an absolute path name will place the file in the explicit path (*see* **Note 13**). The format option allows the file to be stored in formats such as GenBank, FASTA, or others.

4. Notes

1. MySQL is available through a no-cost license for noncommercial use at http://www.mysql.com/.
2. Perl is installed by default on most Linux distributions along with many publicly available modules. See http://www.perl.org/ for stand-alone Perl distributions.

3. The repository for many publicly available Perl modules is http://www.cpan.org/. Both modules for download and their documentation can be found there.

4. See http://www.bioperl.org/.

5. There are several such version-dependent directories. For version 5.x.y of Perl, one is typically /usr/lib/perl5/site_perl. The command to list these directories is "perl –V." Look for the lines following "@INC:."

6. The commands to set environment variables are shell dependent:
   c-shell: setenv SEQUENCE_ARCHIVE /var/genome.
   sh, bash: export SEQUENCE_ARCHIVE=/var/genome.

7. UCSC Genome browser is located at http://genome.ucsc.edu/.

8. A list of genome sequences at the UCSC can be obtained from http://hgdownload. cse.ucsc.edu/downloads.html.

9. The promoter sequences should be a nonredundant set: one promoter per gene. For human promoters, we have analyzed *(4)* a subset of sequences published in **ref. 8**. From the browser at UCSC (http://genome.ucsc.edu/cgi-bin/hgTables), it is possible to retrieve sequences corresponding to promoter regions, predicted with respect to ESTs *(8)*. However, the retrieved sequences should be checked for redundancies. Alternatively, you can create a nonredundant set of accession numbers of cDNA sequences for upload and retrieval of the corresponding promoter sequences. The promoters that you will obtain will be predicted with respect to the 5' end of the uploaded cDNA sequences defined by accession numbers.

10. Standard Query Language (SQL) is an ANSI standard that several relational databases use. Though it does not conform strictly to the standard, MySQL does conform in the relevant commands used here. These commands are well documented in MySQL and other database engine documentation. The MySQL documentation is available at www.mysql.org.

   Some command formats are determined by your operational environment. Note that the database may be remote; it need not reside on the same system as the sequence files and the Toolkit.

11. MySQL has an account scheme independent of the account scheme of the hosting OS; other SQL engines use a more integrated account scheme. Password protection for MySQL user accounts is available. If used, users may set the value of the environment variable MYSQL_PWD to the password so that the user is not prompted for a password each time a program is run (*see* **Note 6**). See the MySQL documentation for more information about securing database accounts.

12. The master9 file contains two columns. Each line provides the ID and the sequence of a 9-mer, for example:
   RF49903f AGGGTTAAG
   RF49903r CTTAACCCT
   The first column provides the IDs, and the second column contains the sequence of the corresponding 9-mer. As shown above, the 9-mers are analyzed as pairs, forward (f) and reverse (r) complement. The sequences are listed from the 5' to 3' direction by convention. The 9-mers are analyzed and collected as pairs in order

to make them independent of their orientation in the DNA *(4)*. This strategy also eliminates problems associated with redundancy and overcounting.

13. An absolute path is one that explicitly states the location of a file or directory among all file systems available to the user, starting at the root file system. (File systems are arranged in a tree structure.) An example would be /export/home/joe/ project1/mydata. If user joe has his home directory (/export/home/joe) set as the current working directory, the relative path would project/mydata.

14. Batch Entrez is located at http://www.ncbi.nlm.nih.gov/entrez/batchentrez.cgi?db =Nucleotide/.

    Follow the second procedure for uploading the file that contains the accession numbers. After uploading the file, click on retrieve. The page will return a listing of files that no longer exist. At the bottom of that page, you will find a link for retrieving the existing sequences. Click on that link. You will obtain a listing of the accession numbers and the corresponding definitions. To download the actual sequences, use the pull-down menu next to summary and select Genbank. You will obtain a large file that contains all sequences in GenBank format. On the top of the page, click on send all to file. You will be shown a form for opening or saving the files. Click on save. It may take a while to complete the download.

15. The grep and sed commands are system commands (installed with the operating system). The former looks for a given pattern and outputs lines containing it. The latter edits this output to strip away the beginning accession number and separating tab.

16. The output is in a format that can be used by map, a program by the Genetics Computer Group (GCG). For details go to http://www.accelrys.com/products/dstudio/ gcg/.

17. Multisequence files are ones that contain multiple, possibly unrelated, sequences, one after the other. In FASTA files, a new sequence starts with the ">" character at the beginning of a line. In GenBank files, a double slash (//) terminates a sequence, and another such sequence may occur afterward. The first sequence of the file is at position 1, the second at position 2, and so on.

18. This is not a particularly important feature for this study, but it is useful for subsequently planned studies.

Table 1
RF_code_hits

Column/field	Data type	Explanation
RF_id	CHARACTER	ID assigned to a 9-mer
RF_9mers	CHARACTER	Nucleotide sequence of a 9-mer
Num_of_hits	INTEGER	Number of times the 9-mer occurs in the RF_code_locate table

Table 2
Genome_count

Column/field	Data type	Explanation
RF_id	CHARACTER	ID of a 9-mer
RF_9mer	CHARACTER	Nucleotide sequence of a 9-mer
Mer_count	INTEGER	Number of times a 9-mer occurs in the genome
Repeat_count	INTEGER	Number of time a 9-mer occurs within the regions classified as repetitive DNA
CDS_count	INTEGER	A count of 9-mers that exist in the coding regions of analyzed cDNA sequences

Table 3
Ranking

Column/field	Data type	Explanation
RF_id	CHARACTER	ID of a 9-mer
Rank_A_Ei	INTEGER	Number of times the 9-mer occurred in the A region
Rank_A	FLOAT	Rank of this 9-mer using the Rank_A_Ei in this record
Rank_B_Ei	INTEGER	Number of times the 9-mer occurred in the B region
Rank_B	FLOAT	Rank of this 9-mer using the Rank_B_Ei in this record
Rank_C_Ei	INTEGER	Number of times the 9-mer occurred in the C region
Rank_C	FLOAT	Rank of this 9-mer using the Rank_C_Ei in this record

Table 4
Sequence_name

Column/field	Data type	Explanation
Sequence_id	INTEGER	Numeric index to uniquely identify this record
Sequence_name	CHARACTER	Name/title/description of the sequence. We used the definition in the GenBank files.
Sequence_labname	CHARACTER	Unique short name given to a promoter sequence (lab-name or promoter1, promoter2, and so on)
Sequence_file	CHARACTER	Name of the file containing the actual promoter sequence. This may be a relative name. The Toolkit uses an external environment variable to specify the absolute path of the relative root (*see* **Note 13**).
Sequence_file_date	INTEGER	Number of seconds past Jan 1, 1970, when the file was last modified. If this time does not match that given by the file system, any data in processed tables referring to this sequence are stale.
Sequence_position	INTEGER	Position of the sequence in a multi-sequence file (*see* **Note 17**)

Table 5
Expt

Column/field	Data type	Explanation
Exp_num	INTEGER	Code for uniquely identifying a record in this table
Exp_comment	CHARACTER	Arbitrary text specified at processing time by experimenter
Sequence_id	INTEGER	The numerical index that appears in the Sequence_name table (*see* **Table 4**)
TI_site	INTEGER	Position of the transcription initiation site within a promoter sequence
Start_position	INTEGER	Specifies the first nucleotide that was used for data collection. The Start_position is specified with respect to TI

(continued)

Table 5 (Continued)

Column/field	Data type	Explanation
End_location	INTEGER	Specifies the last position that was used for data collection. The End_location is specified with respect to TI.
First_mer	CHARACTER	The first 9-mer that was read from a promoter sequence. This is used to verify the integrity of the collected data.
Last_mer	CHARACTER	The last 9-mer that was read from the promoter sequence. This is used to verify the integrity of the collected data.
Name_of_person	CHARACTER	Database user identifier of user inserting this data record into the table (i.e., responsible for this processing run)
Date_of_exp	DATETIME	Time this record was created

Table 6
RF_code_locate

Column/field	Data type	Explanation
RF_id	CHARACTER	ID assigned to a 9-mer
Sequence_id	INTEGER	The numerical index that appears in the Sequence_name table (*see* **Table 4**)
Location	INTEGER	Location of a 9-mer in a promoter sequence with respect to TI (i.e., −400)
Direction	CHARACTER	The 9-mers are collected as a pair of complementary sequences. The direction of a 9-mer was assigned a + if it was found on the same strand as the coding sequence. A − is assigned for the complementary sequence of that 9-mer in the other strand.
Flag	CHARACTER	Single character to flag a 9-mer as special. Specifically used to identify 9-mers that lie in a region of repetitive DNA. N.B.: Flag must be specified in the study_seq command line if this field is to be included in this table.
Exp_num	INTEGER	Refers to a field in the Expt table (*see* **Table 5**)

References

1. International Human Genome Sequencing Consortium. (2001) Initial sequencing and analysis of the human genome. *Nature* **409,** 860–921.
2. Venter, J. C., et al. (2001) The sequence of the human genome. *Science* **291,** 1304–1351.
3. Collins, F. S., Green, E. D., Guttmacher, A. E., and Guyer, M. S. (2003) A vision for the future of genomics research. *Nature* **422,** 835–847.
4. Bina, M., Wyss, P., Ren, W., et al. (2004) Exploring the characteristics of sequence elements in proximal promoters of human genes. *Genomics* **84,** 929–940.
5. Hutchinson, G. B. (1996) The prediction of vertebrate promoter regions using differential hexamer frequency analysis. *Comput. Appl. Biosci.* **12,** 391–398.
6. Marino-Ramirez, L., Spouge, J. L., Kanga, G. C., and Landsman, D. (2004) Statistical analysis of over-represented words in human promoter sequences. *Nucleic Acids Res.* **32,** 949–958.
7. FitzGerald, P. C., Shlyakhtenko, A., Mir, A. A., and Vinson, C. (2004) Clustering of DNA sequences in human promoters. *Genome Res.* **8,** 1562–1574.
8. Trinklein, N. D., Aldred, S. J, Saldanha, A. J., and Myers, R. M. (2003) Identification and functional analysis of human transcriptional promoters. *Genome Res.* **13,** 308–312.
9. Kent, W. J., Sugnet, C. W., Furey, T. S., et al. (2002) The human genome browser at UCSC. *Genome Res.* **12,** 996–1006.
10. Karolchik, D., Baertsch, R., Diekhans, M., et al. (2003) University of California Santa Cruz. The UCSC Genome Browser Database. *Nucleic Acids Res.* **31,** 51–54.
11. Karolchik, D., Hinrichs, A. S., Furey, T. S., et al. (2004) The UCSC Table Browser data retrieval tool. *Nucleic Acids Res.* **32** (Database issue), D493–496.

12

Analysis of Allele-Specific Gene Expression

Julian C. Knight

Summary

The analysis of allele-specific gene expression has been of long-standing interest in the study of genomic imprinting, but there is growing awareness that differences in allelic expression are widespread among autosomal nonimprinted genes. Recent research into *cis*-acting regulatory polymorphisms has utilized the analysis of allele-specific gene expression to identify functionally important regulatory haplotypes and specific genetic polymorphisms. Allele-specific effects are typically of modest magnitude, requiring techniques for analysis of high sensitivity and specificity. Here, strategic approaches to the analysis of allele-specific gene expression are reviewed with protocols for in vivo analysis. These include analysis of the relative allelic abundance of transcribed RNA and of transcription factor recruitment and Pol II loading by chromatin immunoprecipitation.

Key Words: Allele; polymorphism; SNP; gene expression; RNA; transcription.

1. Introduction

Genomic imprinting and X-chromosome inactivation are well-described phenomena in which allele-specific gene expression occurs, although the underlying molecular mechanisms for this remain incompletely understood. A number of recent studies have demonstrated that allele-specific gene expression also occurs among autosomal, nonimprinted genes *(1–4)*. Allele-specific effects appear to be heritable, of relatively modest magnitude (typically 1.5 to 2-fold), highly context specific, and occur relatively commonly. Moreover, differences in gene expression can be mapped as quantitative traits *(4)*. There is growing interest in the phenomenon of allele-specific differences both as evidence to support the existence of *cis*-acting regulatory polymorphisms of the DNA sequence and as a tool for the identification of regulatory DNA variation (reviewed in

From: *Methods in Molecular Biology, vol. 338: Gene Mapping, Discovery, and Expression: Methods and Protocols*
Edited by: M. Bina © Humana Press Inc., Totowa, NJ

refs. *5–7)*. At present the identification of genetic variation modulating gene expression remains highly problematic *(8)*. This is important, as regulatory polymorphisms account for a significant amount of interindividual variation in gene expression *(6)* and are likely to underlie many of the observed associations between genetic variation and susceptibility to disease in population-based genetic association studies *(9)*.

In vitro strategies for allele-specific analysis of gene expression provide important experimental evidence of the effects of DNA sequence diversity. These have typically involved reporter gene analysis and assays of protein-DNA binding *(10)*. In both cases, polymorphisms can be engineered into the DNA sequence and then either incorporated into a reporter gene construct for transfection into a cell of interest or used as probes to compare relative protein-DNA binding affinity. This chapter describes approaches that analyze allele-specific gene expression in living cells. Here it is the naturally occurring sequence diversity that is analyzed in a context of the normal chromatin and regulatory mechanisms controlling gene expression.

To achieve this, the analysis of allele-specific gene expression in vivo of autosomal genes has relied on defining the allelic origin of RNA using a transcribed marker polymorphism and quantifying the relative allele-specific transcript abundance *(1,11)*. If a DNA marker such as a single-nucleotide polymorphism (SNP) is selected in a heterozygous state, with one copy of each allele within the cell, this provides an internally controlled situation in which the relative abundance of the transcript can be assayed (**Fig. 1**). This contrasts with the situation of correlating either total RNA abundance or the resulting translated protein between cells or individuals on the basis of their genotype. Such analyses are potentially confounded at many levels, such as prevailing environmental stimuli, variation in the signaling cascade, and the genetic makeup of the individual.

Chromatin immunoprecipitation affords a further insight into allele-specific effects on gene expression, as proteins bound to DNA can be quantified in an allele-specific manner, again discriminated on the basis of a marker polymorphism but here assaying DNA rather than RNA and thus circumventing the need for the polymorphism to be transcribed. Allele- or haplotype-specific chromatin immunoprecipitation (haploChIP) allows for both discrimination of allele-specific transcription factor binding *(12)* and Pol II loading *(13)*. The latter, when assayed using antibodies to phosphorylated Pol II, may serve as a useful surrogate of levels of gene expression, expanding the number of polymorphisms and haplotypes that can be analyzed for allele-specific effects (**Fig. 1**).

A number of different approaches have been used to quantify the relative allelic abundance of RNA (in allele-specific transcript analysis) or DNA (in allele-specific ChIP), notably primer extension assays. A number of modes of

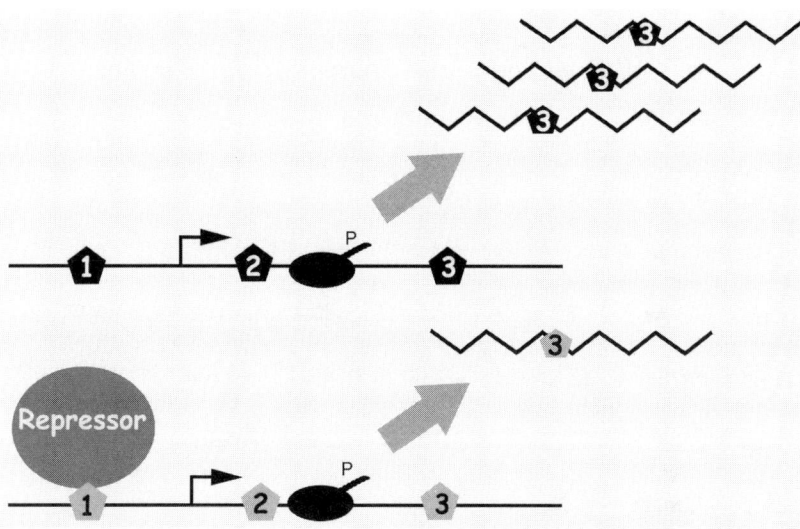

Fig. 1. Allelic discrimination using heterozygous polymorphic markers. In a cell in which there is a relative allelic difference in gene expression seen in transcript abundance, this may arise owing to *cis*-acting regulatory polymorphisms. For example, SNP 1 may act to change a site of DNA-protein binding resulting in recruitment of a transcriptional repressor to one allele. In the presence of a transcribed marker (SNP 3), the relative abundance of RNA can be determined according to its allelic origin. Phosphorylated Pol II loading can also be used to estimate relative gene expression using haplo-ChIP, but here any polymorphic marker within 1 kb of the 5' or 3' end of the gene can be used, as the method does not rely on the marker being transcribed: in this case SNPs 1, 2, or 3 can be used. HaploChIP can also be used to estimate relative allele-specific binding by the transcriptional repressor using SNPs 1, 2, or 3 (if the SNPs lie within 1 kb of the transcriptional repressor binding site).

quantification have been reported with detection using radionucleotides (*11*), fluorescence (*1,14*), and mass spectrometry (*13*). Other strategic approaches to defining relative allelic abundance include the amplification refractory mutation system (ARMS) (*15*) and polymerase chain reaction (PCR) amplification with restriction fragment-length polymorphism analysis. The incorporation of a labeled primer in the final amplification cycle of PCR amplification (hot-stop PCR) offers the advantage of avoiding heteroduplex formation, which may bias observed allelic differences (*16*). High-throughput approaches based on microarray technology have also been used (*3*) together with single-molecule RNA profiling (*17*), but because of space limitations these two approaches are not described here.

2. Materials

2.1. Choice of Cell Type and Ascertainment of Heterozygosity

1. PCR amplification: BioTaq DNA polymerase (Bioline, London, UK), DNA Engine thermal cycler (MJ Research, Waltham, MA).

2.2. Cell Culture and Harvesting

1. Culture medium for lymphoblastoid cell lines (LCLs): RPMI-1640 supplemented with 2 mM glutamine and 10% fetal bovine serum.
2. Phorbol 12-myristate 13-acetate (PMA; Sigma-Aldrich, Gillingham, Dorset, UK) dissolved at 1 mM in dimethyl sulfoxide (DMSO) and stored at –20°C.
3. Ionomycin (Sigma) dissolved at 1 mM in DMSO and stored at –20°C.
4. TRIzol Reagent (Invitrogen, Paisley, UK).
5. Formaldehyde crosslinking buffer (10X): 100 mM NaCl, 1 mM EDTA pH 8, 0.5, mM EGTA pH 8, 50 mM HEPES, pH 8. Autoclave and store at room temperature.
6. 11% Formaldehyde (10X) to be added to 10X formaldehyde crosslinking buffer immediately prior to use in a fume hood.
7. Glycine solution (20X, 2.5 M): autoclave and store at room temperature.
8. Phosphate-buffered saline (PBS, 1X): autoclave and store at 4°C.

2.3. Extraction of RNA and cDNA Synthesis

1. RNase-free DNase I (Ambion [Europe], Huntingdon, Cambridgeshire, UK).
2. AMV reverse transcriptase (Roche Diagnostics, Lewes, East Sussex, UK).
3. RNasein Ribonuclease Inhibitor (Promega, UK, Southampton, UK).

2.4. Chromatin Immunoprecipitation

1. ChIP lysis buffer 1 (1X): 50 mM HEPES, 140 mM NaCl, 1 mM EDTA, 10% glycerol, 0.5% NP-40 (Calbiochem, San Diego, CA), 0.25% Triton X-100 (Sigma). Store at 4°C.
2. ChIP lysis buffer 2 (1X): 200 mM NaCl, 1 mM EDTA pH 8, 0.5 mM EGTA pH 8, 10 mM Tris-HCl, pH 8. Store at room temperature.
3. ChIP lysis buffer 3 (1X): 1 mM EDTA, pH 8, 0.5 mM EGTA, pH 8, 10 mM Tris-HCl, pH 8. Store at 4°C.
4. ChIP lysis buffers 1, 2, and 3 should be supplemented with protease inhibitors immediately before use: complete protease inhibitor tablets (Roche), 1 mM benzamidine (Sigma; 0.1 M stock prepared with water and stored at –20°C), 50 µg/mL TLCK (Roche) (1 mg/mL stock in 0.05 M sodium acetate, pH 5 [Sigma], stored at –20°C), 50 µg/mL TPCK (Roche; stock 3 mg/mL in ethanol, stored at –20°C), 1 µg/mL pepstatin (Roche; stock 1 mg/mL in ethanol, stored at –20°C).
5. Branson 450 Sonifier (Branson Ultrasonics, Branson, CT).
6. The choice of antibody will depend on the application; for analysis of gene expression, antibodies vs phosphorylated serine residues of the CTD of Pol II (Ser5, MMS-134R clone H14; Ser2, MMS-129R clone H5; Covance, Princeton, NJ) may be used.

7. Dynabeads M-280 (Dynal Biotech, Oslo, Norway) with secondary antibody of choice.
8. Magnetic particle concentrator (MPC; Dynal).
9. To wash Dynabeads, prepare a solution of bovine serum albumin (BSA) 5 mg/mL in PBS immediately before use.
10. Nutator (Shelton Scientific, Shelton, CT).
11. 2X RIPA-POL buffer: 20 mM Tris-HCl, pH 8, 2 mM EDTA, 1 mM EGTA, 2% Triton X-100 (Sigma), 0.2% sodium deoxycholate, 0.2% sodium dodecyl sulfate (SDS), 280 mM NaCl, 2X complete protease inhibitor, and 10 µg/mL pepstatin. When required, add 100 µg/mL sonicated herring sperm DNA (Promega, Madison, WI) and 250 mM lithium chloride.
12. Elution buffer: 10 mM Tris-HCl, pH 8, 1 mM EDTA, 1% SDS. Store at room temperature.
13. Qiagen DNA clean-up kit (Qiagen, Crawley, West Sussex).

2.5. Quantification of Relative Allelic Abundance

1. Primer extension quantification by matrix-assisted laser desorption/ionization time-of-flight mass spectrometry (MALDI-TOF MS): massEXTEND platform (Sequenom, San Diego, CA) including SpectroCLEAN (Sequenom) resin, SpectroCHIP (Sequenom) microarray, SpectroPOINT (Sequenom) nanoliter dispenser, and a Spectro-READER (Sequenom) mass spectrometer.
2. SNaPshot Multiplex Kit (Applied Biosystems, Warrington, Cheshire, UK).

3. Methods

3.1. Choice of Cell Type and Ascertainment of Heterozygosity

1. Genomic DNA from the cells or tissue selected for analysis (*see* **Note 1**) is geno-typed using ARMS (*see* **Note 2**). This will define genetic markers (*see* **Note 3**), which can then be used to resolve the allelic origin of RNA or immunoprecipitated DNA as described in **Subheading 3.5.**
2. Primers are designed with a 3' mismatch at the site of the polymorphism and multi-plexed with a control primer set (*see* **Note 4**). PCR amplification is performed using 1X PCR buffer, 0.4 mM dNTPs, 1.9 mM MgCl$_2$, 0.25 U Bioline Taq. Amplification is performed at 96°C for 1 min; 5 cycles of 96°C for 35 s, 70°C for 45 s, 72°C for 35 s; 23 cycles of 96°C for 25 s, 65°C for 50 s, 72°C for 40 s; 6 cycles of 96°C for 35 s, 55°C for 1 min, 72°C for 1.5 min. Products are visualized by agarose gel electrophoresis.

3.2. Cell Culture and Harvesting

1. LCLs (*see* **Note 5**) are grown in RPMI-1640 medium supplemented with penicillin and streptomycin, with 2 mM L-glutamine and 10% heat-inactivated fetal bovine serum. Cells are harvested in mid log phase.
2. For RNA preparation, 10 to 50 × 10^6 cells are typically harvested for each time point of interest (*see* **Note 6**).

a. Cell suspensions are centrifuged at 500*g* for 5 min at 4°C and then placed on ice; culture media are removed by aspiration.

b. The cell pellet is lysed in TRIzol (*see* **Note 7**) by repetitive pipeting using 1 mL TRIzol per 10×10^6 cells.

c. The lysed material can be stored at this point at −80°C or the RNA isolated directly.

3. For chromatin preparation 100 to 500×10^6 cells are used per time point.

a. Crosslinking buffer (10X) containing formaldehyde is added directly to cells in growing media, gently mixed, and left to incubate for 45 min at room temperature (*see* **Note 8**).

b. Then 2.5 *M* glycine (20X) is added to stop crosslinking, and cells are pelleted by centrifugation at 500*g* for 5 min at 4°C.

c. The cells are washed in cold PBS and then either used directly to isolate nuclei or stored as a cell pellet at −80°C (*see* **Note 9**).

3.3. Extraction of RNA and cDNA Synthesis

1. RNA is isolated using TRIzol reagent according to the manufacturer's instructions.

a. Briefly, the homogenized sample is incubated for 5 min at room temperature and then 0.2 mL chloroform is added per 1 mL TRIzol reagent.

b. Following shaking by hand for 15 s and incubation at room temperature for 2 min, samples are centrifuged at 12,000*g* for 15 min at 4°C.

c. The colorless upper aqueous phase is transferred to a new tube and precipitated using isopropyl alcohol (0.5 mL per 1 mL TRIzol used for initial homogenization).

d. The sames are incubated at room temperature for 10 min and centrifuged at 12,000*g* for 10 min at 4°C.

e. Supernatant is removed, and the RNA pellet is washed with 75% ethanol.

f. Purified total RNA is then treated with DNase I (for a 50-μg aliquot of total RNA, 8 U DNase I for 40 min at 37°C) followed by phenol-chloroform extraction and reprecipitation (*see* **Note 10**).

2. The RNA is annealed to random decamers and first-strand cDNA synthesis is performed using AMV reverse transcriptase (*see* **Note 11**) incorporating appropriate negative controls.

a. The reaction mix comprises 2.5 μg RNA in PCR buffer with 1 m*M* dNTP, 2.5 m*M* MgCl$_2$, 10 U/sample RNasin, 5 U AMV RT, and 10 μ*M* random decamer.

b. Reactions are incubated in a thermal cycler at 25°C for 10 min and then 42°C for 60 min and placed on ice before purification using a spin column.

3.4. Chromatin Immunoprecipitation

1. Nuclear material is prepared from cell pellets by resuspending cells on ice in 20 mL ChIP lysis buffer 1 per 500×10^6 cells (*see* **Note 12**).

a. Rock suspension at 4°C for 10 min and then pellet in tabletop centrifuge at 2300*g* for 10 min at 4°C.

b. Resuspend pellet in 16 mL ChIP lysis buffer 2 (containing protease inhibitors) at room temperature, and rock gently at room temperature for 10 min.

c. Repeat centrifugation at 4°C, and resuspend the nuclear material on ice in 4 mL ChIP lysis buffer 3 (containing protease inhibitors).

2. The suspension of nuclear material is then sonicated to achive a final average DNA fragment size of 0.5 to 1 kb (*see* **Note 13**).

3. Sonicated material is centrifuged at 12,000*g* at 4°C for 10 min and adjusted to 10% final glycerol concentration prior to storage at −80°C or used directly in the immunoprecipitation reactions described below (*see* **Note 14**) in **steps 4** to **11**.

4. Immunoprecipitation reactions are prepared by combining 50 μg chromatin with 2 μg primary antibody (such as vs phosphorylated Pol II) attached to 50 μL starting volume of Dynabeads. The experimental design should include appropriate negative controls.

5. Dynabead-antibody complexes are prepared by concentrating Dynabeads M280 precoated with secondary antibody in an MPC, removing supernatant, and resuspending Dynabeads in 1 mL PBS/BSA.
   a. Beads are washed twice in PBS/PSA and primary antibody is added, followed by incubation overnight on a nutator at 4°C.
   b. To remove unbound antibody, use MPC and wash in PBS/BSA three times.
   c. Beads are resuspended in a volume of PBS/BSA equal to the starting volume of Dynabeads taken from stock.

6. To set up immunoprecipitation reactions, combine chromatin with Dynabeads bound to primary antibody to achieve a final concentration of 1X RIPA-POL in a total volume of 500 μL. Incubate overnight on a nutator at 4°C.

7. Save aliquots of chromatin from antibody negative control tube to use as input control (typically take 2.5 μL, make up to 100 μL with TE, and store on ice) and then wash immunoprecipitation reactions using MPC (*see* **Note 15**).
   a. Sequentially wash bead complexes twice with 1 mL freshly prepared 1X RIPA-POL buffer.
   b. Proceed to wash with 1X RIPA-POL buffer containing 100 μg/mL herring sperm DNA, 1X RIPA-POL buffer with 100 μg/mL herring sperm DNA plus 300 m*M* NaCl, and finally 1X RIPA-Pol with 250 m*M* LiCl, rocking at room temperature for exactly 5 min with each wash.
   c. Wash once with 1 mL TE, and remove any remaining liquid.

8. Elute from beads.
   a. First, 50 μL of elution buffer is added, vortexed briefly to resuspend beads, and incubated at 65°C for 10 min.
   b. Samples are centrifuged for 30 s at maximum speed in a microfuge and supernatant is transfered to a new tube.
   c. The remaining bead pellet is discarded.
   d. Then 120 μL of elution buffer is added to the supernatant in the new tube, and crosslinks are reversed by incubating at 65°C overnight in water bath.
   e. For input controls, add 11 μL of 10% SDS and reverse crosslink at 65°C overnight in water bath.

9. To remove proteins, 150 μL of proteinase K/glycogen mix is added to each tube and incubated for 2 h at 37°C.

10. DNA extraction.
    a. An equal volume of equilibrated phenol-chloroform, pH 8, is added and then vortexed and spun at maximum speed in a microfuge for 5 min.
    b. Phenol-chloroform extraction is repeated and then DNA is extracted once with an equal volume of chloroform/isoamyl alcohol.
    c. To precipitate DNA fragments, 1/10th vol of 3 *M* sodium acetate, pH 5.2, 2.5X volume of ice-cold 100% ethanol are added and vortexed briefly.
    d. Samples are incubated at −20°C overnight and then centrifuged at maximum speed in a microfuge for 10 min at 4°C.
    e. The resulting pellet is washed with 1 mL ice-cold 70% ethanol, vortexed, centrifuged for 5 min at 4°C at maximum speed, then air-dried.
11. To remove any contaminating RNA, the pellet is resuspended in 30 μL TE containing 10 μg RNase A and incubated for 1 h at 37°C. RNase can be removed by spin column purification according to the manufacturer's instructions and the DNA fragments eluted in 10 m*M* Tris-HCl, pH 8, and stored at −20°C.

3.5. Quantification of Relative Allelic Abundance

1. Non-allele-specific analysis should first be performed to check the specificity and abundance of the gene of interest using either RNA or the products of chromatin immunoprecipitation. Gene-specific PCR primers can be designed for analysis either as a uniplex or by multiplexing with a housekeeping gene to carry out semi-quantitative PCR *(18)* or by quantitative PCR. It is important to include the negative controls (such as AMV negative samples from RT-PCR or mock antibody controls from ChIP) in this analysis.
2. Allele-specific quantification can be achieved by many approaches; primer extension has been most commonly used. A number of approaches are described below in **steps 3** to **6**, as many laboratories may not have the equipment to allow application of techniques such as fluorescent nucleotide detection or MALDI-TOF MS (*see* **Note 16**).
3. For *primer extension with detection by mass spectrometry (PE/MS)*, an appropriate experimental design to minimize variance should be considered (*see* **Note 17**).
    a. A first-round PCR reaction is performed using approx 1/100th of the product of the RT-PCR reaction (*see* **Subheading 3.3.2.**) or 1/25th of the ChIP DNA (*see* **Subheading 3.4.11.**) in a 25 μL reaction vol using 0.5 U BioTaq with 0.8 m*M* dNTPs, 1.9 m*M* MgCl$_2$, and 0.2 μ*M* each primer.
    b. Thermal cycling parameters should be optimized to ensure that the cycle number remains in the linear phase of amplification with annealing temperatures dependent on the primer design, for example, 96°C for 1 min followed by 6 cycles of 94°C for 45 s, 56°C for 45 s, 72°C for 30 s; then 30 cycles of 94°C for 45 s, 65°C for 45 s, 72°C for 30 s; followed by final extension at 72°C for 10 min.
    c. The PCR product is then subaliquoted onto a 384-well plate, and nonincorporated dNTPs are removed using shrimp alkaline phosphatase by incubating at 37°C for 20 min followed by 85°C for 5 min.

d. Primer extension is performed using a homogeneous MassEXTEND reaction comprising a cocktail of 100 μ*M* extension primer, 0.576 U MassEXTEND enzyme, buffer, and an appropriate deoxy and dideoxy nucleotide termination mix.

e. A typical primer extension reaction would comprise 94°C for 2 min then 40 cycles of 94°C for 5 s, 52°C for 5 s, 72°C for 5 s.

f. The products of primer extension are desalted using SpectroCLEAN resin and transferred onto a microarray by SpectroPOINT nanolitre dispenser.

g. MALDI-TOF analysis is performed using a SpectroREADER mass spectrometer.

4. *Primer extension analysis by fluorescence* is based on the single-base extension of an unlabeled oligonucleotide primer at its 3' end with a fluorescently labeled dideoxyterminator followed by quantification using capillary gel electrophoresis.

a. PCR-amplified samples are incubated with shrimp alkaline phosphatase and exonuclease I and then primer extension is performed using the SNaPshot Multiplex kit according to the manufacturer's instructions (*see* **Note 18**).

b. Primer extension cycling conditions for extension primers between 20 and 22 bp are 95°C for 2 min and then 25 cycles of 95°C for 5 s, 43°C for 5 s, 60°C for 5 s.

c. Products are treated with 0.5 U shrimp alkaline phosphatase for 45 min at 37°C and then at 85°C for 15 min.

d. The SNaPshot reaction is analyzed by electrophoresis on an ABI PRISM 310 or 3100 Genetic Analyzer or 3700 DNA Analyzer.

5. With *Allele-specific quantification by hot-stop PCR*, it is possible to resolve relative allelic abundance by PCR amplification followed by restriction enzyme digestion if the marker polymorphism creates a convenient restriction enzyme site. The problem with this and other PCR-based approaches is that heteroduplexes may occur and confound results. For this reason, hot-stop PCR can be a useful approach *(16)*.

a. cDNA or the products of chromatin immunoprecipitation are amplified by PCR in 1X PCR buffer, 0.4 m*M* dNTPs, 1.5 m*M* MgCl$_2$, 1 μ*M* each primer spanning the region of interest containing the marker polymorphism and Taq (94°C for 2 min then 30 cycles of 94°C for 30 s, 60°C for 15 s, 72°C for 20 s followed by 72°C for 10 min)

b. A final cycle using a labeled primer is now performed.

c. The chosen primer is radiolabeled with [γ-^{32}P]ATP using T4 PNK, and unincorporated radioactivity is removed by spin column.

d. A final round of PCR is performed following addition of 1 pmol of radiolabeled primer (94°C 30 s, 60°C 30 s, 72°C for 10 min).

e. The PCR product is digested using a restriction enzyme of choice and visualized by denaturing polyacrylamide gel electrophoresis.

6. *Allele-specific quantification by the amplification refractory mutation system (ARMS)* provides a simple and robust approach, although it may not be as the quantitative as the other approaches outlined above. The technique is described in **Subheading 3.1.** (*see* **Note 19**).

4. Notes

1. A critical decision in designing experiments to interrogate allele-specific gene expression relates to the choice of cell or tissue type for analysis. Allele-specific effects may be highly context dependent, relating not only to cell and tissue type, but also to the nature and kinetics of cell stimulation, and whether cells are of human or animal origin. Pragmatic considerations of the cell numbers required for analysis may limit investigation of primary cells; for approaches such as ChIP, LCLs provide an attractive model.

2. Genotype may also be derived by direct DNA sequencing, restriction enzyme digestion, primer extension, or any other genotyping method of choice.

3. For the approaches described here, the genetic markers chosen should be SNPs. The choice of marker SNP may be derived from databases of publically available SNPs or by *de novo* analysis through, for example, resequencing. The National Center for Biotechnology Information maintains a public SNP database, dbSNP, available at http://www.ncbi.nlm.nih.gov/projects/SNP/. Other resources for SNP analysis and discovery include http://www.ensembl.org/. For transcript analysis, transcribed exonic SNPs are used, although there are reports of using intronic SNPs to analyze heteronuclear RNA *(19)*. For haploChIP, any SNP within 1 kb 5' or 3' to a gene can be used for analysis of Pol II loading. There is an advantage to identifying the haplotypic structure of the gene region of interest at the outset both to interpret observed allelic differences and to have additional SNP markers with which to confirm any observed allelic differences. Many studies have shown that for a given marker SNP, only a minority of individuals show any allelic difference. This highlights both the need to analyze a sufficiently large number of individuals to have statistical power to detect an effect and the importance of understanding the underlying haplotypic structure, which may otherwise confound effects.

4. Additional discrimination can be achieved by incorporating further mismatches into the primer, for example, changing the third or fifth base from the 3' end. In general 20-bp primers are used with a T_m around 60°C.

5. The protocols below describe the analysis of LCLs, but the principles can be applied to cell or tissue types.

6. The experimental design may be modified to only include unstimulated cells looking at constitutive expression or may span a range of time points following stimulation with, for example, mitogen (ionomycin 125 nM + PMA 200 nM).

7. A variety of different commercial kits are available for isolation of RNA.

8. Fresh formaldehyde should be used from an unopened stock, taking appropriate safety precautions for use and disposal. For a given cell type, the time and temperature of exposure to crosslinking with formaldehyde may need to be titrated.

9. The use of 10% fetal bovine serum in the final PBS wash appears to help protect chromatin integrity in freeze-thawing.

10. Many alternative protocols are available for RNA extraction including those based on spin cup technologies that bind RNA in the presence of a chaotropic salt such as the Absoloutely RNA miniprep kit (Stratagene); these allow for DNase I treatment on the spin cup and avoid the need for phenol-chloroform precipitations steps.

11. A number of different reverse transcriptases are available, and some investigators favor oligo(dT) primers for cDNA synthesis.

12. ChIP lysis buffers 1 and 3 should be chilled and kept on ice; 1X complete protease inhibitor, benzamidine, TLCK, TPCK, and pepstatin should be added to buffers immediately before use. For analysis of phosphorylated Pol II, 10 mM (final concentration) sodium pyrophosphate (Sigma) should be used in all ChIP buffers including nuclear isolation buffers and those used subsequently for immunoprecipitation.

13. Effective sonication can be achieved with a 3-mm microtip on a Branson 450 Sonifier using constant power with a 30-s constant burst and allowing the suspension to cool on ice for 1 min between pulses. To minimize foaming during sonication, place the probe at mid-depth in suspension and then turn on the power; start at lower settings and then increase (for example, starting at setting 4 (six times) then increasing to 5 (six times)). Sonication should result in DNA fragments approximately 500 bp to 3 kb in apparent size when assessed by gel electrophoresis at this stage (the actual size will be smaller). To assess this, run 10 to 12 µL of sonicated material on a 1.4% TAE agarose gel. To avoid samples aggregating in the wells of the gel, add Sarkosyl to 0.5% (final concentration) to samples prior to loading on the gel. If the average size of chromatin fragments is greater than several kb, repeat sonication.

14. For some chromatin immunoprecipitation experiments, it may be useful to purify chromatin by cesium chloride ultracentrifugation following sonication. The need for this appears to be dependent on the protein to be immunoprecipitated and the quality of the antibody used for immunoprecipitation. However, the majority of investigators no longer include the cesium chloride centrifugation step.

15. The wash steps are critical to the success of the immunoprecipitation and should be done as carefully as possible to minimize nonspecific background. Experimental conditions may need to be titrated for the individual protein of interest (for example, the stringency of salt and SDS concentration).

16. The choice of method for allelic quantification will be influenced by factors such as cost, availability of technology, and throughput requirements. For all approaches it is essential that appropriate positive and negative controls be included. To assess the accuracy of the technique, the two allelic species can be prepared and mixed in different proportions so that the observed and actual allelic ratios can be compared. Analysis of genomic DNA heterozygous for the marker polymorphism provides a useful baseline for the expected 1:1 allelic ratio, although this may be influenced by other factors such as epigenetic modifications.

17. Primer extension with detection by MALDI-TOF MS is a highly sensitive quantitative approach. To minimize variance, replication should be included at different levels: these could include use of independent LCLs of a given genotype, multiple replicate immunoprecipitation reactions (typically three), replicate PCR amplification of immunoprecipitated DNA fragments (typically four), replicate spotting of PCR products on detection chip (typically four), and independent reads by the mass spectrometer of the spotted PCR amplified material (typically five).

18. An alternative approach is to use TaqMan probes for quantification *(3,14)*.

19. Real-time quantitative ARMS quantitative PCR has also been reported *(20)*.

Acknowledgments

The author's work is supported by a Wellcome Trust Senior Research Fellowship in Clinical Science.

References

1. Yan, H., Yuan, W., Velculescu, V. E., Vogelstein, B., and Kinzler, K. W. (2002) Allelic variation in human gene expression. *Science* **297,** 1143.
2. Cowles, C. R., Hirschhorn, J. N., Altshuler, D., and Lander, E. S. (2002) Detection of regulatory variation in mouse genes. *Nat. Genet.* **32,** 432–437.
3. Lo, H. S., Wang, Z., Hu, Y., et al. (2003) Allelic variation in gene expression is common in the human genome. *Genome Res.* **13,** 1855–1862.
4. Morley, M., Molony, C. M., Weber, T. M., et al. (2004) Genetic analysis of genome-wide variation in human gene expression. *Nature* **430,** 743–747.
5. Buckland, P. R. (2004) Allele-specific gene expression differences in humans. *Hum. Mol. Genet.* **13 Spec No 2,** R255–260.
6. Pastinen, T. and Hudson, T. J. (2004) Cis-acting regulatory variation in the human genome. *Science* **306,** 647–650.
7. Knight, J. C. (2004) Allele-specific gene expression uncovered. *Trends Genet.* **20,** 113–116.
8. Hudson, T. J. (2003) Wanted: regulatory SNPs. *Nat. Genet.* **33,** 439–440.
9. Peltonen, L. and McKusick, V. A. (2001) Genomics and medicine. Dissecting human disease in the postgenomic era. *Science* **291,** 1224–1229.
10. Rockman, M. V. and Wray, G. A. (2002) Abundant raw material for cis-regulatory evolution in humans. *Mol. Biol. Evol.* **19,** 1991–2004.
11. Singer-Sam, J., LeBon, J. M., Dai, A., and Riggs, A. D. (1992) A sensitive, quantitative assay for measurement of allele-specific transcripts differing by a single nucleotide. *PCR Methods Appl.* **1,** 160–163.
12. Knight, J. C., Keating, B. J., and Kwiatkowski, D. P. (2004) Allele-specific repression of lymphotoxin-alpha by activated B cell factor-1. *Nat. Genet.* **36,** 394–399.
13. Knight, J. C., Keating, B. J., Rockett, K. A., and Kwiatkowski, D. P. (2003) In vivo characterization of regulatory polymorphisms by allele-specific quantification of RNA polymerase loading. *Nat. Genet.* **33,** 469–475.
14. Liu, X., Campbell, M. R., Pittman, G. S., Faulkner, E. C., Watson, M. A., and Bell, D. A. (2005) Expression-based discovery of variation in the human glutathione S-transferase M3 promoter and functional analysis in a glioma cell line using allele-specific chromatin immunoprecipitation. *Cancer Res.* **65,** 99–104.
15. Newton, C. R., Graham, A., Heptinstall, L. E., et al. (1989) Analysis of any point mutation in DNA. The amplification refractory mutation system (ARMS). *Nucleic Acids Res.* **17,** 2503–2516.
16. Uejima, H., Lee, M. P., Cui, H., and Feinberg, A. P. (2000) Hot-stop PCR: a simple and general assay for linear quantitation of allele ratios. *Nat. Genet.* **25,** 375–376.
17. Butz, J. A., Yan, H., Mikkilineni, V., and Edwards, J. S. (2004) Detection of allelic variations of human gene expression by polymerase colonies. *BMC Genet.* **5,** 3.

18. Takahashi, Y., Rayman, J. B., and Dynlacht, B. D. (2000) Analysis of promoter binding by the E2F and pRB families in vivo: distinct E2F proteins mediate activation and repression. *Genes Dev.* **14,** 804–816.
19. Pastinen, T., Sladek, R., Gurd, S., et al. (2004) A survey of genetic and epigenetic variation affecting human gene expression. *Physiol. Genomics* **16,** 184–193.
20. Bai, R. K. and Wong, L. J. (2004) Detection and quantification of heteroplasmic mutant mitochondrial DNA by real-time amplification refractory mutation system quantitative PCR analysis: a single-step approach. *Clin. Chem.* **50,** 996–1001.

13

Construction of microRNA-Containing Vectors for Expression in Mammalian Cells

Yoko Fukuda, Hiroaki Kawasaki, and Kazunari Taira

Summary

MicroRNAs (miRNAs) are a class of noncoding RNAs that regulate gene expression by single-stranded RNAs of 18 to 25 nucleotides in length. Hundreds of miRNAs have been found in animals and plants, some of which play important roles in development or differentiation. Increasing attention has thus been paid to their biogenesis and regulation mechanisms and the identification of target genes. We are constructing a comprehensive expression vector library containing predicted human miRNAs. miRNA expression vectors containing human RNA polymerase II or III promoters, and utilizing a flexible vector system, can be useful for functional analysis.

Key Words: MicroRNA; CMV promoter; U6 promoter.

1. Introduction

MicroRNAs (miRNAs) are a class of tiny noncoding RNAs *(1–3)* that regulate the expression of target genes by translational repression or mRNA cleavage *(4,5)*. Since the first discovery of *lin-4* and *let-7* in *Caenorhabditis elegans (6,7)*, hundreds of miRNA have been found in animals and plants, and their functions are continuously being unveiled. At present, mammalian miRNAs are associated with hematopoietic differentiation *(8)*, adipocyte differentiation *(9)*, insulin secretion *(10)*, and some other important mechanisms.

The primary transcripts of miRNA genes form characteristic stem-loop structures *(11)*, which are initially processed by RNase III Drosha *(12)*; the resultant pre-miRNAs are then exported to the cytoplasm. Mature miRNAs are then generated after processing of pre-miRNAs by RNase III Dicer *(12)* and incorporated into micro-ribonucleoproteins (miRNPs) *(13)*. In miRNPs, miRNAs act as the "guide" sequences that recognize complementary mRNA sequences.

From: *Methods in Molecular Biology, vol. 338: Gene Mapping, Discovery, and Expression: Methods and Protocols*
Edited by: M. Bina © Humana Press Inc., Totowa, NJ

Mature or precursor sequences of predicted miRNAs and their genomic positions are available in the public miRNA Registry database (http://www.sanger. ac.uk /Software/Rfam/mirna/index.shtml). For comprehensive analysis of the functions and molecular mechanisms of human miRNAs, we have constructed an expression vector library of human miRNAs from this database.

The transcription of miRNAs is not yet fully understood. They were first believed to be transcribed by RNA polymerase III (pol III); however, some miRNAs are probably transcribed by RNA pol II *(14)*. Although mature miRNAs have a very short sequence of 18 to 25 nucleotides, their pri-miRNAs are relatively long, sometimes over several kilobases, and have four or more uridine residues *(11)*. Because these observations are characteristic of pol II transcripts, unlike most small RNAs that are RNA pol III transcripts, a considerable number of miRNAs are thought to be transcribed by RNA pol II *(4)*. We have constructed two types of miRNA expression vectors. The system mainly introduced here uses a pol II promoter; miRNA genes are under the control of a cytomegalovirus (CMV; pol II) promoter and can easily be transferred to other expression vectors by the TOPO Gateway system. Detailed analysis of time- and tissue-specific expression of miRNAs can be conducted using this flexible vector system. The other system uses a U6 (pol III) promoter, is suitable for expressing short-length RNA, and allows high expression levels of desired sequences in the transfected cells. Since about 30-bp flanking regions of predicted stem-loop sequences can be necessary for maturation of miRNAs when they are constructed in expression vectors *(8)*, all miRNA expression vectors containing predicted miRNA precursor sequences have been constructed with additional 30 to 40-bp genomic sequences.

This chapter describes the construction of miRNA expression vectors driven by human RNA pol II and III promoters (**Figs. 1–3**).

2. Materials

2.1. Reagents and Kits

1. TE buffer, pH 8.0 (Wako).
2. 5X Annealing buffer (Japan Bio Services).
3. 50X TAE buffer: 2 M Tris-acetate, 0.05 M EDTA (pH 8.0).
4. DNA polymerase and 10X buffer: KOD plus (Toyobo).
5. Restriction enzymes: *Eco*RI and *Xho*I (Takara) and reaction buffer: 10X high-salt buffer (100 mM MgCl$_2$, 10 mM dithiothreitol (DTT), 1 M NaCl, 500 mM Tris-HCl, pH 7.5).
6. Restriction enzymes: *Bbr*I (New England Biolabs) and reaction buffer: 10X NE buffer 2 (10 mM MgCl$_2$, 1 mM DTT, 50 mM NaCl, 10 mM Tris-HCl, pH 7.9).
7. Alkaline phosphatase and 10X buffer: CIAP (Takara).

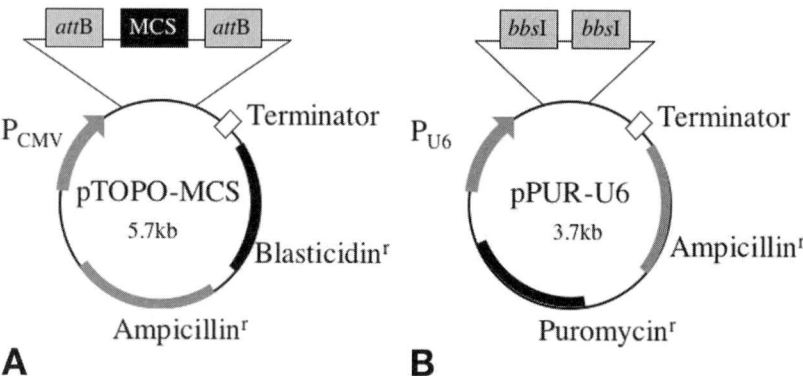

Fig. 1. miRNA expression vector and maturation of miRNA. (**A**) pTOPO-MCS vector containing a CMV promoter and the multiple cloning site, the *att*B site. (**B**) Processing of pre-miRNA sequences transcribed from expression vectors and miRNA maturation.

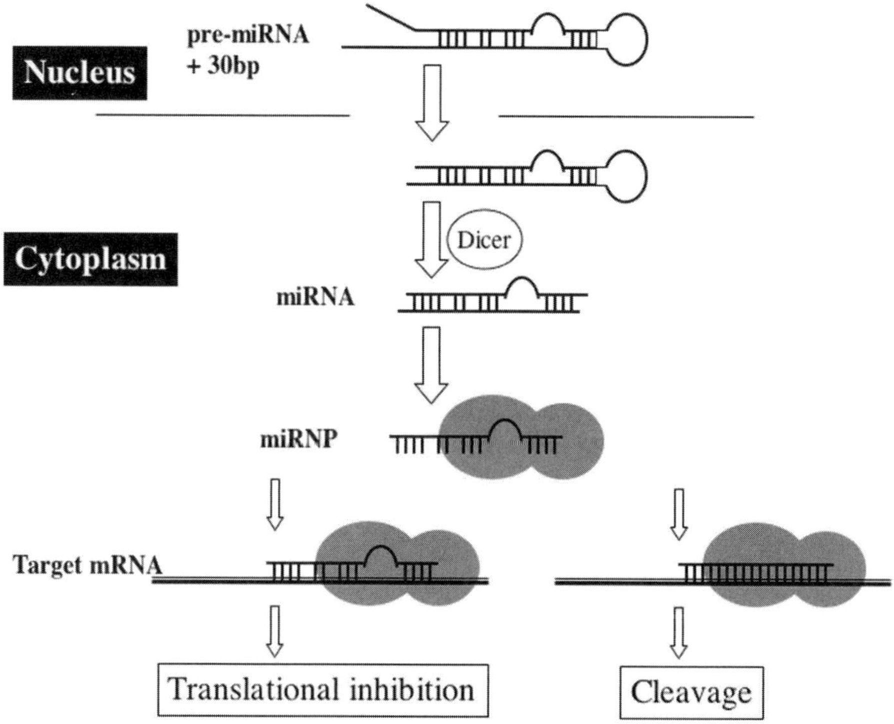

Fig. 2. Model of miRNA maturation and translational inhibition or cleavage of mRNA by an miRNA expression vector.

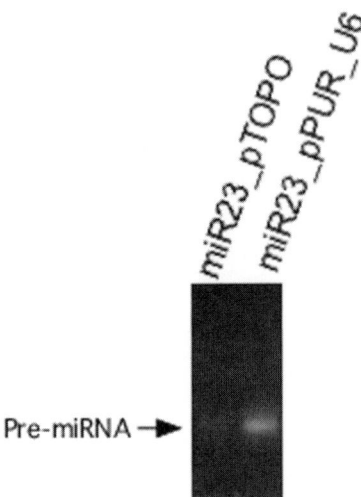

Fig. 3. RT-PCR analysis of pre-miRNA-23 expressed by pTOPO-MCS and pPUR-U6 vectors.

8. Polymerase chain reaction (PCR) purification kit: MinElute (Qiagen).
9. Gel extraction kit: MinElute (Qiagen).
10. Miniprep kit (Qiagen).
11. *E. coli* and SOC medium: One Shot TOP10 Chemically Competent kit (Invitrogen) and DH5α competent cells.
12. DNA ligation solution: Salt Solution of pcDNA™ Gateway Directional TOPO Reagent (Invitrogen) or ligation high (Toyobo).

2.2. Plasmids

1. pcDNA6.2/GW/D-TOPO (Invitrogen).
2. pPUR-U6 (Taira Lab).

3. Methods

3.1. Pol II Promoter-Based miRNA Expression Vectors

3.1.1. Construction of pTOPO-MCS Vector Plasmid

For large-scale cloning of miRNAs, we constructed a pTOPO-MCS vector that contains multiple cloning sites at the cloning site of pcDNA3.2/V5/GW/D-TOPO. Restriction enzyme recognition sites of *Bam*HI, *Eco*RI, *Eco*RV, *Hind*III, *Kpn*I, *Pst*I, and *Xho*I are located at the cloning site from 5' to 3'.

1. Anneal sense (5'-CAC CAT GCT GGG ATC CAA GCT GGA ATT CAG CT GGA TAT CAC GCT GAA GCT TAT ACT GGG TAC CAT CCT GCT GCA

GAT TCT GCT CGA GAT TCT C-3') and antisense sequences. The reaction solution contain sense oligonucleotides (500 pmol), antisense oligonucleotides (500 pmol), annealing buffer in a total volume of 100 µL. Incubate at 90°C for 2 min and 72°C for 10 s and then gradually decrease the annealing temperature to 4°C.

2. Mix 4 µL of annealing products, 1 µL of Salt Solution (pcDNA™ Gateway Directional TOPO Reagent), and 1 µL of TOPO vector. Incubate for about 20 min at room temperature.
3. Transform the reaction mixture to One Shot TOP10 Competent Cells. Mix 6 µL of reaction and 50 µL of competent cells and incubate on ice for 30 min. Place the cells in a 42°C water bath for 30 s and then move immediately and put on ice. Add 250 µL of SOC medium and incubate at 37°C for 1 h. Plate the cells on LB agar plates containing 100 µg/mL ampicillin.
4. After colonies have become apparent, select desired recombinant by colony PCR, and then sequence the positive clones for confirmation.

3.1.2. Cloning of miRNA Precursor Sequences

For the first step of PCR amplification, prepare genomic DNA as the template DNA. Human genomic DNA was extracted from HEK 293 cells using a DNeasy Tissue kit. PCR primers were designed to amplify the predicted miRNA precursor sequence with 30-bp flanking regions and each *Eco*RI, and *Xho*I site in the 5' and 3' ends. If there were the same sequences as *Eco*RI or *Xho*I sites in the miRNA precursor and extended 30-bp sequence, other adequate restriction enzymes were selected from cloning sites in pTOPO-MCS, and PCR primers were designed.

1. Trypsinize adherent cells, centrifuge them for 5 min at 500g, and discard supernatant. Resuspend cells in 10 mL ice-cold PBS, centrifuge for 5 min at 500g, and discard supernatant. Add 100 µL of proteinase K and 200 µL of Buffer AL to the cells, mix by vortexing, and incubate at 70°C for 10 min.
2. Add 200 µL of 99.5% ethanol, mix, and then transfer the mixture to a DNeasy spin column. Centrifuge at 6000g for 1 min and discard the flowthrough.
3. Add 500 µL of buffer AW1, centrifuge at 6000g for 1 min, add AW2 in the same way, and discard the flowthrough at each step.
4. Place the column on a new 1.5-mL tube and add 100 µL of buffer AE to the membrane. Incubate at room temperature for 1 min and then centrifuge at 6000g for 1 min.
5. Amplify the template genomic DNA with specific primers of each miRNA by PCR reaction. The reaction solution contains template 0.2 µg DNA, 50 pmol forward primer, 50 pmol reverse primer, 0.2 mM dNTP mix, 1 mM MgSo$_4$, KOD plus buffer, and 1 U KOD plus in a total volume of 50 µL. The thermal cycle is as follows: denaturing reaction (94°C, 30 s), annealing reaction (50–55°C, 30 s), polymerase reaction (68°C, 1 min); repeat for 30 cycles.
6. After purification of the PCR products using the PCR purification kit, digest 1 µg of the vector plasmid pTOPO-MCS and the PCR products with *Eco*RI and *Xho*I in a high-salt buffer for 2 h at 37°C in a total volume of 20 µL.

7. Isolate the desired DNA fragments by electrophoresis with 1% and 2% agarose gels for the vector plasmids and the insert fragments, respectively. Then, purify them with the gel extraction kit. Elute the DNA fragment in 10 µL of EB buffer.

8. To prevent the vector plasmid from self-ligation, incubate the resulting vector with CIAP in CIAP buffer and in a total volume of 50 µL at 37°C for 30 min. Remove the enzyme by phenol/chloroform extraction and ethanol precipitation, and then resuspend the pellet in 10 µL of TE buffer.

9. The ligation reaction contains 2 µL of the vector plasmid, 3 µL of the insert sequences, and 5 µL of the ligation high. Incubate the reaction mixture at 16°C for 2 h.

10. Transform *E. coli* cells with the ligation mixture. Plate the cells on LB agar plates containing 100 µg/mL ampicillin.

11. After colonies have become apparent, select the desired recombinant by colony PCR, and then sequence the positive clones for confirmation.

3.2. Pol III Promoter-Based miRNA Expression Vectors

The plasmid pPUR-U6 was constructed, which is originally from the pPUR vector (Clontech). A U6 cassette containing the human U6 promoter followed by two *Bbs*I cloning sites and four strings of tyramine residues was inserted into the *Eco*RI and *Bam*HI site of the pPUR vector. pol III promoter-based miRNA expression vectors constructed on pPUR-U6 should follow a procedure similar to that used in the pol II promoter-based miRNA expression vectors. RNA pol III terminates transcription at more than four T tracts; therefore, be sure that there are no T strings of more than 4 nt in the desired sequences of the miRNA precursor.

1. Amplify the template genomic DNA with specific primers of each miRNA by PCR reaction. Our PCR primers were designed to amplify the predicted miRNA precursor sequence with 30-bp flanking regions and *Bbs*I sites in the 5' and 3'ends.

2. After purification of the PCR products by a PCR purification kit, digest 1 µg of the vector plasmid pPUR-U6 and the PCR products with *Bbs*I in NE buffer 2 for 2 h at 37°C in a total volume of 20 µL.

3. After purification and ligation of the vector and the insert fragments, transform *E. coli* cells with the ligation mixture in the same way as set out in **steps 7** to **11** in **Subheading 3.1.2.**

References

1. Lagos-Quintana, M., Rauhut, R., Lendeckel, W., and Tuschl, T. (2001) Identification of novel genes coding for small expressed RNAs. *Science* **294,** 853–858.

2. Lau, N. C., Lim, L. P., Weinstein, E. G., and Bartel, D. P. (2001) An abundant class of tiny RNAs with probable regulatory roles in *Caenorhabditis elegans*. *Science* **294,** 858–862.

3. Lee, R. C. and Ambros, V. (2001) An extensive class of small RNAs in *Caenorhabditis elegans*. *Science* **294,** 862–864.

4. Bartel, D. P. (2004) MicroRNAs: genomics, biogenesis, mechanism, and function. *Cell* **116,** 281–297.
5. Yekta, S., Shih, I. H., and Bartel, D. P. (2004) MicroRNA-directed cleavage of HOXB8 mRNA. *Science* **304,** 594–596.
6. Lee, R. C., Feinbaum, R. L., and Ambros, V. (1993) The *C. elegans* heterochronic gene lin-4 encodes small RNAs with antisense complementarity to lin-14. *Cell* **75,** 843–854.
7. Wightman, B., Ha, I., and Ruvkun, G. (1993) Posttranscriptional regulation of the heterochronic gene *lin-14* by *lin-4* mediates temporal pattern formation in *C. elegans. Cell* **75,** 855–862.
8. Chen, C. Z., Li, L., Lodish, H. F., and Bartel, D. P. (2004) MicroRNAs modulate hematopoietic lineage differentiation. *Science* **303,** 83–86.
9. Esau, C., Kang, X., Peralta, E., et al. (2004) MicroRNA-143 regulates adipocyte differentiation. *J. Biol. Chem.* **279,** 52361–52365.
10. Poy, M. N., Eliasson, L., Krutzfeldt, J., et al. (2004) A pancreatic islet-specific microRNA regulates insulin secretion. *Nature* **432,** 226–230.
11. Lee, Y., Jeon, K., Lee, J. T., Kim, S., and Kim, V. N. (2002) MicroRNA maturation: stepwise processing and subcellular localization. *EMBO J.* **21,** 4663–4670.
12. Lee, Y., Ahn, C., Han, J., et al. (2003) The nuclear RNase III Drosha initiates microRNA processing. *Nature* **425,** 415–419.
13. Mourelatos, Z., Dostie, J., Paushkin, S., et al. (2002) miRNPs: a novel class of ribonucleoproteins containing numerous microRNAs. *Genes Dev.* **16,** 720–728.
14. Lee, Y., Kim, M., Han, J., et al. (2004) MicroRNA genes are transcribed by RNA polymerase II. *EMBO J.* **23,** 4051–4060.

14

Mining Microarray Data at NCBI's Gene Expression Omnibus (GEO)*

Tanya Barrett and Ron Edgar

Summary

The Gene Expression Omnibus (GEO) at the National Center for Biotechnology Information (NCBI) has emerged as the leading fully public repository for gene expression data. This chapter describes how to use Web-based interfaces, applications, and graphics to effectively explore, visualize, and interpret the hundreds of microarray studies and millions of gene expression patterns stored in GEO. Data can be examined from both experiment-centric and gene-centric perspectives using user-friendly tools that do not require specialized expertise in microarray analysis or time-consuming download of massive data sets. The GEO database is publicly accessible through the World Wide Web at http://www.ncbi.nlm.nih.gov/geo.

Key Words: Microarray; gene expression; database; data mining.

1. Introduction

Microarray technology is one of the most important experimental developments in molecular biology in recent years. Microarrays have enabled researchers to conduct large-scale quantitative assessments of gene expression, defining the transcriptome of a multitude of cellular types and states.

The National Center for Biotechnology Information (NCBI) launched the Gene Expression Omnibus (GEO) database in 2000 to support the public use and dissemination of gene expression data generated by high-throughput methodologies *(1,2)*. The database is populated by material supplied by the scientific community. Most researchers submit to GEO in accordance with grant or journal

---

*This chapter is an official contribution of the National Institutes of Health; not subject to copyright in the United States.

From: *Methods in Molecular Biology, vol. 338: Gene Mapping, Discovery, and Expression: Methods and Protocols*
Edited by: M. Bina © Humana Press Inc., Totowa, NJ

requirements stipulating that microarray data be made available through a public repository, in compliance with long-established standards of scientific reporting that allow others to judge or reproduce the results. Consequently, most of the data presented in GEO has been analyzed and published. GEO is not intended or suitable for initial analysis of newly acquired data, which is typically the role taken by laboratory information management systems (LIMS).

The GEO database stores molecular abundance data generated by a wide variety of high-throughput measuring techniques. These include microarray-based experiments that measure gene expression or detect genomic gains and losses (comparative genomic hybridization), as well as genomic tiling arrays that are used to detect transcribed regions or single-nucleotide polymorphisms, or to identify protein-binding genomic regions in conjunction with chromatin immunoprecipitation (ChIP-chip technology). Some non-array-based high-throughput data types are also accepted by GEO, including serial analysis of gene expression (SAGE), massively parallel signature sequencing (MPSS), serial analysis of ribosomal sequence tags (SARST), and some peptide profiling techniques such as tandem mass spectrometry (MS/MS). The data analysis features discussed here are generally applicable to all these technology types, but for the purposes of this chapter the focus is on microarray-generated gene expression data, which currently constitute about 95% of the data in GEO.

At the time of writing, GEO holds over 50,000 submissions, representing approximately half a billion individual molecular abundance measurements, for over 100 organisms, submitted by over 1000 laboratories. These data explore a huge breadth of biological phenomena, for example, mouse models of diabetes, flower development in plants, anthrax sporulation, aging in fruit flies, effect of cigarette smoke on bronchial cells, kidney transplant rejection, toxicological effects of antimalarial drugs, and many others. When one is working with a vast compendium of data, it is important to be able to effectively query the data, focusing on those that are relevant to a specific area of interest. This chapter describes intuitive interfaces and tools that help researchers effectively explore, visualize, and interpret the submitted data. These tools do not require specialized knowledge of microarray analysis methods, nor do they require time-consuming download of large data sets.

2. Organization of the Database

To readily interpret the data in GEO, it helps to have a general understanding of the database structure and content. Researchers provide their data in four sections: Platform, Sample, Series (which receive persistent GPLxxx, GSMxxx, and GSExxx accession numbers, respectively), and raw data. A Platform defines the array template and contains sequence identity tracking information for each feature on the array. A Sample record contains the measured hybridization data, along with a description of the biological source and treatment protocols. A Series record ties together experimentally related Samples. Accompanying raw data (e.g., Affymetrix original probe data or cDNA array.tif images) may be optionally supplied.

The hardware and software packages that generate and process microarray data produce a wide assortment of data styles and formats—Platform and Sample tables can take on many different structures and contain multiple and varying types of ancillary and supporting information. Furthermore, microarray-based technologies and processing strategies are continually evolving. The GEO database was designed with these considerations in mind and has a flexible and open architecture that can accommodate variety. However, data provided in different styles and formats are not readily interpretable or analyzable, even by the experienced user. To address this issue, an upper level of organization is applied. Despite the diversity, a common core of relevant data is supplied to GEO:

- Sequence identity tracking information for each feature on the array.
- Normalized hybridization measurements.
- A description of the biological source used in each hybridization.

These data are extracted from the submitter-supplied records and reassembled by GEO staff into an upper level unit called a GEO DataSet (assigned a persistent GDSxxx accession). A DataSet represents a collection of similarly processed, experimentally related hybridizations. Samples within a DataSet are further organized according to experimental variable subgroups, for example, they are categorized by age, disease state, and so on. A DataSet can be rendered to generate two separate representations of the data:

1. An *experiment-centered* perspective that encapsulates the whole study. This information is presented as a *DataSet record*. DataSet records comprise a synopsis of the experiment, a breakdown of the experimental variables, access to several data display and analysis tools, and download options (**Fig. 1**).
2. A *gene-centered* perspective that provides quantitative gene expression measurements for individual genes across the DataSet. This information is presented as a *GEO Profile*. GEO Profiles comprise gene identity annotation, the DataSet title, and a chart depicting value and rank measurements for that gene across all Samples in that DataSet (**Fig. 2**).

DataSets represent a standardized format for the data in GEO. All the data analysis and mining tools described in this chapter are based on DataSets.

3. Retrieving and Analyzing GEO Data

GEO DataSets may be browsed at http://www.ncbi.nlm.nih.gov/projects/geo/gds/gds_browse.cgi. Original submitter-supplied records may be browsed at http://www.ncbi.nlm.nih.gov/projects/geo/info/print_stats.cgi. A specific

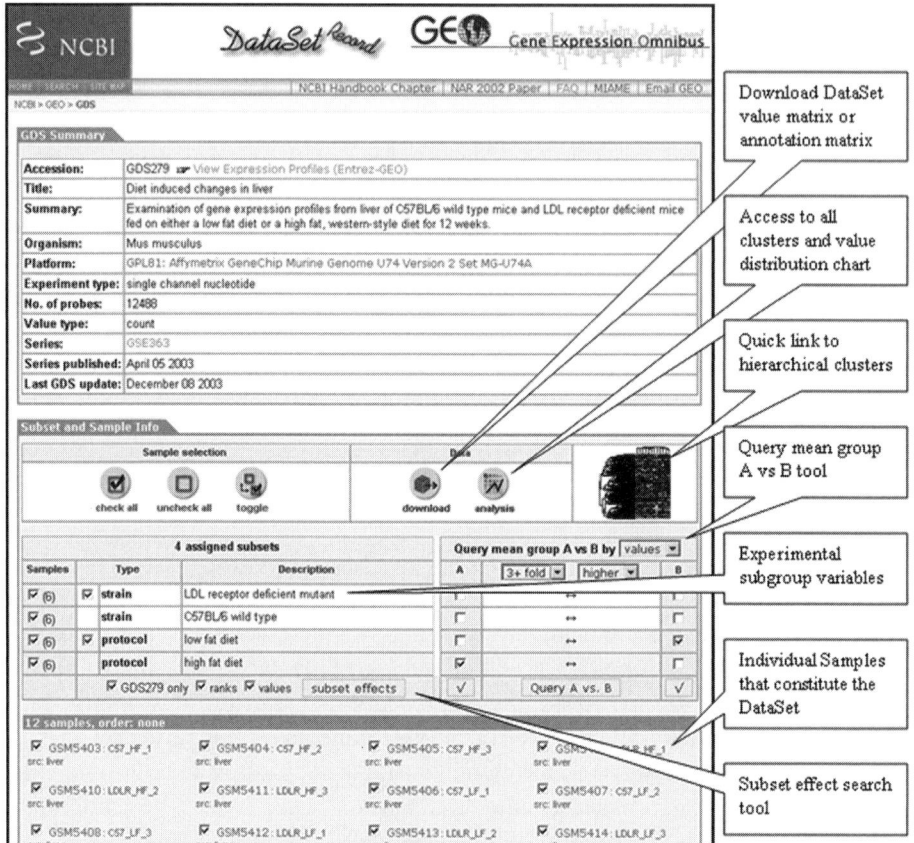

Fig. 1. GEO DataSet record. A screen shot of a typical DataSet record, GDS279, which investigates the effect of a high-fat diet on liver tissue in wild-type and LDL receptor-deficient mice *(6)*. The locations of the main DataSet features and tools are indicated.

record may be accessed directly using its assigned GEO accession number. Additionally, all data (DataSets, original submitter-supplied data, and raw data) are freely available for bulk download via FTP at ftp://ftp.ncbi.nih.gov/pub/geo/data/.

The utility of vast quantities of data such as these is increased greatly if there are effective methods to query the database, allowing reduction to data that are relevant to a specific area of interest. Several different query approaches have been developed, including standard text-based searches, nucleotide sequence-based searches, queries based on characteristics of the expression patterns themselves, or combinations of these parameters.

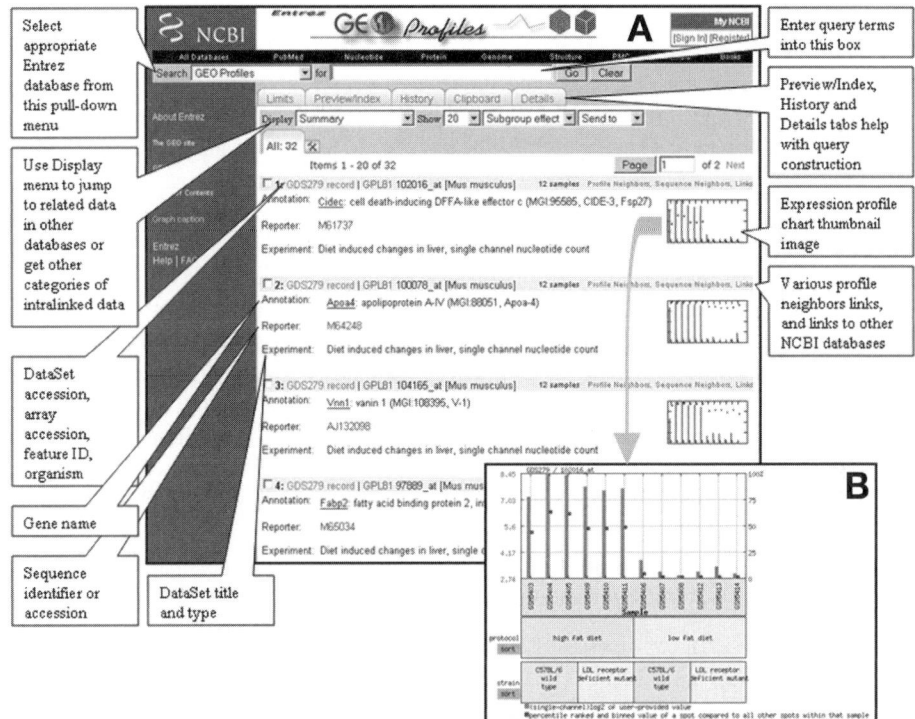

Fig. 2. Screenshot of GEO Profiles retrievals and expanded profile chart. (**A**) Screen shot of GEO Profiles retrievals for GDS279, a DataSet that investigates the effect of a high-fat diet on liver tissue in wild-type and LDL receptor-deficient mice (*6*). Each retrieval represents an individual gene on the array. The locations of various features are indicated. (**B**) Expanded chart for the Cidec gene. The chart bars represent relative gene expression levels in each Sample across the DataSet (*see* **Note 4**). The squares represent rank information; ranks are calculated by rank-ordering expression values for all the genes that are present on the array for each Sample and then placing them into percentile bins. Thus, ranks give an indication of the level of expression of that gene compared with all other genes on the array (or they can be an indication of when data are not properly normalized; *see* **Notes 5** and **6**). The bars at the foot of the chart reflect experimental design (*see* **Note 7**). From the chart, it is clear that expression of Cidec is upregulated in both strains of mice fed a high-fat diet.

3.1. Searches Using NCBI's Entrez Search Engine

Most biological scientists are familiar with Entrez, using it routinely to search NCBI databases like PubMed and GenBank (*3,4*). It has a straightforward interface by which users can simply type in key words and terms to locate relevant material, as well as capabilities to perform complex searches across multiple databases. The Entrez system currently comprises 25 interlinked databases, two of which store GEO data:

1. *Entrez GEO DataSets* contains DataSet definitions. Researchers can search for DataSets using text key word terms. Retrievals display the DataSet title, a synoptic description, the organism, and the experimental variables, as well as links to the complete GDS record (**Fig. 1**), parent Platform, and reference Series records. A user can quickly scan through the retrievals and identify DataSets that look relevant to the area of study. Entrez GEO DataSets is searchable using the "DataSets" query box on the GEO home page, or at http://www.ncbi.nlm.nih.gov/entrez/query.fcgi?db=gds.

2. *Entrez GEO Profiles* contains individual, normalized, DataSet-specific gene expression profiles. Researchers can query for specific genes by name, symbol, accession number, and clone identifier, or genes of interest based on characteristics of the expression profiles. Retrievals display the mapped gene name (determined by the sequence reference provided on the array), the DataSet title, and a thumbnail chart of the gene expression profile generated from the normalized values for an individual gene reporter in each Sample of the DataSet. Clicking on the thumbnail image will enlarge the chart to display the full profile details and the Sample subset partitions that reflect experimental design (**Fig. 2**). Entrez GEO Profiles is searchable using the "Gene profiles" query box on the GEO home page, or at http://www.ncbi.nlm.nih.gov/entrez/query.fcgi?db=geo.

It is usually sufficient to simply enter text key words to retrieve data of interest. For example, searching GEO Profiles with the term "Klk13" will retrieve all profiles that match that gene name. However, information is indexed into several categories (**Table 1**). These indices can be used to refine and restrict Entrez queries. To perform such a search, specify the search terms, their fields, and the Boolean operations to perform on the term using the following syntax:

term [field] OPERATOR term [field]

where term(s) are the search terms, the field(s) are the search fields and qualifiers, and the OPERATOR(s) are the Boolean operators (uppercase AND, OR, NOT). The Preview/Limits link on the Entrez tool bar assists greatly in construction of complex queries. Alternatively, complex search statements can be written and executed directly in the search boxes. The indices (available on the Preview/Limits page) may be used to browse and/or select the terms by which the data are described.

3.1.1. Identifying Experiments of Interest

In many cases, a researcher will begin by looking for DataSets that are pertinent to the area of study. For example, to locate experiments that investigate spermatogenesis or testis development in mice using Affymetrix GeneChip technology, search Entrez GEO DataSets with "(spermatogenesis OR testis development) AND Affymetrix AND mouse."

Table 1
Entrez Qualifier Fields[a]

Field name	Field description
GEO DataSets	
Author	Authors associated with the experiment
Experiment Type	The experiment type, e.g., cDNA, genomic, protein, SAGE
GDS Text	DataSet description text
GEO Accession	The GEO accession number (GPLxxx, GSMxxx, GSExxx, GDSxxx)
GEO Description/Title Text	Text provided in the description/title of original records
Number of Samples	The number of Samples in the DataSet*
Number of Platform Probes	The number of Platform reporters in the DataSet*
Organism	The organism from which the reporters on the array were derived/designed
Reporter Identifier	The identifier for the array reporter (GenBank accession, gene name, and so on)
Sample Source	The source biological material of the Sample
Sample Title	Sample Title
Submitter Institute	Submitter institute
Subset Description	The description of the experimental variable
Subset Variable Type	The type of experimental variable, e.g., age, strain, gender

(*continued*)

Table 1 (Continued)

Field name	Field description
GEO Profiles	
Experiment Type	The experiment type, e.g., cDNA, genomic, protein, SAGE
Flag Information	Specific experimental variable flags, e.g., age, strain, gender
Flag Type	Flag types, e.g., rank and value subset effects
GDS Text	DataSet description text
GEO Accession	The GEO accession number (GPLxxx, GSMxxx, GSExxx, GDSxxx)
GEO Description/Title Text	Text provided in the description/title of original records
GI	Mapped GenBank identifier
Gene Description	Gene description, symbol, alias
ID_REF	The unique identifier for a reporter as given on the array
Max Value Rank	The maximum value rank*
Max value in profile	The maximum value in profile*
Median value in GDS	The median value in DataSet*
Median value in profile	The median value in profile*
Min Value Rank	The minimum value rank*
Min value in profile	The minimum value in profile*
Number of Samples	The number of Samples in the DataSet*
Organism	The organism from which the Samples were derived
Ranked Standard Deviation	The ranked standard deviation
Reporter Identifier	The identifier for a reporter
Sample Source	The source biological material of the Sample

[a]Useful qualifier fields for performing restricted GEO DataSets and GEO Profiles queries. (*) indicates possible range operation, e.g., 20:50[Number of Samples] will find DataSets containing 20 to 50 Samples.

3.1.2. Identifying Gene Expression Profiles of Interest

Once the researcher has located relevant DataSet(s), he or she can use the DataSet accession number (GDSxxx) to restrict searches to that experiment. For example, to view the profiles of all heat shock genes in DataSet GDS181, he or she could query with "GDS181 AND heat shock."

If a DataSet accession is not specified, then the search will be performed across all GEO data. For example, to view profiles of kallikrein family genes in any DataSet that investigates progesterone, enter "kallikrein AND progesterone[GDS Text]."

Often the gene name will contain words that can be found in the DataSet description, and vice versa. In these cases, the only way to specifically retrieve data would be to restrict the search using the appropriate qualifier. For example, a researcher trying to locate profiles for the glycine receptor gene would have to restrict the search to "glycine receptor[Gene Description]"; otherwise, he or she would also retrieve all profiles from GDS967, a DataSet investigating a "Glycine receptor beta subunit mutant model for hyperekplexia."

Several fields are available for refining a search to help identify interesting or significant profiles based on the characteristics of the expression pattern. For example, the expression measurements of each Sample in a DataSet are rank ordered. It is possible to refine searches to identify genes that fall within a specified abundance bracket. To view profiles that fall into the top 5% abundance rank bracket in at least one Sample in DataSet GDS182, search with "GDS182 AND 96:100[Max Value Rank]." Alternatively, to search for profiles in which the median value level across the DataSet is high (approx 12–14 for this DataSet), query with "GDS182 AND 12:14[Median value in profile]."

Typically, researchers seek genes that vary their expression depending on experimental factors. As described in **Subheading 2.**, GEO DataSets are partitioned into subsets that reflect experimental design. Profiles are flagged if they display a significant effect in relation to subsets, that is, if the expression values or ranks pass a threshold of statistical difference between any non-single experimental variable subset and another. These flags assist in the identification of candidate genes as follows:

- Users can restrict their searches to find any profile that exhibits an effect within specific DataSets. For example, to view profiles showing interesting value subset effects in either DataSet GDS186 or GDS187, query with "(GDS186 OR GDS187) AND "value subset effect" [Flag Type]." A convenient way to run this search is from the DataSet record page (*see* subset effect box in **Fig. 1**).
- Users can search across the whole GEO for genes that show an effect with respect to a particular experimental variable type. For example, to search for any gene that shows an effect with respect to gender, query with "gender [Flag Information]."

- Standard GEO Profile retrievals are default ordered according to subset effect flags, bringing potentially significant and interesting profiles to the fore.

The Entrez search system is a powerful tool that interlinks many diverse data domains. Regular users of NCBI resources are well advised to familiarize themselves with the advanced mining features available though Entrez (*see* **Note 1**).

3.2. Tools Available Within Entrez GEO Profiles Results Page

After identification of profile(s) of interest, there are several features on the profile records that link to various types of related profiles, assisting in identification of more genes of interest, or to related information in other NCBI Entrez databases:

Profile neighbors connects genes that show a similar or reversed profile shape within a DataSet, as calculated by Pearson correlation coefficients. Profile neighbors may suggest that those genes have a coordinated transcriptional response, possibly inferring some common function or regulatory elements (*see* **Note 2**).

Sequence neighbors retrieves profiles related by nucleotide sequence similarity across all DataSets, as determined by BLAST *(5)*. Retrievals are shown in decreasing order of similarity to the selected sequence and can provide valuable insights into the possible function of the original sequence if it has not yet been characterized, or they may be useful in identifying related gene family members or for cross-species comparisons.

Homolog neighbors retrieves profiles of genes belonging to the same Homolo-Gene group. HomoloGene is a system for automated detection of homologs among the annotated genes of several completely sequenced eukaryotic genomes.

Links menu allows users to easily traverse from the GEO databases to associated records in other Entrez data domains including GenBank, PubMed, Gene, UniGene, OMIM, HomoloGene, Taxonomy, SAGEMap, and MapViewer.

3.3. Query Mean Group A vs B

The "Query mean group A vs B" feature is available on DataSet records (**Fig. 1**). This tool enables researchers to identify genes that have specific profile characteristics with regard to experimental factors within a DataSet. The user assigns one or multiple subsets to group A and other subset(s) to group B. He or she can then specify that he wants to retrieve all profiles in which the mean value or rank is, for example, 4+-fold higher in group A compared with group B and is directed to Entrez GEO profiles that match those criteria.

For example, for DataSet GDS279, if a researcher wants to locate genes that display a 3+-fold increase in expression between mice that were fed high-fat diet compared with mice that were fed a low-fat diet, he or she would check boxes, as indicated in **Fig. 1**. Hitting the "Query A vs B" button would retrieve profiles that meet these specifications.

3.4. Cluster Heat Maps

One of the most powerful methods to mine and visualize high-dimensional data is through cluster analyses. Details on the mathematical basis of these clustering algorithms are not within the scope of this chapter, but simply speaking, cluster analyses attempt to detect natural groups in data using a combination of distance metrics and linkages. GEO provides nine classic varieties of precomputed unsupervised hierarchical clusters, as well as user-defined K-means and K-median clustering (**Fig. 1**). Columns (Samples), and independently, the rows (genes) are rearranged to place rows with similar response patterns near each other and columns with similar response patterns near each other. Cluster results are graphically represented as "heat maps," whereby high through low expression levels are presented as a two-color spectrum that allows the user to easily identify groups of interesting genes through visual pattern recognition. Each distinct colored "island" in the heat map represents a coordinated transcriptional response, based on the assumption that genes having similar expression profiles across a set of conditions are likely to be involved in the same biological processes. Such biologically relevant clusters can lead to the formulation of testable predictions and can infer functional roles for previously uncharacterized genes.

The GEO cluster heat map images are interactive; using a moveable box, users can select a region, or regions, of interest. This region can be enlarged and the raw data downloaded, plotted as line charts, or linked out to the corresponding profiles in Entrez GEO Profiles (**Fig. 3**; *see* **Note 3**).

3.5. GEO BLAST

The GEO BLAST feature is available on the GEO home page and from NCBI's BLAST page (http://www.ncbi.nlm.nih.gov/BLAST/). This tool allows retrieval of expression profiles on the basis of BLAST nucleotide sequence similarity. Researchers paste in a nucleotide sequence, or specify a sequence accession number, and a BLAST query is performed against all GenBank identifiers represented on microarray Platforms or SAGE libraries in GEO. The output resembles conventional BLAST output with each alignment receiving a quality score; each retrieval has an expression "E" icon that links directly to corresponding GEO Profiles. This interface is helpful in identifying sequence homologs, e.g., related gene family members or for cross-species comparisons, or for providing valuable insight into possible roles of the original sequence if it has not yet been functionally characterized.

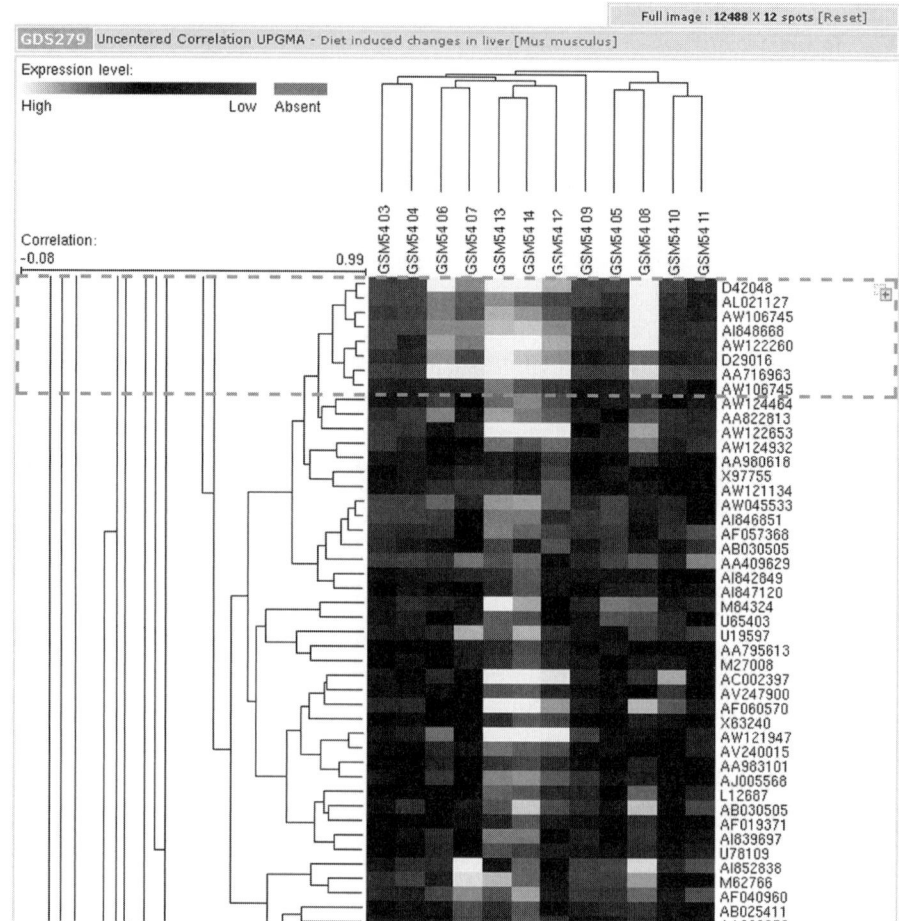

Fig. 3. DataSet cluster analysis. Section of DataSet GDS279 uncentered correlation UPGMA hierarchical cluster analysis. Each column represents an individual Sample, or hybridization; each row represents a gene, identified by a GenBank accession number. The light color indicates high expression and the darker color low expression. The dashed box can be moved and resized to select regions of interest, the data for which may be downloaded, or exported to Entrez GEO Profiles.

3.6. Conclusions

The GEO database archives large volumes of gene expression data generated by the scientific community. Several different approaches to mining GEO data are outlined in this chapter; each of these methods assists biologists to drill down through inherently noisy expression data to genes that are relevant, or behave in a way that is relevant, to their particular area of study.

Making such a large collection of data accessible and analyzable using common interfaces adds a valuable investigative dimension not attained when considering isolated experiments. Through analyzing multiple, independently generated DataSets that examine similar phenomena, it is possible to substantiate interesting gene expression trends that may have been overlooked, or are borderline, in one experiment alone. Researchers can look to see what the preponderance of evidence indicates about the behavior of a gene, or group of genes *(7,8)*. Users can mine GEO for evidence that corroborates laboratory findings, or they may look to GEO for candidate genes worthy of further study in the laboratory. Having sequence information together with expression information can help in the functional annotation and characterization of unknown genes, or in finding novel roles for characterized genes. These data are also valuable to genome-wide studies, allowing biologists to review global gene expression in various cell types and states, to compare with orthologs in other species, and to search for repeated patterns of coregulated groups of transcripts that assist formulation of hypotheses on functional networks and pathways *(9–11)*.

Additionally, integration of GEO data into NCBI's Entrez search engine greatly expands the utility of the data. Entrez is a powerful tool that enables disparate data in multiple databases to be richly interconnected. This can lead to inference of previously unidentified relationships between diverse data types, facilitating novel hypothesis generation, or assisting in the interpretation of available information. Such opportunities for discovery will only increase as the database continues to grow.

The GEO database is under continuous development, so the examples and data presentation strategies described in this chapter may become outdated over time. To keep informed of the latest GEO developments, subscribe to the GEO mailing list at geo@ncbi.nlm.nih.gov.

Acknowledgments

The authors unreservedly acknowledge the expertise and dedication of the GEO curation and development team—Carlos Evangelista, Pierre Ledoux, Dmitry Rudnev, Alexandra Soboleva, Tugba Suzek, Dennis Troup, and Steve Wilhite.

4. Notes

1. Advanced mining tips using Entrez searches.
   a. Use "History" in the Entrez tool bar to see your previous queries. Each search is assigned a number and is stored for up to 8 h. Previous queries can be combined to form a new search query.
   b. Use the "Display" pull-down menu to find related data in other Entrez resources. For example, let us say that your GEO Profiles search has narrowed down to a

list of 100 candidate genes. You next want to check the Gene Expression Nervous System Atlas (GENSAT) database for complementary expression evidence for those genes. (GENSAT is another NCBI resource that contains expression mapping information for genes in the mouse brain at various stages of development.) Instead of checking each of the 100 genes individually in GENSAT, you would simply select "GENSAT Links" from the Display menu in GEO Profiles, and you are immediately directed to GENSAT data that corresponds to your 100 candidate genes.

  c. GEO mining tools, together with the Entrez features described in **steps a** and **b** above, can be combined to form very powerful searches. For example, consider DataSets GDS214 and GDS563—these are independent experiments performed on different arrays that compare normal muscle tissue with muscle tissue from patients affected with Duchenne muscular dystrophy. A user could perform the following set of maneuvers to identify genes that are upregulated in Duchenne patients, in both DataSets.

    i. Use cluster analysis in GDS214 to visually select clusters of genes that are highly expressed in Duchenne Samples compared with control.

    ii. Use the "Get Profiles" button to export these genes to Entrez GEO Profiles.

    iii. Select "Gene Links" from the "Display" pull-down menu—this retrieves a list of corresponding curated genes from NCBI's "Gene" database. From the History tab, you can see that this search is assigned #1.

    iv. Repeat the above three steps for GDS563—from the History tab, you can see that this search is assigned #2.

    v. Combine these two searches by querying Entrez Gene with "#1 AND #2." This retrieves a list of common genes that are found to be upregulated in Duchenne patients in two separate DataSets. The fact that these genes appear to be similarly regulated in both DataSets lends confidence to the results. This also demonstrates a way to effectively perform cross-platform analyses.

  d. The "MyNCBI" feature allows users to save searches and retrieve them in a later session, or monitor how a prior saved search is modified in the context of the current, updated database content. To use the many features of MyNCBI, the user must first establish a login name in that system.

2. Profile neighbor links are subject to cutoff limit. Thus, if this limit is reached, bear in mind that there are probably more genes in the DataSet that demonstrate similar behavior. In this case, you might consider utilizing cluster analyses or the "Query mean group A vs B" tool, which are not subject to such limitations.

3. It is important to realize that different cluster methods will generate different results. An underlying assumption of clustering is that genes with similar expression patterns are more likely to have similar biological function. Clustering does not provide proof of this relationship, but it does provide suggestions for data interpretation.

4. The gene expression value bars are plotted on the left y-axis. Note that this scale slides to fit the values of a particular profile. This sliding scale allows subtle dif-

ferences in values to be more clearly visualized. The ranks are plotted on the right *y*-axis and are always scaled from 0 to 100%.

5. Binned rank information is provided as complementary indication of the relative abundance of a gene compared with all other genes on that array. A rank profile that follows the trend of the corresponding value profile provides additional assurance that the data are properly normalized. Keep in mind that cross-gene rank assessments are made with the assumption that all probes are detecting their target with the same efficiency, which may not always be true.

6. The Samples within any comparable DataSet are assumed to have been processed similarly. You can verify that Sample values are well distributed and normalized with respect to each other (and thus comparable) by viewing the "value distribution" chart that is provided on each DataSet record under the "analysis" button. This presents a box and whisker plot for each Sample within the DataSet, allowing easy visualization of the value median, spread, and overall range.

7. The "Sort" button on GEO Profile charts lets users resort the Samples in the DataSet according to a particular experimental variable. This can assist in clearer visualization of an expression trend in experiments with multiple variables.

References

1. Barrett, T., Suzek, T. O., Troup, D. B., et al. (2005) NCBI GEO: mining millions of expression profiles—database and tools. *Nucleic Acids Res.* **33,** (Database issue) D562–566.
2. Edgar, R., Domrachev, M., and Lash, A. E. (2002) Gene Expression Omnibus: NCBI gene expression and hybridization array data repository. *Nucleic Acids Res.* **30,** 207–210.
3. Wheeler, D. L., Barrett, T., Benson, D. A., et al. (2005) Database resources of the National Center for Biotechnology Information. *Nucleic Acids Res.* **33,** (Database issue) D39–45.
4. Schuler, G. D., Epstein, J. A., Ohkawa, H., and Kans, J. A. (1996) Entrez: molecular biology database and retrieval system. *Methods Enzymol.* **266,** 141–162.
5. Altschul, S. F., Gish, W., Miller, W., Myers, E. W., and Lipman, D. J. (1990) Basic local alignment search tool. *J. Mol. Biol.* **215,** 403–410.
6. Recinos, A. 3rd, Carr, B. K., Bartos, D. B., et al. (2004) Liver gene expression associated with diet and lesion development in atherosclerosis-prone mice: induction of components of alternative complement pathway. *Physiol. Genomics* **19,** 131–142.
7. Wu, X., Li, Y., Crise, B., and Burgess, S. M. (2003) Transcription start regions in the human genome are favored targets for MLV integration. *Science* **300,** 1749–1751.
8. Zerbini, L. F., Wang, Y., Czibere, A., et al. (2004) NF-kappa B-mediated repression of growth arrest- and DNA-damage-inducible proteins 45alpha and gamma is essential for cancer cell survival. *Proc. Natl. Acad. Sci. USA* **101,** 13618–13623.
9. Rodwell, G. E., Sonu, R., Zahn, J. M., et al. (2004) A transcriptional profile of aging in the human kidney. *PLoS Biol.* **2,** e427.

10. Scott, M. S., Perkins, T., Bunnell, S., Pepin, F., Thomas, D. Y., and Hallett, M. T. (2005) Identifying regulatory subnetworks for a set of genes. *Mol. Cell. Proteomics* Feb 18 Epub.
11. Haverty, P. M., Frith, M. C., and Weng, Z. (2004) CARRIE web service: automated transcriptional regulatory network inference and interactive analysis. *Nucleic Acids Res.* **32,** (Web Server issue) W213–216.

15

The Stanford Microarray Database

A User's Guide

Jeremy Gollub, Catherine A. Ball, and Gavin Sherlock

Summary

The Stanford Microarray Database (SMD) is a DNA microarray research database that provides a large amount of data for public use. This chapter describes the use of the primary tools for searching, browsing, retrieving, and analyzing data available for SMD. With this introduction, researchers and students will be able to examine and analyze a large body of gene expression and other experiments. Additional tools for depositing, annotating, sharing, and analyzing data, available only to registered users, are also described. SMD is available for installation as a local database.

Key Words: DNA; cDNA; microarray; gene expression; database; repository; data analysis; expression profiling; clustering.

1. Introduction

The Stanford Microarray Database (SMD; http://smd.stanford.edu/; *1*) is a web-based, research-oriented database for DNA microarray data. As of April 2005, SMD contained data for over 50,000 hybridizations, generated by over 1000 users. In addition, data for nearly 9000 hybridizations, including the data underlying over 200 published manuscripts, were freely available from SMD at that time, for public use without restriction.

Most researchers, educators, and students will find SMD useful as a repository of a very large quantity of publicly available data, together with analysis tools suitable for exploratory, unsupervised analysis and discovery. At the time of this writing, most of the data available to the public were generated from two-color assays on printed cDNA microarrays *(2)*. However, some data from experiments using Affymetrix GeneChips™ (single-channel data from high-density photolithographic arrays) are also available. SMD also has prepublication

From: *Methods in Molecular Biology, vol. 338: Gene Mapping, Discovery, and Expression:*
Methods and Protocols
Edited by: M. Bina © Humana Press Inc., Totowa, NJ

data from Agilent, NimbleGen, and CombiMatrix platforms, so it is likely that data from those platforms will eventually be made public.

This chapter concentrates primarily on the tools for browsing, selecting, downloading, and analyzing these public data.

Providing a vehicle for disseminating the results of published microarray data is only a secondary function of SMD. SMD was primarily designed and is used as a research-oriented database for ongoing, prepublication experiments. Registered users have access to tools to deposit data, annotate them to full compliance with the MIAME standards *(3)*, share them with collaborators, and perform certain analyses that are too resource intensive for general public use. These functions are also briefly covered in this chapter, since most are of interest to any researcher conducting experiments or data analysis utilizing microarray data.

Additionally, there are many tools available to the database curators, for entering array designs, maintaining and updating annotations for genes, managing user accounts, and so forth; these will not be discussed here.

SMD may be found at http://smd.stanford.edu/. SMD's source code and database schema are also available for local installation, under the MIT license (http://www.opensource.org/licenses/mit-license.php), at http://smd.stanford. edu/download/.

2. How To Read This Chapter

This chapter is organized by broad topics, including selecting data by directed queries and browsing; retrieving, filtering, transforming, and analyzing data; assessing data quality; entering, organizing, and annotating data; and additional analysis tools for registered users. The first few topics are treated in more detail, since they are of more interest to "public" (i.e., unregistered) users.

Each tool described is prefaced with instructions for navigating to the tool via the Web interface available from SMD as of April 2005. **Figure 1** shows the critical navigational tools in the context of the SMD home page.

3. Methods

The major functions and tools of SMD are described below.

3.1. Finding Data

The most common approach to dealing with microarray data is to analyze a set of hybridizations that together form an experiment, the individual arrays representing various time points, conditions, biological or surgical samples, or other combinations of experimental factors. The Publication List and the Basic and Advanced Search tools discussed in this section, facilitate identification and selection of microarrays for analysis. After the initial selection of arrays, reporters (e.g., clones, oligos, or genes) can be winnowed down based on their mea-

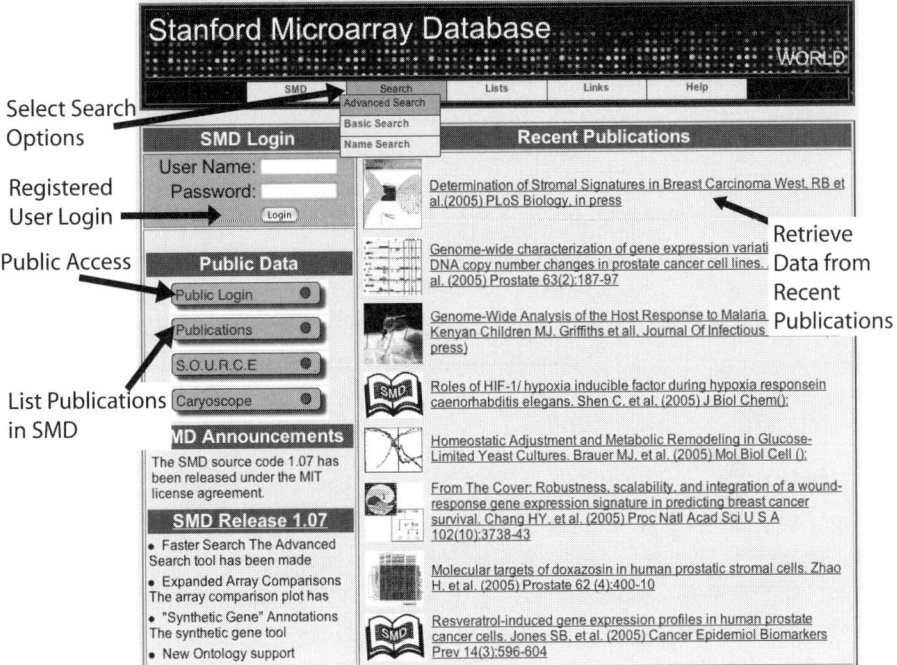

Fig. 1. The SMD home page. Important navigational elements are indicated.

surement quality, expression pattern, or other criteria as described in **Subheading 3.2.**

Alternatively, investigators may start with an individual gene and identify experiments in which that gene was affected. The "Name Search" and "Expression History" tools support this approach.

3.1.1. Access Control

SMD is a research database, containing both published and unpublished data. By default, data are only visible to the owner, members of the owner's group (usually a laboratory), and the database curators. Access may be assigned to other users and groups, or to the public, at the owner's discretion. Therefore:

- All users, including unregistered or "public" users, may view and retrieve data that have been made public upon publication or at the owner's discretion. It is SMD's policy that all data supporting a publication must be made public.
- Registered users may view and retrieve their own data, data belonging to members of their group, and any other data to which they have been explicitly granted access.
- Registered users may edit and delete only their own, nonpublic data. Once data have been made public, the owner may no longer delete it, although editing is still possible.
- Curators may view, edit, and delete all data.

This system permits very flexible collaborations. Unpublished data may be kept private or shared with as many collaborators as required. At the appropriate time, it is easy to make data public, and thousands of microarrays have been published and/or released to the public through SMD.

Some functions of SMD require users to log in before use. Registered users may provide their username and password to gain access to unpublished data. Public users may instead activate a "world session" directly from SMD's home page, gaining access to all published data and most of the analysis tools provided by SMD.

3.1.2. The Publication List

Navigation (no login required):

> Home page: "Publications" button.
> List menu: "Publications" option.

The publication list organizes data that support manuscripts or other published materials. This is usually the simplest way for "public" users to find data of interest. Citations are listed along with links to the data in SMD, a PubMed link if appropriate, the full text of the article if it is available online, and any supplemental web site for the publication. The list may be sorted and searched by organism studied, date of publication, citation, and authors.

Clicking on the "SMD" icon leads to a page with the title, citation, and abstract of the article, along with the links listed above. In addition, there are links to the "raw" data files for the experiment(s) (*see* **Subheading 3.2.2.**) and to machine-readable metadata describing the microarrays included. There are also options to display (*see* **Subheading 3.2.1.**) or retrieve (*see* **Subheading 3.2.4.**) and cluster (*see* **Subheading 3.2.5.**) the data, organized by experiment sets (*see* **Subheading 3.4.3.**).

3.1.3. Basic Search

Navigation (login required):

> Search menu: "Basic Search" option.

The "Basic Search" tool permits users to browse organized data sets. Users may select an organism and then browse publications (*see* **Subheading 3.1.2.**), experiment sets (*see* **Subheading 3.4.3.**), or all arrays annotated to a given category (*see* **Subheading 3.4.1.**). Options are provided to display (*see* **Subheading 3.2.1.**) or retrieve (*see* **Subheading 3.2.4.**) and cluster (*see* **Subheading 3.2.5.**) the selected data.

Basic Search is frequently the best way to find public, as yet unpublished data, although not all data that a user is allowed to see can be found through this interface. The arrays that are shown in the Basic Search tool are only those that have been organized by the experimenters or curators into experiment sets (and to which the user has access), so hybridizations of interest will not appear if they have not been so organized. In that case, the "Advanced Search" tool is required (*see* **Subheading 3.1.4.**). (In the case of browsing by "category," all accessible arrays are presented by Basic Search, but in a manner determined by the annotation provided by the experimenter.)

Basic Search is most powerful in allowing access to experiment sets (*see* **Subheading 3.4.3.**), which are the primary means of collaborative communication and data organization in SMD. By using Basic Search rather than the "Advanced" option, collaborators can easily view data as organized and annotated by their coworkers, eliminating confusion and redundancy.

3.1.4. Advanced Search

Navigation (login required):

Search menu: "Advanced Search" option.

The "Advanced Search" tool provides several ways to identify, browse, and select data for analysis. All data to which the user has access may be found with this utility. Hybridizations may be identified by their owner (listed by username, which may be looked up in the User List), print run or array design, or key words (i.e., Category and Subcategory). Registered users may additionally use personal "array lists" for sets of hybridizations that they routinely revisit (*see* **Subheading 3.4.2.**). Data may be displayed (*see* **Subheading 3.2.1.**) or retrieved (*see* **Subheading 3.2.4.**) and then clustered (*see* **Subheading 3.2.5.**). Advanced Search is also the jumping-off point for creation of experiment sets (*see* **Subheading 3.4.3.**) and array lists.

This is the most powerful search tool and usually the most suitable for experimenters working with their own current data or assembling data from various sources. However, it can be difficult to identify data of interest other than your own, since that depends on the quality of annotations assigned by the data owners (frequently for their own use). Basic Search or the publication list are much more convenient when the data of interest have already been organized and annotated.

3.1.5. Reporter/Gene-Centric Search

Navigation (login required):

Search menu: "Name Search" option.

The "Name Search" tool finds reporters (clone, oligos, or other molecules placed or synthesized on a microarray), rather than experiments or hybridizations. Only those reporters found on microarrays in the database may be found in this way—this is not a general replacement for NCBI's Entrez, or Stanford's SOURCE *(4)*. Reporters may be identified by organism and gene name, description, identifier, and so on (wildcards are allowed). All identifiers, annotations, and so forth will be returned for all matches to the search term.

If data for a so-found reporter are available, a link to the reporter's "expression history" will be presented. This will lead to a histogram of all data for the reporter (from arrays to which the user has access). The graph is interactive: clicking on a bar in the histogram will produce a list of the hybridizations in which the reporter had that value, with options to display (*see* **Subheading 3.2.1.**) or retrieve (*see* **Subheading 3.2.4.**) and cluster (*see* **Subheading 3.2.5.**) all data from those hybridizations. This can serve to identify experiments in which a particular gene was affected. Note, however, that microarray data may not be easily comparable across different experimental conditions, technical protocols, and reference RNA mixtures, so the histogram may be somewhat misleading.

3.2. Analysis

SMD provides several methods for unsupervised analysis, primarily hierarchical clustering (*see* **Subheading 3.2.5.**), and supervised analysis methods such as Gene Ontology (GO) enrichment analysis, using GO::TermFinder *(5)*. However, SMD is primarily a platform for data storage, quality assessment, and collaboration. There are many excellent software packages for data analysis, such as the R statistical programming language, MatLab, and specialized tools such as Significance Analysis of Microarrays (SAM) to name a few. SMD currently supports analysis using these packages by allowing users to easily download the entire raw data for an array or to filter and download selected data from multiple arrays, in a convenient text-based format. Importing into other software packages is usually a matter of making simple changes to the data format, using a spreadsheet program such as Excel. Several analysis tools, such as the BioConductor (http://www.bioconductor.org) packages for R, have facilities for reading the data files available from SMD with no editing required.

3.2.1. Display Data

Navigation (login required unless entering via the publication list):

Any search tool: "Display Data" button.

The first task in analysis, of course, is to identify the data to be analyzed. The search tools described in **Subheading 3.1.** are the entry to this process. For detailed information on the hybridizations found by the search tools, click the "Display Data" button. This appears next to each experiment set contained in a publication (*see* **Subheading 3.1.2.**) on the Basic (*see* **Subheading 3.1.3.**) and Advanced (*see* **Subheading 3.1.4.**) Search pages and with the list of arrays selected through the "expression history" tool (*see* **Subheading 3.1.5.**).

The display page presents each array selected or found by the search, along with a number of options for examining the data (**Fig. 2**). Most of these options are discussed below, in **Subheading 3.3.** (Quality Control and Other Tools). Most relevant here is the "view details" option, invoked by clicking on the "View" icon for an array of interest. This page presents all descriptive information provided by the experimenter, including channel and general descriptions, and any procedural information entered, as well as links to various quality control utilities. This information indicates the role of the hybridization in an experiment (e.g., the time point or the tissue type or disease state of the sample, and so on), and is thus critical for proper data analysis and understanding of the experiment.

The display page provides additional information if entered from Basic Search or the publication list. In this case, an entry for the experiment set appears at the top of the page. The "view" icon for this entry leads to summary information for the experiment as a whole, including the experimenter's description of the overall experimental design and a listing of experimental factors and their values for each array (e.g., incubation time, age, disease state, or whatever factors were deemed critical by the data owner). If provided by the experimenter, this summary serves as a guide to supervised analysis with other data analysis packages and to understanding the experiment.

3.2.2. Raw Data Files

Navigation (login required unless entering via the publication list):

> Any search tool: "Display Data" button: "Raw Data" icon (single array).
> Publication list: SMD "book" icon: "Raw Data" icon (all arrays in set).
> Lists menu: "All Programs" option: "Get Public Data by Organism" link (public users only).

All measured data, array layout, manufacture information, and reporter annotations for a single array are combined in the "raw data" file for each hybridization. These files can be easily examined and edited in a spreadsheet program and contain all available information to support analysis in other software.

Files in SMD's format are provided for two-color arrays and for summary (at the gene or "probe set" level) data for Affymetrix-style single-channel arrays. Affymetrix probe-level data are provided in the original .cel files, for better compatibility with the many analysis packages that use the .cel format. When requested by registered users, these files (other than .cel files) are dynamically

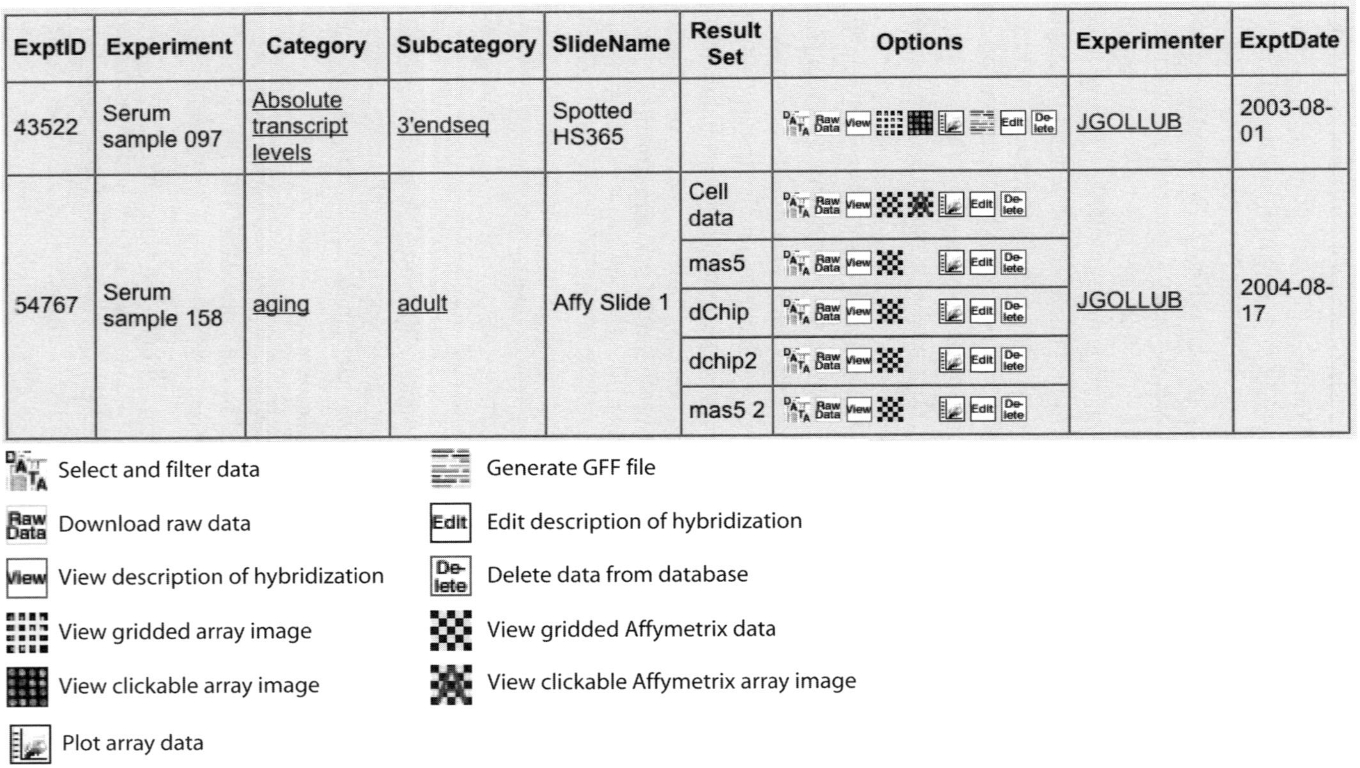

ExptID	Experiment	Category	Subcategory	SlideName	Result Set	Options	Experimenter	ExptDate
43522	Serum sample 097	Absolute transcript levels	3'endseq	Spotted HS365		DATA RawData View [grid] [clickable] Edit Delete	JGOLLUB	2003-08-01
54767	Serum sample 158	aging	adult	Affy Slide 1	Cell data	DATA RawData View [grid] Edit Delete	JGOLLUB	2004-08-17
					mas5	DATA RawData View [grid] Edit Delete		
					dChip	DATA RawData View [grid] Edit Delete		
					dchip2	DATA RawData View [grid] Edit Delete		
					mas5 2	DATA RawData View [grid] Edit Delete		

DATA — Select and filter data

RawData — Download raw data

View — View description of hybridization

[icon] — View gridded array image

[icon] — View clickable array image

[icon] — Plot array data

[icon] — Generate GFF file

Edit — Edit description of hybridization

Delete — Delete data from database

[icon] — View gridded Affymetrix data

[icon] — View clickable Affymetrix array image

Fig. 2. The "Display Data" view of search results. This page presents many options for examining data, many of which are described in **Subheading 3.3.** The icons are for various tools, as shown.

198

generated, in order to present the most current annotations for the reporters (cDNA clones, and so on) on the array. Public users will receive a static file from SMD's ftp site, which is refreshed periodically with current annotations.

3.2.3. View Data

Navigation (login required unless entering via the publication list):

Any search tool: "Display Data" button: "Data" icon.

This tool returns a subset of the data in a raw data file (*see* **Subheading 3.2.2.**). The user may select any or all of the measured data fields, array layouts, and manufacture information (e.g., block, row, and column, or polymerase chain reaction [PCR] quality code), and reporter annotations (e.g., gene symbol or UniGene cluster ID). The data may be filtered for spot quality according to metrics chosen by the user (of which there are several dozen to choose from) and sorted by any data field. For example, to list the well-measured spots with the highest ratio of red to green signal, the user could filter for spots with a measured foreground signal more than twice the locally measured background and sort by red to green ratio in descending order. This tool may be used interactively, or the user may specify the fields and filters and download a file of all results.

3.2.4. Data Retrieval from Multiple Hybridizations

Navigation (login required unless entering via the publication list):

Any search tool: "Data Retrieval and Analysis" button.

Most forms of analysis require data from multiple arrays. SMD provides a "preclustering" (pcl) file of data selected and filtered as specified by the user (*see* http://smd.stanford.edu/help/formats.shtml#pcl for details of the pcl format). These files can be used for hierarchical clustering, either within SMD (*see* **Subheading 3.2.5.**) or using a variety of external tools. They are also suitable for use with other analysis tools, generally with only simple modifications in a spreadsheet program.

To generate a pcl file, a user selects arrays using any of the search tools (*see* **Subheading 3.1.**) and then clicks on the "Data Retrieval and Analysis" button. This leads to an interactive list of the search results, which may be refined further by selecting specific arrays. Clicking again on the "Data Retrieval and Analysis" button proceeds with retrieval, whereas clicking on the "Display Data" button allows examination of the selected arrays (*see* **Subheading 3.2.1.**). Registered users will also see options to create experiment sets or array lists (*see* **Subheading 3.4.**).

Proceeding with data retrieval, the user is presented with a series of pages for setting retrieval parameters. Detailed help documents, describing the various options, are available within SMD. Briefly, the major choices to be made are: which arrays to include (determined at the previous step); which reporters to include (all, or a restricted list); which field to retrieve (generally log ratio for two-color data, but any measured value may be selected); what spot quality filters to impose; what if any filters to impose on the expression pattern of each reporter (e.g., rank or variance filters); and whether to transform the data (by centering, and/or by log transformation or simple variance stabilization for single-channel data).

The pcl file may be downloaded upon retrieval from the database, or after filtering for expression profile and/or transformation. A summary of the retrieval and filtering options is also provided for download, making it a simple matter to record the procedure. Alternately, the user may stay within the SMD system and use the hierarchical clustering facilities on the data (*see* **Subheading 3.2.5.**).

3.2.5. Hierarchical Clustering

Navigation (login required unless entering via the publication list):

> Any search tool: "Data Retrieval and Analysis" button: data retrieval: cluster options page.

Hierarchical clustering is a primary tool for discovery and data exploration in microarray research (and many other fields). Clustering is an iterative process of grouping items, in this case expression patterns or vectors, according to their similarity. There are many examples of cluster analysis in the microarray literature; *see* in particular **ref. 6** for an early demonstration of the utility of clustering of microarray data. The technique itself is well described as a means for finding patterns and subgroups within data—*see* **ref. 7** for detailed descriptions of clustering methods.

In SMD, the clustering tool is available as the final step in the data retrieval process described in **Subheading 3.2.4.** Both reporters (genes) and arrays (conditions) may be clustered by centroid linkage, using either a Euclidean distance metric, or a centered or uncentered Pearson correlation metric. SMD also supports data partitioning, using self-organizing maps (SOMs; *8,9*).

The clustering results may be examined using several tools within SMD, including GeneExplorer *(10)*, and the Java TreeView applet *(11)*. Alternately, users may download the result files for examination with other tools. SMD supports GO *(12)* term enrichment analysis (*see* **Subheading 3.2.6.**) for nodes of the gene cluster tree, currently for human, mouse, and yeast data, via the Web-based interactive cluster viewer. To access this tool, users click on the miniature image of the cluster heat map displayed at the end of the clustering process.

3.2.6. GO Term Analysis

Navigation (login required—registered users only):

Lists: All Programs option: "Ontology Term Finder" link.

GO term analysis is frequently an illuminating way to explore the biological significance of clustering results or of any procedure that produces a list of genes as its result. Rather than the researcher needing to know the functions of all the gene products, and the pathways in which they perform those functions, this type of analysis provides a robust way of determining whether there is a biological theme to a group of genes, and whether that theme is present at a rate higher than would be expected by chance.

SMD's GO term analysis tool is not limited to reporters found in the database —any human, mouse, or yeast genes may be analyzed, although not all genes have been annotated to the GO ontology at this time. The pcl file that was clustered, or another list of "background" genes, may be supplied to improve the accuracy of the calculation. All terms to which any gene in the list is annotated are returned, along with p values and estimated false discovery rates. A detailed help document is available within SMD.

3.3. Quality Control and Other Tools

Microarray data are subject to many sources of bias, error, and noise. All data should be filtered for quality before analysis, as briefly described in **Subheading 3.2.** Some hybridizations are simply too low quality and should be excluded from analysis. A full discussion of microarray quality control is beyond the scope of this writing.

Generally, experimenters will police the quality of their data before publication. However, public users may also want to use the quality assessment tools described in this section before reaching any conclusions about the results of their analyses.

3.3.1. Images

Navigation (login required unless entering via the publication list):

Any search tool: "Display Data" button: image icon.

Microarray data are derived from images of the microarray slide, taken by a fluorescent scanner. SMD generates lower resolution, two-color (for two-color data) versions of these images and makes them available for visual inspection. These images alone can frequently reveal problems with background fluorescence, poor hybridization or array manufacture, and other issues. The image is interactive: clicking on a feature will lead to a page with detailed information on the reporter and measured data for it.

3.3.2. Single-Array Plots

Navigation (login required unless entering via the publication list):

Any search tool: "Display Data" button: scatter plot icon.

There are a variety of standard plots for examining the quality of data obtained from a hybridization, e.g., a scatter plot of log ratio (M) versus overall intensity (A); red vs green intensities; a histogram of channel intensities or log ratios. These and many other plots can be generated with SMD's plotting utility. Any one or two measured data fields (e.g., channel intensities, log ratios, background intensity values, and so on) may be plotted against one another in a scatter plot or histogram. The data may optionally be filtered, in order to examine only "good" spots.

3.3.3. Array Comparison Plots

Navigation (login required unless entering via the publication list):

Any search tool: "Display Data" button: "View" icon: "Compare…" link.

Replicate hybridizations may be compared using the array comparison tool. This utility will plot values from one array against another. A variety of fields may be selected for plotting. (Usually log ratios are the most appropriate for two-color data, or summary signal for single-channel data.) Features may be filtered for quality; all data are plotted, but spots that failed the filters are shown in red. A simple linear regression is fit to the data that pass the filters, giving an indication of the concordance between arrays.

3.3.4. Spatial Bias Plot

Navigation (login required unless entering via the publication list):

Any search tool: "Display Data" button: "View" icon: "Ratios on Array" link.

Actual images of microarrays (*see* **Subheading 3.3.1.**) are very large and contain a vast amount of information. As such, they can be very difficult to assess by eye for any but the most blatant defects. One very common problem in microarray data, spatial or print-tip bias, is easier to see in the simplified heat map presented by this tool. Each spot is presented as a blue (green), yellow (red), or white square against a black background, as determined by threshold values set by the user. The entire image can be assessed at a glance. Any obvious pattern to the data (e.g., red predominant in the center and green around the edges) indicates a systematic bias owing to uneven drying or other problems during hybridization.

The spatial bias tool also performs a simple ANOVA calculation, indicating the strength of the relationship between log ratio and print tip or printing plate.

(This tool is not available for single-channel data.) These values provide a statistically meaningful assessment of the apparent bias or lack thereof. Generally speaking, a statistical strength of 10% or more ($R^2 > 0.1$) in the print-tip ANOVA indicates a significant problem with the data.

3.4. Entering and Organizing Data

Registered, unrestricted SMD users may enter and edit their own experimental data. This section very briefly covers the high points; extensive help documentation is available within SMD. Note that the design (layout) of a microarray must be entered, by a curator, before experimental data from the corresponding arrays may be entered.

3.4.1. Entering Experimental Data

Navigation (unrestricted account and login required):

> Data menu: "Enter my data" option.

SMD currently accepts two-color data from the Agilent Feature Extraction, GenePix, ScanAlyze, and SpotReader feature extraction software packages. Single-channel data are accepted from the Affymetrix MAS 5 and GCOS software, and from DNA Chip Analyzer (dChip). Original TIFF images, data files, and grid files are required for two-color data; image (.DAT), primary (.CEL), and summary (.txt) data files are required for Affymetrix/dChip data.

The primary data may be entered via a Web form, or in batch mode using a text file prepared by the user. Procedural information and annotations are entered as a separate step, again either interactively (by editing the hybridization record), or in batch mode.

The primary data from GenePix, ScanAlyze, and SpotReader are automatically normalized on data entry, using a simple total-intensity normalization calculation. At the experimenter's option, the data may be renormalized at any time, using the same simple calculation, or using more sophisticated loess normalization options provided by the marray package *(13)* for BioConductor *(14)*. There are currently no facilities for renormalizing Agilent or Affymetrix/dChip data, although Agilent's FeatureExtraction software provides a number of normalization options that may be employed before entering data into SMD. SMD does provide some simple options for normalizing and transforming data during retrieval, prior to clustering (*see* **Subheading 3.2.4.**).

3.4.2. Array Lists

Registered users may create "array lists," which are lists of arrays/hybridizations, optionally with specifications for specific data filters for each one (*see* **Subheading 3.2.4.**). These lists may be created by hand, or using online tools within SMD, and are stored within SMD. They are used for a variety of purposes, most commonly to specify a group of hybridizations that the user will frequently analyze. Array lists are typically used by a single researcher in the course of active analysis. When the analysis has matured or must be shared with collaborators, or published, an "experiment set" (*see* **Subheadings 3.1.3.** and **3.4.3.**) is generally a better tool.

3.4.3. Experiment Sets

Navigation (unrestricted account and login required):

> Advanced Search tool: "Data Retrieval and Analysis" button:
> "Create Experiment Set" button

Registered users may create "experiment sets," which are the primary organizational tool for analysis, publication, and collaboration in SMD. An experiment set is an ordered list of hybridizations, together with experimental factor values for each and an overall description of the experiment. Experiment sets are created using interactive tools within SMD. The owner/creator of the set may assign access to other users, making it a simple matter to share annotations and descriptions along with the data.

3.4.4. Publication Records

Navigation (curator account and login required):

> Lists menu: "All Programs" option: "Create Publication" link.

Database curators may create publication records (*see* **Subheading 3.1.2.**). This requires the creation of an experiment set or sets to associate with the publication (*see* **Subheading 3.4.3.**) and a grant of public access on all data to be included in the publication. Given a PubMed ID number, the publication creation tool will download all relevant information (authors, citation, abstract, and so on), or the information may be entered by hand. The curator may also enter URLs for web sites containing the full text of the article, supplemental information, and so forth.

3.5. The Repository

Navigation (registered account and login required):

> "My Data" menu: "My Repository" option.

Registered users may save retrieved data and/or cluster analyses in their "repository." Preclustering files, or the output of the cluster analysis, may be saved at any stage of data retrieval or analysis (*see* **Subheadings 3.2.4.** and **3.2.5.**), in which case a summary of all the options selected in the retrieval process is also saved for later viewing. Alternately, preclustering and cluster files may be uploaded from the user's desktop computer, for use with SMD's tools. Users may control access to items in their repositories, making this a useful tool for collaborative research as well as for storing intermediate or final analysis results. **Figure 3** shows a view of one example repository, with the various options for examining or analyzing the data.

From the repository, users have access to a variety of tools.

3.5.1. Reentry into Data Retrieval and Analysis Process

Navigation (registered account and login required):

"My Data" menu: "My Repository" option: "Filter" or "to-cluster" icon.

Preclustering files stored in the repository may be filtered or clustered as described in **Subheadings 3.2.4.** and **3.2.5.** This makes it convenient to try clustering with several different metrics, or to use and compare multiple filters for reporter (gene) response to experimental conditions, and so forth.

3.5.2. View Clustering Results

Navigation (registered account and login required):

"My Data" menu: "My Repository" option: Clusterheat-map or "TreeView" icon.

Clustered data may be displayed using either the GeneExplorer *(10)* or Tree View *(11)* software.

3.5.3. KNN Impute

Navigation (registered account and login required):

"My Data" menu: "My Repository" option: Two-headed arrow "impute" icon.

Many statistical analysis methods require a complete data set, with no missing values. (One example is singular value decomposition; *see* **Subheading 3.5.4.**). However, it is very common for values to be missing in microarray data, owing to poor measurement quality or the combination of different array platforms with disjoint reporters. The K-nearest neighbors (KNN Impute) algorithm replaces missing values by estimating them from the values that are present for the most similar reporters (genes) to the one with the missing value *(15)*. The original file is preserved, and the new one (with imputed values) may be analyzed, downloaded, or saved in the repository.

3.5.4. Singular Value Decomposition

Navigation (registered account and login required):

"My Data" menu: "My Repository" option: "SVD" icon.

Name	Organism	Date	Type	Genes	Expts.	Size	Options
aCGH -- prostate samples	Homo sapiens	09/27/04	PCL	41293	7	5962 kB	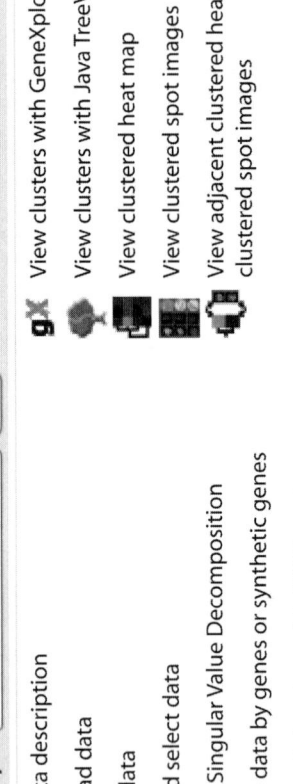
Amplification and labeling trials	Homo sapiens	09/27/04	PCL	17848	9	2196 kB	
Breast cancer - basal cells, new gene list	Homo sapiens	09/27/04	PCL	267	10	55 kB	
Breast cancer - basal cells, new gene list cluster	Homo sapiens	09/27/04	CDT	267	10	1803 kB	
Breast tumors, luminal A and B	Homo sapiens	09/27/04	PCL	8389	13	1737 kB	
cell line -- stress response	Homo sapiens	09/27/04	PCL	45290	226	112001 kB	
Hypoxia short course	Homo sapiens	09/27/04	PCL	6160	13	434 kB	
Prostate samples -- centered and clustered	Homo sapiens	09/27/04	CDT	8389	13	35902 kB	

1 to 8

MY REPOSITORY | UPLOAD

View the repository of JGOLLUB (Jeremy Gollub) [::] Submit

View — View data description

↑ — Download data

Cluster data

Filter — Filter and select data

SVD — Perform Singular Value Decomposition

Synth — Collapse data by genes or synthetic genes

↕ — Estimate missing data with KNN Impute

g^X — View clusters with GeneXplorer

— View clusters with Java TreeView

— View clustered heat map

— View clustered spot images

— View adjacent clustered heat map and clustered spot images

Fig. 3. The Repository. Researchers may store analyses at various stages. The various icons lead to analysis tools, as shown.

Singular value decomposition (SVD) is an unsupervised method for finding underlying patterns in data. It may be used for discovery, or for removing systematic biases from the data (e.g., *see* **refs.** *16* and *17*). Complete data are required; if KNN Impute or some other imputation method has not been employed, SMD's SVD tool will offer to use simple imputation by row averaging. Extensive help documentation is available within SMD.

4. To Learn More

Extensive online documentation, as well as PowerPoint and video copies of SMD beginner and advanced tutorials, is available at SMD (http://smd.stanford. edu/). In addition, those interested in installing a local version of SMD may find useful information on the SMD developers' forum, http://smdforum.stanford. edu/smdforum/.

References

1. Ball, C. A., Awad, I. A., Demeter, J., et al. (2005) The Stanford Microarray Database accommodates additional microarray platforms and data formats. *Nucleic Acids Res.* **33**, (Database issue) D580–582.
2. Schena, M., Shalon, D., Davis, R. W., and Brown, P. O. (1995) Quantitative monitoring of gene expression patterns with a complementary DNA microarray. *Science* **270**, 467–470.
3. Brazma, A., Hingamp, P., Quackenbush, J., et al. (2001) Minimum information about a microarray experiment (MIAME)—toward standards for microarray data. *Nat. Genet.* **29**, 365–371.
4. Diehn, M., Sherlock, G., Binkley, G., et al. (2003) SOURCE: a unified genomic resource of functional annotations, ontologies, and gene expression data. *Nucleic Acids Res.* **31**, 219–223.
5. Boyle, E. I., Weng, S., Gollub, J., et al. (2004) GO:TermFinder—open source software for accessing Gene Ontology information and finding significantly enriched Gene Ontology terms associated with a list of genes. *Bioinformatics* **20**, 3710–3715.
6. Eisen, M. B., Spellman, P. T., Brown, P. O., and Botstein, D. (1998) Cluster analysis and display of genome-wide expression patterns. *Proc. Natl. Acad. Sci. USA* **95**, 14863–14868.
7. Everitt, B. (1974) *Cluster Analysis 122*, Heinemann, London.
8. Kohonen, T. (1995) *Self-Organizing Maps*. Springer, Berlin.
9. Tamayo, P., Slonim, D., Mesirov, J., et al. (1999) Interpreting patterns of gene expression with self-organizing maps: methods and application to hematopoietic differentiation. *Proc. Natl. Acad. Sci. USA* **96**, 2907–2912.
10. Rees, C. A., Demeter, J., Matese, J. C., Botstein, D., and Sherlock, G. (2004) GeneXplorer: an interactive web application for microarray data visualization and analysis. *BMC Bioinformatics* **5**, 141.
11. Saldanha, A. J. (2004) Java Treeview—extensible visualization of microarray data. *Bioinformatics* **20**, 3246–3248.

12. Ashburner, M., Ball, C. A., Blake, J. A., et al. (2000) Gene ontology: tool for the unification of biology. The Gene Ontology Consortium. *Nat. Genet.* **25,** 25–29.

13. Wang, J., Nygaard, V., Smith-Sorensen, B., Hovig, E., and Myklebost, O. (2002) MArray: analysing single, replicated or reversed microarray experiments. *Bioinformatics* **18,** 1139–1140.

14. Gentleman, R. C., Carey, V. J., Bates, D. M., et al. (2004) Bioconductor: open software development for computational biology and bioinformatics. *Genome Biol.* **5,** R80.

15. Troyanskaya, O., Cantor, M., Sherlock, G., et al. (2001) Missing value estimation methods for DNA microarrays. *Bioinformatics* **17,** 520–525.

16. Alter, O., Brown, P. O., and Botstein, D. (2000) Singular value decomposition for genome-wide expression data processing and modeling. *Proc. Natl. Acad. Sci. USA* **97,** 10101–10106.

17. Nielsen, T. O., West, R. B., Linn, S. C., et al. (2002) Molecular characterisation of soft tissue tumours: a gene expression study. *Lancet* **359,** 1301–1307.

16

Detecting Nucleosome Ladders on Unique DNA Sequences in Mouse Liver Nuclei

Tomara J. Fleury, Alfred Cioffi, and Arnold Stein

Summary

Nucleosome arrangements and possibly chromatin higher order structures differ in different regions of genomic DNA, and these differences could be functionally important. Nucleosome arrangements are reflected by the nucleosome ladders they give rise to upon micrococcal nuclease digestion of the chromatin. Here we describe how Southern hybridization can be used to detect the nucleosome ladders arising from different regions of DNA in mouse liver nuclei.

Key Words: Nucleosome ladders; chromatin; mouse liver nuclei; genomic DNA; Southern hybridization.

1. Introduction

Chromosomes in the cells of higher organisms consist of DNA-protein complexes, referred to as chromatin. A mass of histone proteins approximately equal to the mass of DNA is largely responsible for the high degree of compaction of the very long DNA molecule in a chromosome. About 150 bp of DNA is tightly wrapped around an octamer consisting of two copies each of the core histones H2A, H2B, H3, and H4 forming the nucleosome, the fundamental building block of chromatin. Nucleosomes are separated from each other by approximately 45 bp of linker DNA, which is associated with the linker histone H1. This "string of beads" or nucleosome array is further condensed into an irregular approx 30-nm fiber to form interphase chromatin, in which the overall level of DNA compaction is about 40-fold (*1*). The irregularities and the variability among chromatin fibers viewed by cryoelectron microscopy (*2,3*) or scanning force microscopy (*4,5*) are considerable, with some fibers appearing significantly more condensed than others. Moreover, some fibers

From: *Methods in Molecular Biology, vol. 338: Gene Mapping, Discovery, and Expression: Methods and Protocols*
Edited by: M. Bina © Humana Press Inc., Totowa, NJ

exhibit abrupt bends, and others possess loops that bring distal regions of the DNA molecule into close proximity. The possibility exists that such variations might not simply arise from the random coiling of the samples prepared for microscopy, rather, different regions of DNA might tend to form different chromatin higher order structures, and these structural differences might have functional significance *(6,7)*.

Variations also exist among the nucleosome arrangements in different regions of DNA *(6,8)*. Some regions of DNA have more ordered nucleosome arrangements than others, and the value of the nucleosome repeat length often differs from that of the bulk chromatin, sometimes over large DNA regions *(9)*. In the very plausible models of chromatin higher order structure in which the linker DNA is straight *(10–12)*, small changes in linker DNA length can generate large differences in the appearance of the chromatin fiber *(2)*. This sensitivity is a consequence of the structural fact that the histone-DNA contacts in the nucleosome occur across the minor groove of DNA *(13)*, which rotates 360 degrees every 10 bp for straight DNA. Thus, for each base pair change in the length of linker DNA between nucleosomes, the plane of one nucleosome should be expected to rotate by an angle of 36 degrees with respect to the other. Because all the nucleosomes in an array are connected, this effect will propagate, unless constrained, throughout the chain of nucleosomes, with each linker DNA variation contributing, and could thus generate many different higher order structures. Therefore, it is of interest to be able to specifically probe different regions of DNA to assess the nucleosome arrangement in a particular region. Recent work strongly suggests that computationally recognizable long-range periodic sequence motifs in genomic DNA are responsible for the variations in nucleosome arrangements that have been observed *(9)*.

2. Materials

All chemicals were obtained from Mallinckrodt Chemicals, except where noted.

2.1. Mouse Liver Nuclei Prep

1. HB buffer: 0.34 M sucrose (Ultra Pure, Gibco BRL), 15 mM Tris-HCl, pH 7.5, 15 mM NaCl, 60 mM KCl, 0.2 mM EDTA, 0.2 mM EGTA, 0.15 mM spermine, 0.5 mM spermidine *(14)*.
2. 20% (v/v) Triton X-100 Stock made by dilution of Triton X-100 with water.
3. Cheesecloth.
4. Surgical scissors.
5. Frozen mouse livers.
6. VirTis mechanical homogenizer.
7. Dounce homogenizer with loose-fitting pestle B (Bellco).

8. 2X Pro-K Stop Solution: 20 m*M* EDTA, pH 8.0, 0.2% sodium dodecyl sulfate (SDS), 0.5 mg Pro-K/mL.
9. 15-mL Corex glass centrifuge tubes.
10. Sorvall SS-34 rotor.
11. Adaptors for use of 15-mL tubes in the SS-34 rotor.
12. Micrococcal nuclease (Worthington): dissolved in water at 10 U/mL; small portions are stored frozen.
13. 25:24:1 Phenol/chloroform/isoamyl alcohol.
14. 24:1 Chloroform/isoamyl alcohol.
15. Ethanol (200 proof).

2.2. DNA Marker Labeling

1. 5 µg Double-stranded DNA marker with suitable ends for labeling.
2. 10X DNA Polymerase I, Large (Klenow) Fragment buffer, 10 mg bovine serum albumin (BSA)/mL.
3. 5 U DNA Polymerase I, Large (Klenow) Fragment.
4. 50 µCi [α-^{32}P]dATP (3000 Ci/mmol).
5. DEAE Sephadex A-50.
6. 1000-µL Blue pipet tip plugged with glass wool.
7. 0.2 TE: 0.2 *M* NaCl, 10 m*M* Tris-HCl, pH 8.0, 1 m*M* EDTA.
8. 1.0 TE: 1.0 *M* NaCl, 10 m*M* Tris-HCl, pH 8.0, 1 m*M* EDTA.

2.3. PCR Amplification of Mouse Genomic DNA

1. 0.4 mg/mL PCR oligonucleotide primers, forward and reverse (IDT).
2. 10 m*M* dNTP mix.
3. 25 m*M* MgCl$_2$.
4. 10X Taq Polymerase reaction buffer, MgCl$_2$-free.
5. 5 U/µL Taq polymerase.
6. Approx 30 µg/mL Mouse genomic DNA.

2.4. Blotting

1. Gel preparation: 200 mL of 0.25 *N* HCl; 200 mL of 0.4 *N* NaOH/0.6 *M* NaCl; 200 mL of 1.5 *M* NaCl/0.5 *M* Tris-HCl pH 8.0.
2. Blotting by capillary transfer.
   a. 10X SSC: 1.5 *M* NaCl, 0.15 *M* sodium citrate.
   b. Glass dish, cellulose sponge.
   c. GeneScreen Plus charged nylon hybridization transfer membrane (Perkin-Elmer).
   d. Whatman 3 MM filter paper, cut and stacked paper towels.
   e. Glass or Plexiglas plate of size slightly larger than that of the nylon membrane.
   f. 500 g weight arranged to provide even pressure over the membrane.
   g. Plastic wrap.
   h. Small level.

2.5. Southern Hybridization

1. Prehybridization buffer: add 1 g dextran sulfate and 1 mL 10% SDS to 8.46 mL dH$_2$O and heat at 65°C for 30 min (until everything is dissolved); then add 0.58 g NaCl and heat for an additional 15 min at 65°C, until dissolved.
2. Denatured sheared salmon sperm DNA: dilute 50 µg sheared salmon sperm DNA to 2 mg/mL in dH$_2$O in a microfuge tube. Boil for 10 min in a beaker on a hot plate. Immediately chill in wet ice for 15 min.
3. Glass capillary tubes (Microcaps, 50 µL size, Drummond).
4. Diamond tip scriber.
5. Probe template (gel purified using Qiagen Qiaex II Gel Extraction Kit).
6. 10 mg/mL BSA.
7. LS solution: 440 mM HEPES-NaOH, pH 7.6, 44 µM each of dTTP, dCTP, and dGTP, 110 mM Tris-HCl, pH 7.5, 11 mM MgCl$_2$, 22 mM 2-mercaptoethanol, 300 µg/mL random oligodeoxyribonucleotide hexamers (IDT).
8. 50 µCi [α-^{32}P]dATP.
9. DNA Polymerase I, Large (Klenow) Fragment.
10. Kodak Biomax MR film.
11. Kodak Biomax MS intensifying screen.
12. Autoradiogram exposure cassettes.

2.6. Stripping Blots

1. 0.1X SSC/1% SDS.
2. Hot plate with magnetic stirrer for boiling and stirring.
3. Large beaker or glass dish.

3. Methods

Nucleosome arrangements can be assessed from the appearance and the analysis of the nucleosome ladder, which is also called the MNase ladder after the nuclease used to cleave the chromatin, micrococcal nuclease (MNase). Partial MNase digestion of nuclei cleaves the chromatin with a very high preference for linker DNA, rather than for the DNA within the (approx 150-bp) nucleosome core particle. After purification of the partially cleaved DNA and gel electrophoresis, a ladder pattern is produced reflecting the DNA sizes associated with each of the nucleosome oligomers excised, generally extending from monomer to 10-mer DNA lengths. To assess the ladder in a particular region of the DNA, a Southern blot is performed. More ordered nucleosome arrays have ladders with more DNA bands than less ordered nucleosome arrays, and the bands are more distinct over the continuum of background DNA fragments produced. Additionally, in highly ordered nucleosome arrays, the bands of the ladder are multiples of a unit repeat. The nucleosome repeat length can be measured with a precision of about ±5 bp by carefully determining the sizes (bp) of the centers of each nucleosome oligomer band on the autoradiogram, using an adjacent size marker lane to calibrate the gel and then plotting the nucleosome oligomer size vs the nucleosome oligomer number. The slope of the best straight-line fit gives the nucleosome repeat length *(15)*.

3.1. Preparation of Nuclei

1. Defrost four frozen mouse livers in HB buffer on ice (*see* **Note 1**).
2. Drain buffer from livers. Mince livers well with surgical scissors.
3. Add minced livers and 50 mL HB buffer containing 0.1% Triton X-100 to the (pre-chilled VirTis) homogenization vessel. Homogenize the livers at half speed for 2 min.
4. Filter the homogenized livers through two layers of cheesecloth into 15-mL glass Corex centrifuge tubes (usually takes four tubes).
5. Centrifuge the nuclei in an SS-34 Sorvall rotor at 4°C, 4500*g* for 15 min.
6. Carefully decant supernatant and discard.
7. Resuspend the pellets in about 10 mL HB buffer per tube (*see* **Note 2**).
8. Homogenize the resuspended pellets using a (prechilled) Dounce homogenizer with a loose-fitting pestle, using about four strokes of the pestle.
9. Transfer the homogenized nuclei to clean prechilled Corex tubes and centrifuge again at 2300*g* for 10 min.
10. Again discard the supernatant, resuspend pellets in HB Buffer, and Dounce homogenize as above.
11. Centrifuge again as before. Pellets should appear to be a creamy off-white to light tan color. The wash with HB buffer can be repeated once more if necessary.
12. After the final wash and homogenization, resuspend the nuclei very gently in 0.1 *M* TE to a total volume of 20 mL; do not Dounce homogenize (*see* **Note 3**).
13. To measure the approximate DNA concentration of the nuclear suspension:
    a. Remove 10 μL of the suspension to a new tube, immediately after dispersal by gentle agitation, using a cut-off pipet tip.
    b. Add 990 μL of 0.1 *N* NaOH to the 10 μL of nuclei, heat to 95°C for 5 min, mix well, and measure the A_{260} value.
    c. Calculate the approximate concentration (mg DNA/mL) of the undiluted suspension using an A_{260} value of 28 for denatured DNA, and assuming that about 50% of the A_{260} arises from RNA.

3.2. MNase Digest of Mouse Liver Nuclei and DNA Purification

First, perform a small-scale test MNase digest of the nuclei:

1. Add 20 μL 2X Pro-K Stop Solution to each of five microfuge tubes (labeled time points 0, 1, 2, 3, and 5 min).
2. Add 0.10 *M* $CaCl_2$ to a 100-μL aliquot of the nuclei for a final concentration of 2 m*M* (*see* **Note 4**).
3. Warm the nuclei to 37°C for 5 min.
4. Withdraw a 20-μL aliquot at the 0-min time point, and add it to the tube of Pro-K marked 0.

5. Add 4 U of MNase to the remaining 80 μL of nuclei and mix well by pipeting up and down several times. Incubate at 37°C and withdraw 20-μL aliquots at 1, 2, 3, and 5 min, plunging each aliquot immediately into the corresponding tube of 2X Pro-K.
6. Incubate in the Pro-K for about 30 min to 1 h in a heating block at 50°C.
7. Add 1/10 vol of 1 *M* Tris-HCl, pH 8.0, to each tube, and extract with equal volumes of 25:24:1 phenol/chloroform/isoamyl alcohol, and then with 24:1 chloroform/isoamyl alcohol.
8. Adjust to 0.2 *M* NaCl (using a 5 *M* stock), and ethanol precipitate with 2 vol of ethanol. Microfuge to collect the precipitated DNA.
9. Dissolve each DNA pellet in 100 μL 1X gel loading buffer (plus RNase) and incubate for about 30 min at room temperature.
10. Run 10 μL of each time point, along with appropriate DNA size markers, on a 1.5% agarose minigel at 100 V for about 1 h to check extent of digestion of each time point.

Once the MNase digestion conditions have been optimized using small-scale digests, perform the large-scale MNase digests of the nuclei using those conditions scaled up proportionately. The large-scale digests should be incubated in the Pro-K Stop Solution overnight at 37°C.

3.3. Radiolabeling of Appropriate DNA Markers

Nucleosome ladders contain DNA fragment lengths ranging from about 150 to 2000 bp. Thus, useful size markers should contain fragment sizes that are spread roughly evenly over this size range. It is sufficient to label only the 3' ends of the fragments. To label the 3' ends with [α-^{32}p]dATP and Klenow enzyme, there should be an A at the 3' terminus of either a blunt-end or a recessed 3' end. The following procedure has been used routinely for λ *Afl*III fragments and Bio-Rad 100-bp ladder *Hind*III fragments.

1. Prepare a 50-μL reaction vol containing: 1X Klenow reaction buffer, 0.1 μg BSA, 5 μg double-stranded DNA marker, 50 μCi [α-^{32}P]dATP, and 5 U Klenow enzyme. Incubate at 25°C for 30 min. Stop reaction with 1 μL of 0.5 *M* EDTA, pH 8.0. Adjust to approx 0.2 *M* NaCl for column purification.
2. Remove unincorporated [α^{32}-P]dATP by making an anion exchange column in a (blue) 1-mL pipet tip plugged with glass wool.
   a. Make a slurry of 0.05 mg/mL DEAE Sephadex A-50 in 1 *M* NaCl TE, pH 8.0.
   b. Pipet approx 0.5 mL of the DEAE Sephadex slurry into the plugged pipet tip.
   c. Equilibrate the column by three washes with 0.5 mL of 0.2 TE.
   d. Load the salt-adjusted labeled marker reaction onto the DEAE Sephadex column.
   e. Wash the column with 1-mL portions of 0.2 TE, monitoring radioactivity of each wash.
   f. When the signal of the washes has reached background level, elute the labeled marker DNA with 1.0 TE using 0.5-mL washes; usually two washes are sufficient to recover most of the radioactivity from the column (the labeled DNA).

g. Precipitate labeled marker with two volumes of ethanol.
h. Dissolve DNA pellet in 100 µL 1X gel LB.
3. Run loads of 1, 2, and 5 µL on an agarose gel, dry the gel, and expose to autoradio-gram film to check the signal strength. Make appropriate dilutions for use as markers on blots, which are exposed for long times using an intensifying screen. Generally a 200-fold or greater dilution is required (when loading 10 µL).

3.4. Agarose Gel Electrophoresis and Blotting

1. A 1.5% gel is best for separation of fragments without compromising transfer of DNA from gel to blotting membrane.
   a. Load no more than 15 µg genomic DNA per lane. Using more will cause altered mobility of the DNA fragments in a complex fashion owing to the migration of the ladder DNA fragments through more viscous higher molecular weight fragments usually present in the digest.
   b. **Figure 1** shows that at DNA loads equal to or less than about 12 µg, most of the DNA fragments run correctly for the 8-min digest and can therefore be accurately sized using markers present in an adjacent lane.
   c. In contrast, the 28-µg load exhibits a pronounced retardation of fragments less than about 1500 bp and increased mobilities for larger fragments for the 2-min digest; smaller mobility aberrations occur for the 8-min digest (*see* **Fig. 2A** for an example of the mouse liver DNA fragment sizes resulting from a 2-min and an 8-min digestion).
   d. Running a 13-cm-long gel for 3.5 to 4 h at 100 V gives a good separation of the fragments.
2. To prepare gel for Southern blotting, we follow Sambrook and Russell *(17)*.
   a. Cut the nylon membrane to fit the gel exactly.
   b. Prewet the membrane with distilled water, and then equilibrate in 10X SSC for 15 min.
   c. Agitate gel in 0.25 *N* HCl for 10 to 15 min to depurinate the DNA.
   d. Rinse gel in dH$_2$O, and then agitate in 0.4 *N* NaOH/0.6 *M* NaCl for 15 min to denature the DNA.
   e. Then agitate gel in 1.5 *M* NaCl/0.5 *M* Tri-HCl, pH 8.0, for 15 min for neutral-ization.
3. Using a glass dish, add 10X SSC to about half full.
   a. Soak a cellulose sponge that is bigger than the gel, allowing it to saturate with the 10X SSC.
   b. Lay on top of the sponge a piece of Whatman 3MM paper covering the sponge.
   c. Lay the gel onto the Whatman paper face down (so that the smooth under side of the gel will contact the membrane).
   d. Carefully layer the equilibrated membrane onto the gel. Do not reposition the membrane once it is in place on the gel.
   e. Use a Pasteur pipet to gently roll along the membrane to remove any air bubbles between the gel and the membrane. The DNA will not transfer in areas where there are air bubbles.

Unlabeled DNA load: 6 µg   12 µg   18 µg   28 µg

Digestion time: 2' 8' M 2' 8' M 2' 8' M 2' 8'

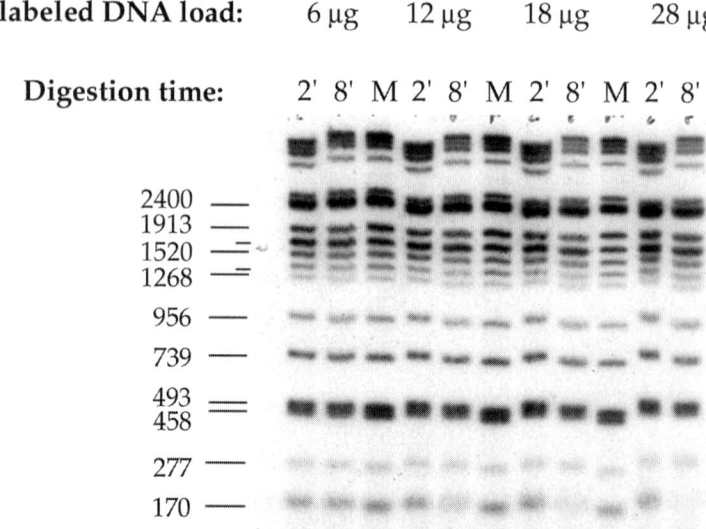

Fig. 1. Effects of the presence of different amounts of unlabeled MNase ladder DNA on the gel mobilities of a tracer amount of labeled size marker fragments. A constant tracer amount of labeled DNA size markers was mixed with 6, 12, 18, or 28 µg of purified MNase ladder DNA from 2-min (2') or 8-min (8') time points of a digestion of mouse liver nuclei. After electrophoresis, the gel was dried and autoradiographed. Lanes labeled M contained only marker, with no mouse liver DNA. Marker fragment sizes (bp) are indicated at the left; the shorter tick marks correspond to the positions of the 1399-bp fragment and the 1712-bp fragment (not listed). The lane containing 28 µg of mouse liver DNA (2' digest) exhibited the most severe effects on the mobilities of the size marker fragments.

    f. Finally, lay two more pieces of Whatman 3MM paper over the membrane.

    g. Stack about 2 inches (compressed) of dry paper towels or other blotting paper over the Whatman paper.

    h. Top the blotting towels or paper with a glass or Plexiglas plate that is as large as or slightly larger than the gel and about 500 g of weights, so that pressure is applied evenly over the membrane.

    i. Cover the glass dish with plastic wrap, and ensure that capillary transfer will occur by solvent flow through the gel, rather than around the gel. Only a few paper towels should become wet during the first hour. Place small levels on the plate to make sure that the arrangement is level.

    j. Check to relevel every 1/2 h or as often as possible, especially for the first 1 1/2 h. Unleveled blotting can result in uneven transfer of DNA.

4. After transfer overnight, remove membrane from gel, and cut the lower right-hand corner of the membrane with the DNA face side up.

A

D M 2' 4' 8' M

B

D M 4' M' 8' M

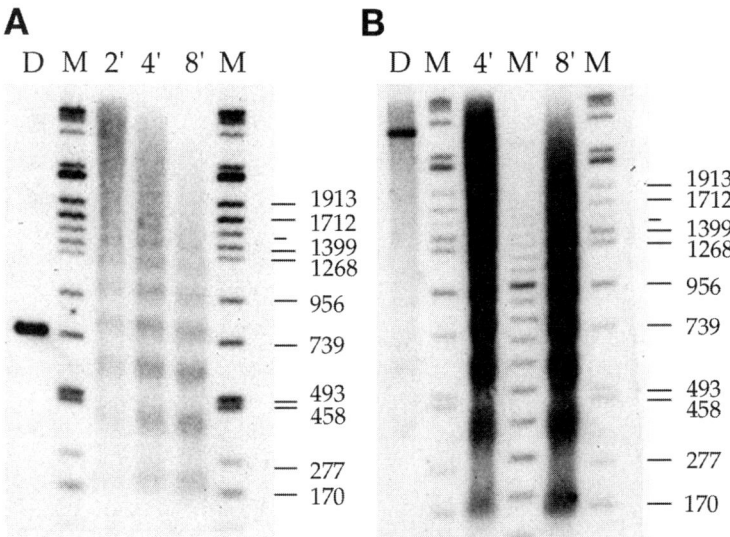

Fig. 2. Nucleosome ladders from mouse liver nuclei revealed by two different hybridization probes that differed in their specificities. (**A**) The probe was highly specific, as evidenced by the specific detection of the predicted 740-bp restriction fragment obtained by *Hind*III and *Pst*I digestion of purified mouse liver DNA (lane D). Nucleosome ladders from this locus are shown for 2-, 4-, or 8-min digests of mouse liver nuclei (lanes labeled 2', 4', or 8'). Lanes labeled M contained DNA size markers of the sizes (bp) indicated at the right; the short tick mark corresponds to the 1520-bp fragment (not listed). (**B**) This probe was not specific for the locus intended, as evidenced by the background detected in lane D in addition to the predicted 3-kb restriction fragment. Analysis of lane D (which was underloaded relative to the 4' and 8' lanes) indicated that the 3-kb fragment contributed less than 50% to the total intensity in this lane. Lanes are labeled as in (A). The nucleosome ladders detected by this probe (lanes labeled 4' and 8') are very intense because the signal is coming from hundreds of loci.

a. Agitate the membrane in 0.4 *N* NaOH for 1 min, and then in 0.2 *M* Tris-HCl, pH 8.0/1X SSC for 1 min.
b. Lay the membrane on a wet filter paper, and fix the DNA to the membrane in a UV crosslinker following the manufacturer's instructions for a wet membrane.
c. Proceed directly to prehybridization or air-dry and store between pieces of dry Whatman filter paper for future use.

3.5. Hybridization Probe Selection

The hybridization probe must hybridize only with the unique DNA region of interest. The size of the hybridization probe selected should be between 300 and about 800 bp. Probes shorter than about 300 bp may require several day-long film exposures to achieve a suitable signal. The larger the probe is, the more likely it will pick up repetitive DNA elements, which will give nonspecific hybridization. A BLAST (blastn, Use MegaBLAST) alignment with the mouse genome can be used to determine whether the chosen probe sequence is sufficiently unique, before oligonucleotides for PCR are ordered or before subcloning is performed. Ideally, the sequence chosen should give a single BLAST hit (red line) with the following parameters: Expect = 0.01, Filter = none, Descriptions = 500, Alignments = 500. Sometimes the presence of a few low-score alignments can be tolerated. However, when thousands of hits occur, the sequence should not be used. Hence, this computational method at least eliminates potential probe sequences containing repetitive DNA elements that are extremely abundant in genomic DNA.

It is essential to experimentally assess the probe specificity under the exact hybridization conditions used to detect the nucleosome ladders. Nonspecific hybridization will still detect ladders, but they will represent the superimposed ladders arising from hundreds or thousands of loci in addition to the locus of interest. Most likely, the ladders detected with nonspecific hybridization will resemble those detected by ethidium bromide staining of the gel. Unfortunately, nearly half of the probes that we make do not hybridize with high specificity, despite use of the BLAST test. To assess the probe specificity; we include a lane on the gel (and the blot) containing a restriction enzyme digest of purified mouse genomic DNA. The probe should ideally detect only the predicted restriction fragment(s), without detecting a large number of other fragment sizes or a background signal consisting of a continuum of fragments. **Figure 2** shows an example of a specific hybridization (**Fig. 2A**) and a nonspecific hybridization (**Fig. 2B**). For **Fig. 2A** the probe was 740 bp, and the BLAST test gave a unique alignment. For **Fig. 2B** the probe was 608 bp, and the BLAST test detected 9039 hits. The restriction digest of purified mouse genomic DNA (lanes D) shows detection of the unique predicted 740-bp fragment for **Fig. 2A**, but a high background continuum of fragments over the predicted 2.2-kb fragment for **Fig. 2B**. Additionally, the intensities of the ladder bands for **Fig. 2B** are unusually high for an overnight film exposure (compare with **Fig. 2A**). In this case, the BLAST predictions were confirmed experimentally.

3.6. PCR Amplification of Mouse Genomic DNA for Use as Probes

1. Set up a master mix PCR reaction of 175 μL vol by mixing together 118 μL deionized water, 17.5 μL of 10X reaction buffer, 14 μL of 25 mM MgCl$_2$, 3.5 μL of 10 mM dNTP mix, 7 μL mouse genomic DNA, and 1 μL of the 5 U/μL Taq polymerase.
2. Mix well, and divide evenly between two tubes. Add 7 μL of the 0.4 mg/mL forward primer to one tube and 7 μL of the 0.4 mg/mL reverse primer to the second tube.
3. Set the thermal cycler for a prewarming time of 5 min at 94°C; a cycling set of 94°C denaturing, 61 to 62°C annealing, and 72°C extension, 35 cycles; and a final cold soak of 4°C. For 24-bp primers designed with IDT's online program, the cal-

culated T_m is usually roughly 60.0 to 60.5°C, and an annealing temperature of 62°C will usually work well for these primers.

4. Let the two halves of the reaction, in their separate tubes, reach 94°C. Then quickly withdraw the contents (the aqueous portion only when an oil layer is used) of the reverse primer tube and pipet them into the bottom of the forward primer tube, mixing briefly. Let the thermal cycling of the now complete reaction mixture begin.
5. After the cycling is finished, add 5X gel loading buffer to the PCR reaction and run it on a 1.5% agarose gel at 100 V for about 1 h. Stain with ethidium bromide to visualize DNA.
6. Excise the target band of interest and purify from the gel with the Qiaex II extraction kit.
7. Estimate the concentration of the recovered DNA by running a dilution series on an agarose gel along with a lane of DNA size markers of known quantity.

3.7. Southern Hybridization

3.7.1. Prewarm the Prehybridization Buffer to 65°C

1. Use 5 mL in a small (150 × 35-mm) hybridization bottle.
2. Place the blot in a prewarmed hybridization bottle and rinse with 2X SSC (prewarmed to 65°C), making sure that the DNA side of the membrane faces in and that all air bubbles between the blot and the bottle are removed.
3. Discard the 2X SSC from the bottle.
4. Add the prehybridization buffer (prewarmed to 65°C) to the bottle.
5. Add denatured sheared salmon sperm DNA to the prehybridization buffer for a final concentration of 100 µg/mL.
6. Prehybridize at 65°C with medium speed rotation in a hybridization oven for at least 1 h.

3.7.2. Radiolabeling of the Gel-Purified Probe is Adapted from the Guide to Molecular Cloning Techniques *(16)*

1. Denature a 10-ng probe template diluted in dH$_2$O to 10 µL vol by first sealing in a glass capillary tube.
2. To do this, touch the tip of the tube to the template solution: the solution should be taken up into the tube by capillary action.
3. Next, center the solution in the tube by tipping the tube to one end.
4. Seal the ends of the tube by holding each end, sequentially, in the hottest part of a Bunsen burner flame for about 3 s.
5. Let the sealed ends cool; then test the seals by applying negative pressure to each with a pipet bulb.
6. Finally, boil the sealed tube of probe template for 5 min submerged in a small beaker, and then chill in an ice water bath for 5 min.
7. Dry off the capillary tube and carefully score it about 2 cm away from the centrally contained liquid on both sides using a diamond tip scriber; then snap the capillary at each score line.

8. Use the supplied pipet bulb to apply a small amount of pressure to eject the denatured probe template solution into a chilled microfuge tube containing: 1 μL of 10 mg/mL BSA, 11.5 μL of LS solution (440 mM HEPES-NaOH, pH 7.6, 44 μM each of dTTP, dCTP, and dGTP, 110 mM Tris-HCl, pH 7.5, 11 mM MgCl$_2$, 22 mM 2-mercaptoethanol, and 300 μg/mL random deoxyribonucleotide hexamers).
9. Add 50 μCi [α-^{32}P]dATP (3000 Ci/mmol) and 2.5 U Klenow fragment for a total volume of 30 μL.
10. Mix well.
11. Incubate at 23°C for 1/2 h.
12. Add 1 μL 0.5 M EDTA, pH 8.0, to stop the reaction.
13. This will yield approximately 35 ng of radiolabeled probe.

3.7.3. Testing the Specific Activity of the Labeled Probe

1. Make a 1/16 dilution of the reaction mixture in 0.1 N NaOH.
2. Spot 1 μL three times on a piece of dry blotting membrane.
3. Let spots dry, and then UV crosslink the membrane.
4. Wash the membrane three times in 2X SSC in a hybridization bottle at 65°C to remove unincorporated [α-^{32}P]dATP.
5. Each spot should be no lower than 200 counts/s to get a useable signal on the hybridized blot.
6. Make a record of the test spots by exposing to Kodak Biomax MR X-ray film for 15 min at room temperature.
7. **Figure 3** shows the results obtained from several probe labeling reactions.

3.7.4. Hybridization

1. Add 5 μL of 10 mg/mL sheared salmon sperm DNA to the labeled probe reaction and mix well in a microfuge tube.
2. Add dH$_2$O to a 250 μL vol. Make sure to poke a small hole into the lid of the tube to prevent the lid from bursting open upon heating.
3. Boil the reaction for 10 min in a beaker using a hot plate, taking care to not dilute the contents of the tube.
4. Chill on wet ice for 15 min.
5. Spin down briefly in a microfuge tube.
6. Add the denatured labeled probe reaction mix to the prehybridization buffer with the prehybridized blot for a final concentration of 5 ng/mL.
7. Higher probe concentrations may lead to nonspecific hybridization.
8. Hybridize from 4 h to overnight at 65°C.
9. After hybridization, discard the hybridization buffer from the bottle (into the liquid radioactive waste).
10. Wash the blot once for 15 min at 65°C with maximum speed rotation in prewarmed 2X SSC (use at least 10 times the hybridization buffer volume for all washes).
11. Wash twice under the same conditions in prewarmed 2X SSC containing 1% SDS, once in prewarmed 0.1X SSC/0.1% SDS at the same speed, 63°C for 30 min (the

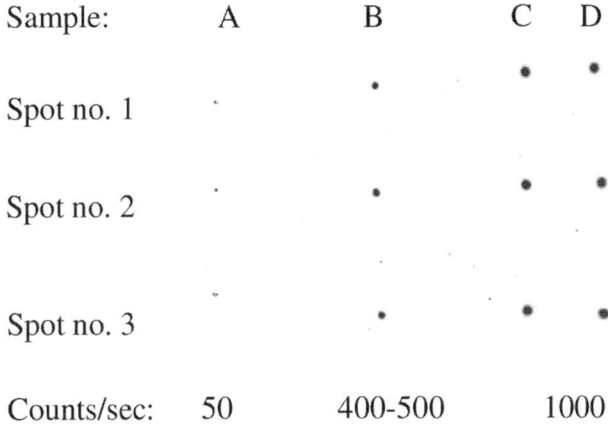

Fig. 3. Testing the specific activity of the labeled probe. Small portions of each probe-labeling reaction mixture (A, B, C, or D) were diluted 16-fold with 0.1 N NaOH, and 1 µL of each diluted mixture was spotted on a piece of nylon hybridization membrane in triplicate (Spot nos. 1, 2, and 3). The spots were dried and UV-crosslinked, and the membrane was washed three times with 2X SSC in a hybridization bottle at 65°C to remove unincorporated [α-^{32}P]dATP. The specific activities of the probes were estimated using a Geiger counter; the 1000 Counts/s spots correspond to approx 10^9 counts/min per µg of DNA. The membrane was then exposed to X-ray film for 15 min.

high-stringency wash) and once with vigorous shaking on a shaker at room temperature in 0.1X SSC for 15 to 30 min (*see* **Note 5**).

12. The final wash is to remove radioactive particles trapped on the blot that may give black spots on the autoradiogram.

13. Monitor the activity of the washes after each wash. The activity should decrease to almost background levels by the third wash. If activity increases significantly in the fourth wash, this is usually a sign of a nonspecifically hybridized probe.

14. Also monitor the activity of the blot background and edges, where no probe should be hybridizing. Additional washes may be necessary to further reduce background.

15. Wrap the washed blot wet in plastic wrap. Try to avoid creases and bubbles in the plastic that can affect the exposure.

16. Radioactivity levels of 5 to 10 counts/s usually need an overnight (16-h) exposure at −70°C using a Kodak Biomax MS intensifying screen.

17. Keeping the blot wet allows for additional washes of the blot if necessary or stripping of the probe from the blot and subsequent rehybridization of the blot. Drying the blot will irreversibly fix the probe to the membrane.

3.8. Stripping Southern Blots for Reuse with Another Probe (see *Note 6*)

1. Fill a beaker or glass dish about half to two-thirds full with 0.1X SSC/1% SDS using a heated magnetic stirrer and heat the buffer to boiling.

2. Add the blot to be stripped and reduce heat to a low, steady boil. Cover with a glass plate, or other cover.
3. Boil for about 30 min with gentle stirring. Check the reduction of radioactive signal strength of the blot with a Geiger counter. If necessary, boil again in fresh buffer.
4. Expose the stripped blot wet, wrapped in plastic, to film under the same conditions that it will be exposed to for the next hybridization to check for any residual probe signal left.
5. Before hybridizing the stripped blot, agitate in 0.4 N NaOH for 1 min and then in 0.2 M Tris-HCl, pH 8.0/1X SSC for 1 min to make sure the DNA fixed on the membrane is denatured.

4. Notes

1. Make sure all buffers, rotors, beakers, tubes, and so on are prechilled to 0°C to 4°C.
2. Break up the pellet well with a 1-mL plastic pipet before Dounce homogenization.
3. Nuclei are very fragile in this buffer and will lyse upon harsh treatment such as high g-force or Dounce homogenization.
4. This will give an effective Ca^{2+} concentration of 1 mM.
5. All washes should be discarded into the ^{32}P liquid radioactive waste.
6. There may be a loss of up to 50% of the DNA fixed to the blot after stripping.

Acknowledgments

This work was supported by NIH grant GM62857, NIGMS to A.S.

References

1. van Holde, K. E. (1989) *Chromatin.* Springer Verlag, New York.
2. Woodcock, C. L., Grigoryev, S. A., Horowitz, R. A., and Whitaker, N. (1993) A chromatin folding model that incorporates linker variability generates fibers resembling the native structures. *Proc. Natl. Acad. Sci. USA* **90,** 9021–9025.
3. Woodcock, C. L. and Horowitz, R. A. (1995) Chromatin organization re-viewed. *Trends Cell Biol.* **5,** 272–277.
4. Zlatanova, J., Leuba, S. H., Yang, G., Bustamante, C., and van Holde, K. (1994) Linker DNA accessibility in chromatin fibers of different conformations: a reevaluation. *Proc. Natl. Acad. Sci. USA* **91,** 5277–5280.
5. van Holde, K. and Zlatanova, J. (1995) Chromatin higher order structure: chasing a mirage? *J. Biol. Chem.* **270,** 8373–8376.
6. Sun, F.-L., Cuaycong, M. H., and Elgin, S. C. R. (2001) Long-range nucleosome ordering is associated with gene silencing in *Drosophila melanogaster* pericentric heterochromatin. *Mol. Cell. Biol.* **21,** 2867–2879.
7. Collins, F. S., Green, E. D., Guttmacher, A. E., and Guyer, M. S. (2003) A vision for the future of genomics research. *Nature* **422,** 835–847.
8. Liu, K. and Stein, A. (1997) DNA sequence encodes information for nucleosome array formation. *J. Mol. Biol.* **270,** 559–573.

9. Dalal, Y., Fleury, T., Cioffi, A., and Stein, A. (2005) Long-range oscillation in a periodic DNA sequence motif may influence nucleosome array formation. *Nucleic Acids Res.* **33,** 934–945.

10. Smith, M. F., Athey, B. D., Williams, S. D., and Langmore, J. P. (1990) Radial density distribution of chromatin: evidence that chromatin fibers have solid centers. *J. Cell Biol.* **110,** 245–254.

11. Bednar, J., Horowitz, R. A., Grigoryev, S. A., et al. (1998) Nucleosomes, linker DNA, and linker histone form a unique structural motif that directs the higher-order folding and compaction of chromatin. *Proc. Natl. Acad. Sci. USA* **95,** 14173–14178.

12. Dorigo, B., Schalch, T., Kulangara, A., Duda, S., Schroeder, R. R., and Richmond, T. J. (2004) Nucleosome arrays reveal the two-start organization of the chromatin fiber. *Science* **306,** 1571–1573.

13. Luger, K., Mäder, A. W., Richmond, R. K., Sargent, D. F., and Richmond, T. J. (1997) Crystal structure of the nucleosome core particle at 2.8Å resolution. *Nature* **389,** 251–260.

14. Hewish, D. R. and Burgoyne, L. A. (1973) Chromatin sub-structure. The digestion of chromatin DNA at regularly spaced sites by a nuclear deoxyribonuclease. *Biochem. Biophys. Res. Commun.* **52,** 504–510.

15. Thomas, J. O. and Thompson, R. J. (1977) Variation in chromatin structure in two cell types from the same tissue: a short DNA repeat length in cerebral cortex neurons. *Cell* **10,** 33–640.

16. Cobianchi, F. and Wilson, S. H. (1987) Enzymes for modifying and labeling DNA and RNA. *Methods Enzymol. Guide to Molecular Cloning Techniques* **152,** 94–110.

17. Sambrook, J. and Russell, D. W. (2001) *Molecular Cloning: A Laboratory Manual,* 3rd ed., Cold Spring Harbor Laboratory Press, Cold Spring Harbor, NY.

17

DNA Methyltransferase Probing of DNA–Protein Interactions

Scott A. Hoose and Michael P. Kladde

Summary

Effective methods of probing chromatin structure without disrupting DNA-protein interactions and associations are necessary for creating an accurate picture of chromatin and its processes in vivo. Expression of cytidine-5 DNA methyltransferases (C5 DMTases) in *Saccharomyces cerevisiae* provides a powerful noninvasive method of assaying relative DNA accessibility in chromatin. DNA MTases are occluded from protein-associated DNA based on the strength and span of the DNA-protein interaction. Ectopic regulation of C5 DMTase expression systems allows for minimal disruption of yeast physiology. Methylated sites are detected by bisulfite genomic sequencing, which leads to a positive signal corresponding to modified cytidine residues. High-resolution C5 DMTases with dinucleotide recognition specificity are shown to provide sufficient coverage to map interactions spanning a relatively short distance.

Key Words: Chromatin; footprinting; methyltransferase; nucleosome; transcription; yeast.

1. Introduction

DNA-protein interactions are central to the regulation of living systems. Thus, effective methods for mapping and characterization of these interactions are crucial to a better understanding of how both chromatin structure and function contribute to genome function. We have found cytidine-5 DNA methyltransferases (C5 DMTases) to be highly effective probes of chromatin structure in living cells *(1–3)*. When expressed in vivo, DMTases bind to and methylate accessible DNA target sequences but are obstructed by DNA-bound and other DNA-associated proteins to a degree dictated by the spans and strengths of their respective interactions. The DNA remains otherwise unaltered, resulting in minimal impact to the physiology of the cell during the short induction times commonly applied. In addition, no cell permeabilization or nuclei isolation is required

From: *Methods in Molecular Biology, vol. 338: Gene Mapping, Discovery, and Expression: Methods and Protocols*
Edited by: M. Bina © Humana Press Inc., Totowa, NJ

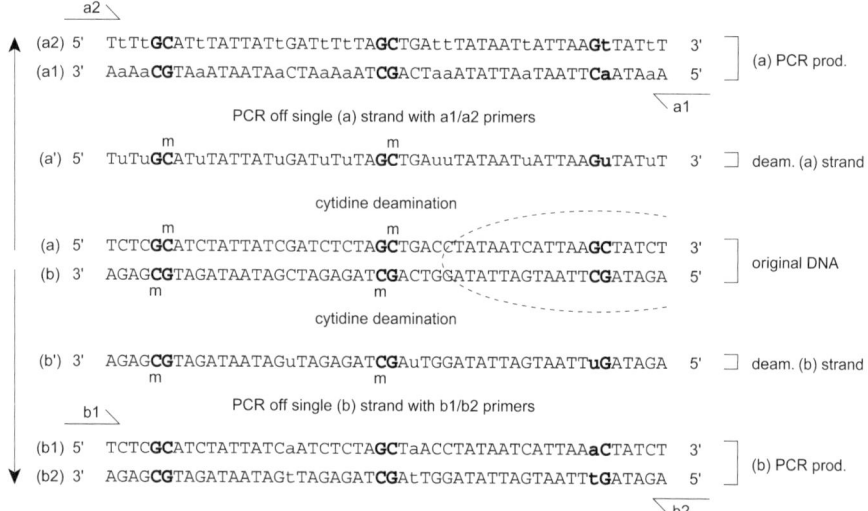

Fig. 1. Detection of methylated cytidine by bisulfite genomic sequencing. M.CviPI sites (GC) are shown in bold. An obstructing DNA-bound protein is shown as a dashed partial ellipse. Unmodified cytidine residues are deaminated to uridine. Cytidine residues that have been methylated (m) resist deamination. Note that the a' and b' DNA strands are no longer complementary following deamination and thus can and must be amplified separately with dedicated primer pairs (a1/a2 or b1/b2). PCR products are then sequenced using ddGTP (a1 or b1 primer) or ddCTP (a2 or b2 primer) terminators.

since the only necessary cofactor, the methyl donor *S*-adenosyl methionine, is naturally present in the cell. This approach allows for accurate analysis of the true in vivo chromatin state by allowing detection not only of nucleosomes but also of other more easily disrupted nonhistone protein-DNA interactions *(1,2)*.

5-Methyl-cytidine (m^5C) yields a positive signal when isolated DNA containing the modification is subjected to the polymerase chain reaction (PCR)-based process of bisulfite genomic sequencing; the strength of the sequencing signal is directly proportional to the fraction of templates methylated at the site of interest across the population *(4,5)*. Deamination by bisulfite ion chemically converts unmethylated cytidine residues in the DNA to uridine *(6)*. Subsequent PCR amplification of a region of interest yields product in which uridine has been replaced by thymidine, resulting in a net unmethylated C to T transition. Because methylated cytidines resist deamination, corresponding residues are propagated as cytidine in subsequent steps of the process. The location and relative amount of cytidine, and hence DNA methylation, are determined by sequencing using 2',3'-dideoxyguanosine terminator with dATP, dCTP, and dTTP. A diagram of the deamination/amplification/sequencing process is shown in **Fig. 1**. By comparing the relative strength of signals within a region in experimental vs control populations (e.g., intact chromatin vs naked DNA, inducing vs repressive conditions, mutant vs wild type, and so on), conclusions can be drawn about the respective chromatin states. DMTases with short and/or degenerate recognition sequences will yield higher resolution owing to the higher frequency of naturally occurring sites.

Many common yeast plasmid vectors can be used for cloning the appropriate coding and regulatory sequences for expression of DMTases in vivo in yeast. Such vectors should include an antibiotic resistance gene (such as the β-lactamase gene, *bla*, conferring amp^r) for *E. coli* cloning purposes and an appropriate yeast-selectable marker (such as *LYS2*) for selection of yeast transformants. We have constructed a series of yeast integrating plasmids (YIps) for use in DMTase expression, comprising estrogen-, doxycycline-, and galactose-inducible systems, with a variety of C5-DMTases including M.CviPI (Gm^{5}C) *(7)*, M.SssI (m^{5}CG) *(8)*, and M.HhaI (Gm^{5}CGC) *(9)*. In addition, fusions of M.CviPI and M.HhaI have been constructed that confer additional functional and regulatory characteristics, such as a Pho4-M.CviPI fusion expressed from the endogenous *PHO4* promoter *(10)*. Ongoing work in our and other laboratories suggests that these approaches can probably be applied across a variety of experimental host organisms as well as to in vitro analysis of DNA-protein complexes *(2)*. The Materials and Methods sections address the preparation and execution of an experiment using pSH1052, a *LYS2*-marked *HO*-targeted YIp that codes for all necessary exogenous components of an estrogen-inducible free M.CviPI expression system.

2. Materials

1. Equipment and software.
   a. Micropipets.
   b. 30°C and 37°C incubators.
   c. Incubator-shakers (New Brunswick Scientific Series 25D).
   d. Water baths.
   e. Hot/stir plate.
   f. Heat block.
   g. Multitube vortexer (VWR 58816-115).
   h. Benchtop vortexer.
   i. Refrigerated superspeed centrifuge (Sorvall RC-5B).
   j. Swinging-bucket rotor with tube holders (Sorvall SH3000).
   k. Microcentrifuge (Eppendorf 5415C).
   l. Spectrophotometer.
   m. Thermocycler with an ALD-1234 dual 30/48 position block alpha unit with heated lids (MJ Research PTC-200).
   n. pH meter.
   o. Vac-Man vacuum manifold (Promega).

    p. Sequencing gel box and mini-gel boxes (Owl Model S3S).
    q. Gel dryer.
    r. UV transilluminator.
    s. Light imaging system and software (Alpha Innotech AlphaImager 950).
    t. Phosphor screen (Molecular Dynamics).
    u. Phosphorimager (Molecular Dynamics Storm 860).
    v. ImageQuant software.
    w. Comparable equipment/software may be substituted.
2. Sterile consumables.
    a. 1250-µL Micropipet tips (VWR 53508-924).
    b. 200-µL Micropipet tips (VWR 30128-376).
    c. 50-mL Conical centrifuge tubes (VWR 21008-242).
    d. 15-mL Conical centrifuge tubes (VWR 21008-216).
    e. 1.7-mL Microcentrifuge tubes (Costar 3620).
    f. 1.5-mL Screw-cap microcentrifuge tubes (VWR 20170-215).
    g. 0.65-mL microcentrifuge tubes (Costar 3208).
    h. 0.5-mL Thin-wall PCR tubes (MJ Research TBI-0521).
    i. 1.2-mL Cryogenic storage vials (Corning 430658).
    j. 10-mL Pipets (BD Falcon 357551).
    k. 5-mL Pipets (BD Falcon 357543).
    l. 0.45-µm Bottle-top filters (Nalgene 295-3345 or Corning 430625).
    m. Wood applicator sticks (Puritan 807).
    n. 1-mL Slip-tip syringe barrels (BD 309602).
    o. 0.45-µm syringe filters (Whatman 6780-2504).
    p. Nonsterile disposable cuvetes (VWR 58017-847).
    q. Comparable consumables may be substituted.

Unless otherwise specified, solutions are prepared using water that has been purified to a resistance of 18.2 MΩ-cm (dH$_2$O).

Caution: A number of hazardous compounds are used in these protocols, including but not limited to 17β-estradiol, chloroform, phenol, sodium metabisulfite, sodium hydroxide, hydroquinone, [γ^{32}P]ATP, unpolymerized acrylamide, and TEMED. Personnel should be trained in the use of the materials listed here and should take appropriate precautions as set forth by the manufacturer and institutional policy.

2.1. Construction of DMTase-Expressing Yeast Strains

2.1.1. Bacterial Strain Transformation, Storage, and Plasmid Isolation

1. Chemically competent or electrocompetent *E. coli* cells that are deficient for restriction of methylated DNA (Mrr⁻ Hsd⁻ Mcr⁻). We have used competent DH10B™ cells (Invitrogen) with success.
2. 2X yeast extract tryptone + ampicillin medium (2X YT + amp): 16 g/L tryptone (BD Bacto™ 211705), 10 g/L yeast extract (BD Bacto™ 210929), 5 g/L sodium

chloride (NaCl, JT Baker 4058-05), pH to 7.5 with 10 N sodium hydroxide (NaOH, JT Baker 3722-01), 20 g/L bacteriological agar (US Biological A0930) for plates. Autoclave to sterilize. Add sterile-filtered 100 mg/mL ampicillin sodium salt (amp, Fisher BP1760-25) in dH$_2$O to a final concentration of 100 µg/mL *after* autoclaving and cooling. 2X YT liquid medium may be stored at room temperature (RT); plates containing amp must be stored at 4°C and should not be prepared more than 1 mo prior to use. Add amp to liquid medium immediately before use. Store amp stock at −20°C for up to 6 mo.

3. pSH1052.
4. Sterile 50% glycerol (Fisher BP229-4): dilute with dH$_2$O. Filter sterilize. Store at 4°C.
5. Materials for plasmid miniprep procedure.
6. Sterile 0.1X TE, pH 8.0: 1 mM Tris-HCl, pH 8.0 (JT Baker 4109-06), 0.1 mM EDTA, pH 8.0 (JT Baker 4040-04). Autoclave to sterilize.

2.1.2. Yeast Transformation, Selection, and Storage

1. pSH1052.
2. R.AatII (New England Biolabs R0117S) and R.AscI (New England Biolabs R0558S), with accompanying 10X reaction buffers and supplements.
3. A yeast strain with a *lys2* auxotrophic marker.
4. Yeast extract peptone dextrose medium (YPD): 20 g/L dextrose (Fisher BP350-1), 20 g/L peptone (BD Bacto™ 211677), 10 g/L yeast extract (BD Bacto™ 210929), 20 g/L bacteriological agar (US Biological A0930) for plates. Autoclave for 20 min to sterilize.
5. Sterile dH$_2$O. Autoclave to sterilize.
6. Sterile 0.1 M lithium acetate (LiOAc; Sigma L6883). Autoclave to sterilize.
7. Sterile 50% (w/v) poly(ethylene glycol) average molecular weight 3350 (PEG4000, Sigma P4338): dissolve PEG4000 in dH$_2$O. Filter sterilize.
8. Sterile 1.0 M LiOAc (Sigma L6883). Autoclave to sterilize.
9. 5 mg/mL Carrier DNA: dissolve 0.05 g salmon testes DNA sodium salt (Sigma D1626) in 10 mL sterile 1X TE, pH 8.0 (10 mM Tris-HCl, pH 8.0, 1 mM EDTA, pH 8.0) with vigorous stirring or agitation. Aliquot 200 µL into 0.65-mL tubes and store at −20°C. The aliquots must be placed in a boiling water bath for 5 min and quick-chilled on ice prior to first use.
10. Complete supplement mixture minus lysine selective medium (CSM-Lys): 20 g/L dextrose (Fisher BP350-1), 6.7 g/L yeast nitrogen base (US Biological Y2025), 0.74 g/L CSM-Lys drop out mix (BIO101 4510-622), 20 g/L bacteriological agar (US Biological A0930) for plates. Autoclave for 20 min to sterilize.
11. Sterile 50% glycerol: (*see* **Subheading 2.1.1., item 4**).

2.1.3. Screening for Yeast Strains With Functional M.CviPI Expression Systems

2.1.3.1 Rapid Screening for M.CviPI Activity by Culture Density

1. YPD (*see* **Subheading 2.1.2., item 4**).

2. 100 µ*M* 17β-estradiol (E₂): dissolve 13.62 mg E₂ (Sigma E8875) in 10 mL 95% ethanol (EtOH, AAPER Alcohol and Chemical Co. absolute ethyl alcohol USP) to yield a 5 m*M* solution. Mix 1470 µL 95% EtOH with 30 µL 5 m*M* E₂ to dilute to 100 µ*M*. Store at –20°C.

2.1.3.2. TOTAL YEAST DNA ISOLATION

1. 1X TE, pH 8.0: 10 mM Tris-HCl, pH 8.0 (JT Baker 4109-06), 1 m*M* EDTA, pH 8.0 (JT Baker 4040-04).
2. 425 to 600-µm Diameter acid-washed glass beads (Sigma G8772).
3. "Smash buffer": 10 m*M* Tris-HCl, pH 8.0 (JT Baker 4109-06), 1 m*M* EDTA, pH 8.0 (JT Baker 4040-04), 2% Triton X-100 (JT Baker X198-05), 1% sodium dodecyl sulfate (SDS, JT Baker 4095-02), 100 m*M* NaCl (JT Baker 4058-05).
4. 24:1 Chloroform/isoamyl alcohol (24:1 CHCl₃:IAA): add 20.8 mL IAA (Fisher BP 1150-500) to a new 500-mL bottle of CHCl₃ (JT Baker 9180-01) and mix thoroughly.
5. Equilibrated phenol: to a 400-mL bottle of saturated phenol (Fisher BP1750I-400) add the entire contents of the accompanying small bottle of equilibration buffer to raise the pH from 6.6 to 7.9. Mix the two phases thoroughly and allow them to separate. The aqueous buffer will lie on top of the equilibrated phenol. Store at 4°C.
6. 10 *M* Ammonium acetate (NH₄OAc, EMD AX1220-1).
7. Isopropanol (Fisher A416-500).
8. 70% EtOH:30% 1X TE, pH 8.0: aliquot 13 mL 1X TE, pH 8.0 to a 50-mL centrifuge tube. qs to 50 mL with 95% EtOH (AAPER Alcohol and Chemical Co. absolute ethyl alcohol USP) and mix thoroughly.
9. 0.1X TE, pH 8.0 (*see* **Subheading 2.1.1.**, item 6).

2.1.3.3. R.MCRBC DIGESTION OF ISOLATED DNA

1. R.McrBC (New England Biolabs M0272S) with accompanying 10X NEB2 buffer and 100X BSA/GTP supplements.
2. 50% Glycerol (*see* **Subheading 2.1.1.**, item 4).
3. Electrophoretic gel: 1% agarose (Bio-Rad 162-0138), 1X Tris-acetate-EDTA (TAE), 0.5 µg/mL ethidium bromide (EtBr).
4. Electrophoretic buffer: 1X TAE, 0.5 µg/mL EtBr.
5. 10X loading buffer for agarose gel electrophoresis.

2.2. DMTase Mapping of DNA-Protein Interactions

2.2.2. Bisulfite Deamination of Isolated Yeast DNA

1. 1X TE (*see* **Subheading 2.1.3.2.**, item 1).
2. 20-mL Glass scintillation vial.
3. Wizard® Minicolumns (Promega A7211).
4. 3-mL Luer-lock syringe barrels (BD 309585).
5. Unopened vial containing 5 g aliquot of sodium metabisulfite (Sigma 255556) (*see* **Note 1**).

6. 3 mg/mL Sheared calf thymus DNA (sh.CTDNA): dilute from stock of calf thymus DNA (Sigma D8661; lot concentration is given on packaging). Shear by passing 15 times through a 21-gage needle/syringe.

7. Degassed distilled water (dg.dH$_2$O): boil >200 mL dH$_2$O for 20 min in a glass beaker. After 20 min, carefully pour the boiling water into a 125-mL bottle until it is completely full (above the lip) and screw the cap on tightly. Cool overnight on the benchtop.

8. 3 N NaOH: make fresh shortly before use. Weigh out approximately 0.4 g NaOH pellets (JT Baker 3722-01) into a 15-mL centrifuge tube. Add the appropriate volume of dg.dH$_2$O (8.333 × g of NaOH = mL of dg.dH$_2$O). Dissolve the pellets by gently rocking the tube rather than by vortexing, in order to minimize aeration.

9. 100 mM Hydroquinone (HQ): make fresh shortly before use. Weigh out approximately 0.04 g HQ (Sigma H9003) into a 15-mL centrifuge tube. Add the appropriate volume of dg.dH$_2$O (90.827 × g of HQ = mL of dg.dH$_2$O). Dissolve the HQ by gently rocking the tube rather than by vortexing, in order to minimize aeration.

10. 0.5 M EDTA, pH 8.0.

11. Sample denaturation buffer (SDB): make fresh shortly before use. Mix buffer components in the following ratios (10 µL total volume per sample to be deaminated, plus a little excess): 5.8 µL dg.dH$_2$O, 3.0 µL 3 N NaOH, 0.7 µL 3 mg/mL sh.CTDNA, 0.5 µL 0.5 M EDTA, pH 8.0.

12. Saturated sodium metabisulfite solution (SMBS): make fresh shortly before use.

    a. Put a small stir bar in a clean 20-mL scintillation vial and place the vial in the center of a stir plate next to the pH meter.

    b. Take all reagents to be used (100 mM HQ, an unopened 5-g vial of sodium metabisulfite, dg.dH$_2$O, and 3 N NaOH) to the bench.

    c. Take P100 and P1000 micropipets, tips, and a 5-mL pipet and pipet-aid to the bench.

    d. Calibrate and prepare the pH meter.

    e. Prepare appropriate volumes of reagents prior to beginning preparation of SMBS as follows.

    f. Draw 7 mL dg.dH$_2$O into the pipet and set aside.

    g. Draw 1 mL 3 N NaOH using the P1000 and set aside.

    h. Draw 100 µL 100 mM HQ using the P100 and add to the scintillation vial.

    i. Open the 5-g vial of sodium metabisulfite and quickly dump the entire contents into the scintillation vial.

    j. Immediately and in rapid succession add the 7 mL dg.dH$_2$O, begin stirring, and add the 1 mL 3 N NaOH.

    k. Begin taking the pH of the solution. Adjust the pH to 5.0 using 3 N NaOH (it usually requires an additional 200–300 µL). Record the final pH.

    l. Since this is designed to be a saturated solution at RT, there will probably be some undissolved sodium metabisulfite remaining.

13. 0.1X TE (*see* **Subheading 2.1.1., item 6**).

14. Desulfonation solution: make fresh shortly before use. Mix buffer components in the following ratios (8 μL total volume per sample to be desulfonated, plus a little excess): 7.0 μL 3 N NaOH, 1.0 μL 3 mg/mL sh.CTDNA.
15. Wizard® PCR Preps DNA Purification Resin (Promega A7181).
16. 80% Isopropanol (Fisher A416-500).
17. 10 M NH$_4$OAc (EMD AX1220-1).
18. 95% EtOH (AAPER Alcohol and Chemical Co. absolute ethyl alcohol USP).
19. 70% EtOH:30% 1X TE, pH 8.0 (*see* **Subheading 2.1.3.2., item 8**).

2.2.3. PCR Amplification of Deaminated DNA Template and Equalization of Product Concentration

1. JumpStart™ Taq DNA polymerase (Sigma D4184) with accompanying 10X reaction buffer and 25 mM MgCl$_2$.
2. 2.5 mM Deoxynucleoside triphosphates (dNTPs): the solution is prepared by mixing 360 μL dH$_2$O with 10 μL of each 100 mM stock of dNTP (Amersham Biosciences 27-2035-01) to yield a 2.5 mM solution with respect to individual nucleotides. Store at –20°C.
3. 20 μM DNA deamination PCR primers (b1/b2 or a1/a2 pair) (*see* **Note 2**).
4. Electrophoretic gels: 1% agarose (Bio-Rad 162-0138), 1X Tris-acetate-EDTA (TAE), 0.5 μg/mL ethidium bromide (EtBr).
5. Electrophoretic buffer: 1X TAE, 0.5 μg/mL EtBr.
6. 10X loading buffer for agarose gel electrophoresis.
7. Montage Microcon PCR filter units (Millipore UFC7PC250).

2.2.4. ^{32}P End-Labeling of Primer To Be Used for Sequencing Extension

1. T4 polynucleotide kinase (T4 PNK, New England Biolabs M0201S) with accompanying 10X reaction buffer.
2. 20 μM DNA deamination extension primer (b1 or a1) (*see* **Note 3**).
3. 1.7 μM 6000 Ci/mmol (10 mCi/mL) [γ^{32}P]ATP (PerkinElmer BLU502Z).
4. Silane-treated glass wool (Supelco 2-0411).
5. Sephadex G-25-80 (Sigma G2580) equilibrated in sterile dH$_2$O.

2.2.5. Primer Extension Sequencing

1. SequiTherm™ DNA polymerase (Epicentre Technologies S3301K) with accompanying 10X reaction buffer.
2. 5 mM 2′,3′-Dideoxyguanosine triphosphate (ddGTP) terminator, diluted with dH$_2$O from 100 mM stock (Amersham Biosciences 27-2071-01). Store at –20°C.
3. 2.5 mM Deoxyadenosine-deoxycytidine-deoxythymidine triphosphates (d[A,C,T]TP): Mix 92.5 μL dH$_2$O with 2.5 μL of each of the 100 mM nucleotide stocks of dATP, dCTP, and dTTP (Amersham Biosciences 27-2035-01) to yield a 2.5 mM solution with respect to individual nucleotides. Store at –20°C.
4. Approx 1 μM ^{32}P-end-labeled extension primer (*see* **Subheading 3.2.4.**).
5. "Stop dye": 95% super pure formamide (Fisher BP228-100), 20 mM EDTA, pH 9.5 (JT Baker 4040-04), 0.025% xylene cyanol FF (XC, EMD 9710), 0.025% bro-

mophenol blue (BB, JT Baker D293-03). This can be made by mixing 150 µL 1.0 *M* EDTA, pH 9.5, 112.5 µL 1.67% XC, 112.5 µL 1.67% BB (saturated solution will contain undissolved BB; vortex before adding), and QS to 7.5 mL with formamide. Mix thoroughly and separate into 1.5-mL aliquots. Store at –20°C.

2.2.6. Electrophoresis and Imaging of Primer Extension Products

1. 40% 19:1 acrylamide/bis-acrylamide solution (EMD 1300).
2. 10X Tris-borate-EDTA (TBE) electrophoretic buffer.
3. Urea (Fisher BP169-212).
4. 10% (w/v) ammonium persulfate (APS; JT Baker 4030-04). Store at 4°C.
5. *N,N,N′,N′*-tetramethyl-ethylenediamine (TEMED; JT Baker 4098-01).
6. TBE-based electrophoretic buffers (*see* **Note 4**).
7. "Dummy dye": 900 µL dH$_2$O, 500 µL stop dye (*see* **Subheading 2.2.5., item 5**), 100 µL 10X SequiTherm™ reaction buffer (*see* **Subheading 2.2.5., item 1**). Mix thoroughly. Store at –20°C.

3. Methods

3.1. Construction of DMTase-Expressing Yeast Strains

3.1.1. Bacterial Strain Transformation, Storage, and Plasmid Isolation

1. Transform a chemically or electrocompetent strain of *E. coli* that is deficient for restriction of methylated DNA (Mrr⁻ Hsd⁻ Mcr⁻) with pSH1052. Select transformants on a 2X YT + ampicillin (amp) plate with incubation for approx 12 to 16 h at 37°C (*see* **Note 5**).
2. Inoculate approx 5 to 7 mL 2X YT + amp with a single colony. Incubate the cultures for 8 to 14 h at 37°C with shaking at 250 rpm (*see* **Note 6**).
3. Prepare a frozen stock of the transformed strain by mixing 350 µL sterile-filtered 50% glycerol with 700 µL of culture in a cryogenic storage vial or 1.7-mL microcentrifuge tube. Store at –80°C. Future cultures can be prepared by directly inoculating media with a small amount of cells that are sterilely scraped from the surface of the frozen stock.
4. Isolate plasmid DNA using any of the many common techniques or commercially available kits. Dissolve/elute the DNA in 30 to 50 µL 0.1X TE. Store the plasmid solution at –20°C.

3.1.2. Yeast Transformation, Selection, and Storage

1. Digest approx 100 ng pSH1052 with R.AatII and R.AscI in a double digest according to the directions of enzyme supplier (*see* **Note 7**).
2. Yeast cells are prepared for transformation by early log-phase culture growth for approx 3 to 6 h in YPD at 30°C with shaking at 250 rpm. For our purposes, early log is defined as log-phase growth at an optical density at 600 nm (OD$_{600}$) of ≤1.0. About 5 U of cells are used per transformation (1.0 U = 1.0 mL of culture at an OD$_{600}$ of 1.0 = 2.0 mL of culture at an OD$_{600}$ of 0.5, and so on). A negative control

transformation with no plasmid DNA, if performed, counts as one transformation (*see* **Note 8**).

3. Pellet the cells for 5 min at 3300*g* at 25 to 30°C in a centrifuge. Aseptically decant the media. Wash the cells by resuspending in 1.0 mL sterile distilled water (st.dH$_2$O), transferring to a 1.7-mL microcentrifuge tube and vortexing briefly. Centrifuge briefly at 16,000*g* in a microcentrifuge to pellet the cells (tap spin). Draw off and discard the wash.

4. Resuspend the cells in 100 μL 0.1 *M* sterile LiOAc per transformation; if multiple transformations or negative controls are to be done, aliquot equal volumes to new 1.7-mL microcentrifuge tubes. Tap spin to pellet the cells. Draw off and discard the LiOAc.

5. Resuspend the cells in the following solutions in the order listed, vortexing well after the addition of each: 240 μL 50% (w/v) sterile PEG4000 (*see* **Note 9**), 36 μL 1.0 *M* sterile LiOAc, 5 μL 5 mg/mL carrier DNA, and 75 μL approx 100 ng pSH1052 + st.dH$_2$O (i.e., plasmid digest, qs to 75 μL with st.dH$_2$O), for a total of 356 μL.

6. Incubate for 30 min in a 30°C water bath. Transfer to a 42°C water bath, and incubate an additional 45 min.

7. Tap spin. Draw off and discard the transformation solution. Wash the cells by resuspending in approx 500 μL st.dH$_2$O and vortexing briefly. Tap spin. Draw off and discard the wash. Resuspend in 100 μL st.dH$_2$O and plate on CSM-Lys.

8. Incubate the plates at 30°C. Colony growth generally requires 2 to 3 days. Upon obtaining colonies, purify four independent transformants by picking a single colony and streaking the cells out to single colonies on quadrants of a second CSM-Lys plate.

9. When distinct single colonies arise, inoculate approx 5 mL YPD with a single colony. Inoculate a culture from each quadrant (i.e., independent transformants) to increase the odds of obtaining a strain that will express active M.CviPI. Incubate at 30°C with shaking at 250 rpm for 12 to 16 h (*see* **Note 10**).

10. Prepare frozen stocks of the clonal transformed strains by mixing 350 μL sterile-filtered 50% glycerol with 700 μL of culture in a cryogenic storage vial or 1.7-mL microcentrifuge tube. Store at −80°C. The strain can then be prepared for future applications by patching a small amount of cells sterilely scraped from the surface of the frozen stock onto rich or selective media, incubating the plate at 30°C, and inoculating liquid media from the patch once it has grown up (*see* **Note 11**).

3.1.3. Screening for Yeast Strains With Functional M.CviPI Expression Systems (see **Note 12**)

3.1.3.1. RAPID SCREENING FOR M.CVIPI ACTIVITY BY CULTURE DENSITY

1. From a plate, inoculate approx 5 mL YPD with the strains to be screened to an OD$_{600}$ of approx 0.5. Incubate at 30°C with shaking at 250 rpm for several minutes to achieve uniformity.

2. Read the A$_{600}$ and reseed a 10.0-mL YPD culture at an OD$_{600}$ = 0.1 in a 50-mL conical centrifuge tube. Incubate as above to achieve uniformity. From this culture aliquot 5.0 mL into a new tube. Add 5.0 μL 100 μ*M* 17β-estradiol (E$_2$) to this 5-mL

culture to a final concentration of 100 nM, and incubate both tubes for 12 to 16 h (overnight) at 30°C with shaking at 250 rpm.

3. At the conclusion of the incubation, compare the densities of the E_2^+ and E_2^- cultures. Strains with a functional M.CviPI expression system are consistently significantly less dense in the E_2^+ vs. the E_2^- cultures (an $OD_{600} \approx 1.0–2.0$ vs $OD_{600} \approx 4.0–7.0$ is typical for our strains) (*see* **Note 13**).

3.1.3.2. TOTAL YEAST DNA ISOLATION

1. Grow a culture under inducing conditions for approx 12 to 16 h (the E_2^+ culture from **Subheading. 3.1.3.1.** can be used).
2. Pellet cells for 5 min at 3300g at 4 to 30°C. Decant the supernatant, resuspend cells in 700 µL 1X TE, pH 8.0, and transfer to a 1.5-mL screw-cap microcentrifuge tube (*see* **Note 14**).
3. Add approx 0.3 g acid washed 400 to 625-µm-diameter glass beads. Pellet the cells by tap spinning. Draw off and discard the supernatant and add the following in the order indicated: 200 µL 1X TE, pH 8.0, 200 µL smash buffer, 200 µL 24:1 CHCl₃/IAA, and 200 µL equilibrated phenol (*see* **Note 15**).
4. Cap the tubes tightly and vortex at maximum speed for 8 min. Centrifuge for 5 min at 16,000g at RT. Transfer the upper (aqueous) phase to a 1.7-mL microcentrifuge tube, being careful not to transfer any of the debris at the phase interface. Add 200 µL 10 M NH₄OAc to the supernatant and vortex. Incubate the tube on ice for 2 h to overnight.
5. Centrifuge for 5 min at 16,000g at RT. Transfer the supernatant to a new 1.7-mL tube. Add 600 µL isopropanol to the supernatant and invert the tube multiple times until mixed. Allow the tube to sit at RT for 5 min.
6. Centrifuge for 5 min at 16,000g at RT to pellet the nucleic acids. Draw off and discard the supernatant. Tap spin, and then carefully draw off and discard the remainder of the supernatant.
7. Add 400 µL 70% EtOH:30% 1X TE, pH 8.0, and vortex briefly to wash the pellet and tube. Centrifuge for 2 min at 16,000g at RT. Carefully draw off and discard the wash. Tap spin, and then carefully draw off and discard the remainder of the supernatant. Dry the pellet briefly by centrifuging the open tube 1 min at 80g in an open microcentrifuge rotor. Add 45 µL 0.1X TE to the nucleic acid pellet. Incubate for 1 to 24 h at 4°C, and then mix the DNA solution gently by pipeting to ensure homogeneity. Store the DNA at −20°C.

3.1.3.3. R.McrBC DIGESTION OF ISOLATED DNA

1. Make a reaction mixture of the following composition: 13.2 µL dH₂O, 2.4 µL 10X NEB2 (supplied by manufacturer), 2.4 µL 10X BSA (supplied as 100X by manufacturer), 2.4 µL 10X GTP (supplied as 100X by manufacturer), and 2.4 µL isolated DNA solution (*see* **Subheading 3.1.3.2.**), for a total of 22.8 µL (*see* **Note 16**).
2. Aliquot 9.5 µL to two 1.7-mL microcentrifuge tubes. To one, add 0.5 µL (5 U) R.McrBC. To the other, add 0.5 µL 50% glycerol. Incubate both tubes for 1 to 4 h at 37°C.

3. Tap spin both tubes. Add 1.1 μL 10X loading buffer to each and mix thoroughly. Load 10 μL of each sample in adjacent lanes on a 1% agarose-TAE-EtBr mini-gel and electrophorese for 30 min at 100 V. Visualize by UV illumination (*see* **Note 17**).

3.2. DMTase Mapping of DNA-Protein Interactions

3.2.1. Growth of Strain(s) Under Experimental Conditions

The nature of the experiment will determine the details of this portion of the protocol. Variations may include growth of a single DMTase-expressing strain under several different conditions or growth of DMTase-expressing experimental and control strains under a single condition. Depending on the method of regulation, induction of the DMTase may be accomplished at high levels in a short pulse at the desired time, at lower levels over a longer period, or constitutively as in the case of a native DNA binding factor-DMTase fusion expressed from an endogenous constitutive promoter. At the completion of the experiment, total yeast DNA is rapidly isolated as in **Subheading 3.1.3.2.** and subsequently subjected to selective deamination of unmethylated cytidines by bisulfite ion.

3.2.2. Bisulfite Deamination of Isolated Yeast DNA

1. Prepare the degassed water the evening prior to execution of subsequent steps.
2. Prepare the 3 N NaOH and 100 mM hydroquinone solutions. Prepare the sample denaturation buffer (SDB). Aliquot 10 μL of SDB to a 0.65-mL microfuge tube for each sample. Add 20 μL yeast DNA solution to the SDB and mix by pipeting. The samples should be left at RT while subsequent reagents are prepared.
3. Prepare the sodium metabisulfite solution (SMBS) and prewarm it to 50°C. It is not necessary to confirm the temperature prior to use.
4. While the SMBS is warming, denature the samples in a thermocycler for 5 min at 98°C.
   a. At about the 4-min point, bring the SMBS to the thermocycler, uncap it, and stir briefly.
   b. At 5 min, open the thermocycler while maintaining block temperature at 98°C and, working rapidly, open the first tube in the block, add 200 μL SMBS, cap and remove the tube, vortex it immediately, and place it in a rack or float.
   c. Proceed likewise for all the samples, using a new tip for each sample.
   d. After all samples have been prepared, incubate them and the remaining SMBS at 50°C for 6 h in the dark (*see* **Note 18**).
5. During the incubation, label two sets of 1.7-mL microcentrifuge tubes. Attach one labeled Promega minicolumn to a 3-mL syringe barrel for each sample.
6. Near the conclusion of the incubation, attach the minicolumn assemblies to the vacuum manifold. Heat the 0.1X TE to 95°C. Prepare the desulfonation solution (DSS) and aliquot 8 μL to each 1.7-mL microcentrifuge tube of the first labeled set from **step 5**. Read and record the final pH of the unused SMBS (*see* **Note 19**).

7. At the conclusion of the incubation, transfer the sample solutions to the second set of 1.7-mL microcentrifuge tubes from **step 5**.
   a. Add 1 mL Promega Wizard® PCR Preps resin to each sample and vortex.
   b. Transfer the resin to the syringe barrels on the vacuum manifold.
   c. Apply the vacuum to draw the resin into the minicolumns.
   d. Close individual stopcocks as the syringe barrels empty completely.
   e. Reopen the stopcocks, add 2 mL 80% isopropanol to each syringe barrel, and apply the vacuum to draw the wash through the columns.
   f. Repeat the wash with 1 mL 80% isopropanol.
   g. After the washes have completely passed through all columns, reopen the stopcocks and apply the vacuum for 30 s to begin drying them.
   h. Detach the minicolumns from the syringe barrels and press them into 1.7-mL microcentrifuge tubes (use a set of reusable tubes with lids removed that are designated for this purpose, not labeled tubes from **step 5**).
   i. Centrifuge at 16,000g for 2 min to remove residual wash.
8. Transfer the minicolumns to the first set of labeled tubes. Add 52 μL 95°C 0.1X TE to each minicolumn and let sit at RT for 5 min. Centrifuge for 20 s at 16,000g to elute the DNA. Mix the eluate and DSS by vortexing. Tap spin and incubate in a 37°C water bath for 15 min (*see* **Note 20**).
9. Add 18 μL 10 M NH$_4$OAc and mix by pipeting. Add 200 μL 95% EtOH and vortex. Incubate overnight at –20°C.
10. Centrifuge the tubes at 16,000g for 20 min at RT to pellet the DNA. Draw off and discard the supernatant. Add 400 μL 70% EtOH:30% 1X TE, pH 8.0, and vortex to wash. Centrifuge at 16,000g for 2 min, carefully draw off and discard the supernatant, centrifuge again, and carefully draw off and discard the remainder of the wash (*see* **Note 21**).
11. Dry the pellet briefly by centrifuging the open tube 1 min at approx 80g in an open microcentrifuge rotor. Before resuspending the pellet, check to be sure there is no residual EtOH wash left in the tube. Resuspend/dissolve the pellet in 25 μL 0.1X TE. Incubate for 1 to 24 h at 4°C. Mix by pipeting and centrifuge prior to use. Store the deaminate at –20°C.

3.2.3. PCR Amplification of Deaminated DNA Template and Equalization of Product Concentration

1. Use 2 to 4 μL of deaminate as template in the following 50-μL PCR reaction: dH$_2$O to volume, 1X Sigma JumpStart™ PCR buffer, 2.25 mM MgCl$_2$, 0.2 mM dNTPs, 0.8 μM b1 (or a1) primer, 0.8 μM b2 (or a2) primer, 1.25 U Sigma JumpStart™ Taq DNA polymerase. Thermocycle as follows: 1 cycle of 3 min at 94°C, 30 cycles of 45 s at 94°C, 45 s at 5°C below the calculated T_m of the primers, and 1 min/kb at 72°C, and 1 cycle of 5 min at 72°C (*see* **Note 22**).
2. Check for successful amplification of PCR product by electrophoresis of 1 μL of the unpurified reaction on a 1% agarose-TAE-EtBr mini-gel for 30 min at 100 V. Visualize by UV illumination. PCR product is purified with Microcon PCR filters

according to the manufacturer's directions. Dissolve product with and elute off filter in 30 μL st.dH$_2$O. Store at –20°C (*see* **Note 23**).

3. In preparation for sequencing primer extensions, quantify the relative product concentration by electrophoresing 2 μL of each product to be sequenced on a 1% agarose-TAE-EtBr mini-gel for 30 min at 100 V and UV imaging. Obtain a linear digital image of the gel (i.e., not saturated owing to overexposure) under UV trans-illumination. Using image analysis software, calculate the relative intensities of the product bands. Equalize product concentrations by diluting products to the level of the least concentrated of the panel with st.dH$_2$O. Store all dilutions at –20°C (*see* **Note 24**).

3.2.4. ^{32}P End-Labeling of Primer To Be Used for Sequencing Extension

1. 1.0 to 1.2 pmol of primer (1.0–1.2 μL at 1 μ*M*) will be used for each extension. Set up a labeling reaction in the following ratio with some excess in a 1.7-mL microcentrifuge tube: 10.0 μL dH$_2$O, 2.0 μL 10X PNK buffer (supplied by manufacturer), 1.0 μL 20 μ*M* b1 (or a1) primer, 1.0 μL 10 U/μL T4 PNK, and 6.0 μL 1.7 μ*M* 6000 Ci/mmol (10 mCi/mL) [γ^{32}P]ATP, for a total of 20.0 μL.

2. Incubate at 37°C for 1 h.

3. During the incubation, pack two 1-mL syringe barrels with silanized glass wool to approx 0.1 mL graduation. Fill the barrels to the top with Sephadex G-25-80 equilibrated in dH$_2$O. Remove the caps from two 1.7-mL screw-cap tubes and place them into two 15-mL centrifuge tubes. Place the syringe barrels into the tube assemblies such that the slip-tip ends sit in the 1.7-mL tubes and the shoulders rest on the 15-mL tubes. Pack the columns by centrifuging the assemblies 5 min at 1800*g* at RT. Remove the assemblies, decant the water from the collection tubes, replace them, fill the columns again to the top with Sephadex, and centrifuge again under the exact same conditions. Decant the water as before and add 100 μL dH$_2$O to the top of each column. Reassemble the spin column apparatus and centrifuge again under the exact same conditions.

4. At the conclusion of the incubation, centrifuge the reaction microcentrifuge tube for 5 min at 16,000*g*.

5. Using new 1.7-mL screw-cap tubes for collection, assemble the spin column apparatus again. Pipet the entire volume of the primer-labeling reaction onto the top of the first column. Centrifuge exactly as in **step 3**, with a separate balance. Upon completion, pipet the entire volume of the eluate onto the top of the second column and repeat the centrifugation. Note the volume of the eluate from each purification for estimation of primer concentration.

3.2.5. Primer Extension Sequencing

1. Use 2 to 4 μL of equalized PCR product (*see* **Subheading 3.2.3.**) as template in the following 8-μL primer extension reaction: dH$_2$O to volume, 1X Epicentre SequiTherm™ buffer, 150 μ*M* ddGTP, 50 μ*M* d[A,C,T]TP, 1.0 to 1.2 pmol primer (*see* **Subheading 3.2.4.**), 1.25 U Epicentre SequiTherm™ thermostable DNA polymerase. Thermocycle as follows: 1 cycle of 2 min at 94°C, 15 cycles of 30 s

at 94°C, 30 s at 5°C below the calculated T_m of the primers, and 1 min/kb at 70°C, and 1 cycle of 2 min at 70°C. At the conclusion of the cycling, put the tubes on ice, tap spin, add 4 μL stop dye to each, and mix by pipeting. Store at –20°C (*see* **Note 25**).

3.2.6. Electrophoresis and Imaging of Primer Extension Products

1. Cast a 35 × 43 × 0.04 cm polyacrylamide denaturing sequencing gel with the following composition: dH$_2$O to volume, 4 to 6% 19:1 acrylamide/bisacrylamide, 1X TBE, 50% urea, polymerized with APS and TEMED. Insert a 40-well straight-tooth comb, and allow to polymerize overnight (*see* **Note 26**).
2. Carefully remove the comb and assemble the electrophoretic apparatus. Buffer compositions should be determined by the desired separation pattern (*see* **Note 4**).
3. Using a syringe, flush out air bubbles and urea from the wells with buffer. Flush out any bubbles from the bottom edge of the gel with buffer. Prior to loading, pre-electrophorese the gel for approx 30 min at 60 W to warm it.
4. Near the end of the preelectrophoresis, denature the primer extension reactions for 5 min at 70°C in a thermocycler. Quick chill the samples on ice at the conclusion of the incubation; do not allow them to cool gradually. Flush the wells with buffer prior to loading. Load 3 to 4 μL of each sample per well. If empty lanes are to be left between samples, load an equal volume of dummy dye in the corresponding wells. Load two to three wells with dummy dye on both sides of the loaded primer extension reactions (*see* **Note 27**).
5. Electrophorese at 60 W. Under gradient buffer conditions, electrophoresis times vary from 2.5 to 4 h at 60 W and should be determined by the number, lengths, and relative positions of fragments to be resolved.
6. Dry the gel for 2 h under heat and vacuum. Expose the dried gel to a phosphor screen. Image using a Molecular Dynamics Storm860 phosphorimager. A representative image is shown in **Fig. 2**.

4. Notes

1. We package approximately 5-g aliquots of sodium metabisulfite in glass 5-g vials in an H$_2$O- and oxygen-free safety hood. The vials are stored tightly capped in the dark in a sealed bottle with Drierite® dessicant to protect against oxidation. Artifacts owing to poor deamination efficiency have been widely reported (*11*), but we have found that measures taken to protect the reagents from oxidation during storage and use (*6*) eliminate such artifacts or greatly reduce them to nondetectable levels.
2. Deamination primers are designated as a1/a2 or b1/b2 according to the convention of Frommer et al. (*4*). Primers must be designed to account for deamination of unmethylated residues. For best results, primers should discriminate between deaminated and undeaminated template and should contain as many C (a1/b1) or G (a2/b2) residues as possible while avoiding potential methylation sites. It becomes difficult to achieve high levels of product as amplification regions exceed 1 kb. We usually design primers to have a T_m of 55 to 57°C at 0.8 μM primer in 0.1 M NaCl.

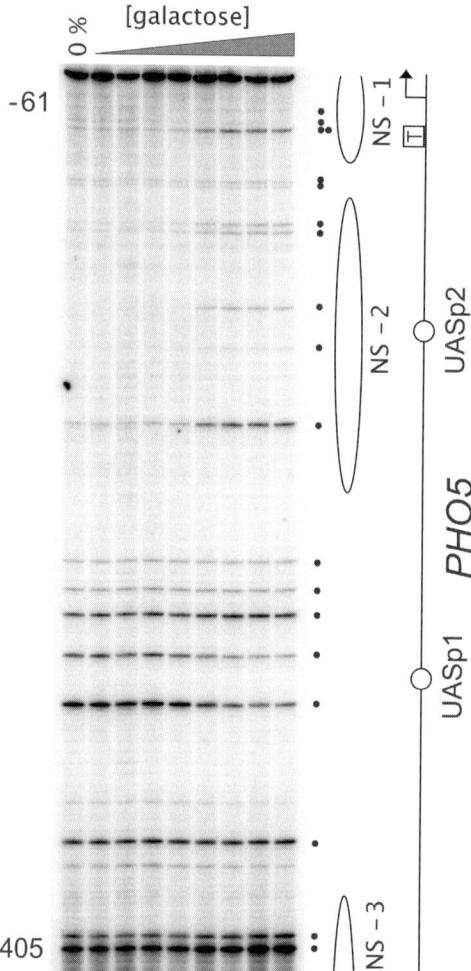

Fig. 2. Accessibility of *S. cerevisiae PHO5* promoter DNA to M.CviPI. The *PHO5* gene, coding for the major yeast acid phosphatase, is induced in conditions of limiting inorganic phosphate (P$_i$). Under these conditions, the trans-activator Pho4 is imported into the nucleus and binds to upstream activation sequence (UAS) elements UASp1 and UASp2, leading to remodeling of several positioned promoter nucleosomes (NS). In this experiment, *PHO4* was placed under the control of the *GAL1* promoter to allow for controlled regulation of Pho4 levels. A single culture was grown in CSM glutamine (gln) raffinose (raf) + 13.4 mM P$_i$ medium overnight and then split and shifted to CSM gln raf-P$_i$ medium with increasing concentrations of galactose (gal) for 3 h. M.CviPI was induced with 100 nM E$_2$ for the last 90 min of growth in gal medium. DNA was then rapidly isolated and processed. Deaminated DNA was amplified with a b1/b2 primer pair, and extension products using a second b1 primer that anneals internally are shown. Filled dots, GC sites; complete and partial ellipses, positioned nucleosomes; open

3. We find that the best results are obtained when a second nested b1 (or a1) primer that anneals downstream of the first is used for sequencing.

4. We commonly use 0.5X TBE in the top reservoir and 0.66X TBE, 1.0 M sodium acetate (Fisher S209-10) in the bottom reservoir to achieve a band compression/stacking effect at the (+) end of the gel *(12)*. For a standard gel, use 1X TBE in both reservoirs.

5. Owing to potential toxicity of foreign DNA methylation, incubation times both on plates and in liquid media should be minimized in order to avoid selection for hosts carrying plasmids with mutations in the DMTase coding sequence and/or regulatory sequences. To help mitigate these effects, we have cloned the strong constitutive *E. coli* promoter conII in inverse orientation immediately downstream of the M.CviPI open reading frame in pSH1052, which leads to high antisense transcription through the DMTase coding sequence in *E. coli*.

6. If the plasmid used in the transformation has not yet been tested for activity, inoculate multiple cultures in order to increase the odds of obtaining a functional expression plasmid.

7. When transforming with a YIp, plasmid DNA is prepared for transformation by restriction with enzyme(s) chosen to place the region(s) of homology at the ends of the linear fragment that is to be integrated. Yeast episomal (YEp) and centromeric (YCp) plasmids require no additional preparation prior to transformation.

8. Culture A_{600} readings are linear within the range of approx 0.1 to 0.5. Therefore, in order to determine the actual density of cultures that are too dense to read accurately, a dilution of a small portion must be done using growth medium to bring the read density into the linear range. The actual culture optical density (OD_{600}) is calculated by multiplying the resulting A_{600} value by the dilution factor.

9. It is very difficult to accomplish the initial resuspension in 50% PEG4000 by pipeting or vortexing, owing to the viscosity of this solution. For this reason it works best to add the PEG4000 to the tube and then use the pipet tip to directly disrupt and stir the pellet. This results in an uneven suspension, which can then be vortexed to uniformity. Make sure that all the cell pellet and resuspension is off the tip before discarding it, and be sure that no residual cell pellet remains in the bottom of the tube before proceeding.

10. Success rates of approx 50% are common.

11. Owing to potential toxicity of foreign DNA methylation, it is advisable to work with strains that have been freshly patched (<1 wk) for screening and experiments. The highly stable epigenetic mark m^{5}C is primarily (if not solely) lost by dilution through the semiconservative replication and segregation of DNA during

---

Fig. 2. *(Continued)* circles, UAS elements; T, TATA box; bent arrow, major transcription start site. The positions of the first and last GC sites relative to the *PHO5* ATG are shown. Note the decreased accessibility of the sites near UASp1 as Pho4 binding increases, as well as increased accessibility where nucleosome remodeling has occurred. Also note that sites near the edges of nucleosomes are more accessible than sites closer to the center (pseudodyad) *(1,2,14)*.

cell growth and division *(13)*. Therefore, patching onto a rich medium such as YPD is recommended when possible.

12. Of the methods detailed here, the most sensitive for determining DNA methylation status is bisulfite genomic sequencing. However, the more rapid and cost-effective screening approaches presented in **Subheading 3.1.3.1.** through **3.1.3.3.** are generally quite sufficient for demonstration of DMTase activity.

13. The slow growth phenotype leading to the low culture density observed in the overnight induction appears to stem from continuous high-level expression of M.CviPI over many hours. No detectable effect on culture growth or cell physiology has thus far been observed when M.CviPI is induced for 1 to 2 h at 100 nM E$_2$, a time frame that generally allows for excellent discrimination between accessible and inaccessible sites.

14. Low temperatures can be used to help rapidly stop cellular processes during the initial steps of the isolation and are recommended when early or closely spaced time points are taken during an experiment.

15. A simple device for aliquoting glass beads can be fashioned by measuring 0.3 g of beads into a 1.7-mL microcentrifuge tube and cutting the point of the tube off at the level of the beads to form a small measuring cup. A handle for fingers or forceps can be made from a loop of lab tape. As an exact mass of 0.3 g is not required; the indefinitely reusable cup can simply be used to scoop a level amount and pour.

16. When digests are to be performed on multiple samples, make a cocktail of the common constituents to ensure homogeneity among reactions.

17. R.McrBC cleaves DNA between two half-sites of purine-m^{5}C. DNA that has been heavily methylated by M.CviPI, M.SssI, or M.HhaI will be digested, resulting in diminution of the intensity of the large molecular weight DNA and/or a visible smear of resulting low molecular weight fragments of variable length in the R.McrBC vs the control reactions.

18. Incubation for 4 to 6 h is probably sufficient to achieve complete deamination of unmodified cytidine residues.

19. Significant deviation from the initial pH of 5 of the SMBS may indicate oxidation of the reactive sulfite to inert sulfate ion, which could lead to poor deamination efficiency.

20. There will likely be a pellet in the bottom consisting of silica resin that has passed though the minicolumn filter; this silica is of no concern as it will remain insoluble in the tube throughout the remainder of the procedure and can simply be pelleted by centrifugation to remove it from suspension prior to use of the final deaminated DNA solution as template for PCR.

21. Pellets will generally be small and sometimes cannot be seen if there is no silica residue, so it is helpful to position the tubes in the microcentrifuge with the tabs facing out so that pellet position is known regardless of pellet size.

22. When primers have different melting temperatures (T_m), use the lower of the two when determining thermocycling parameters. When amplifying multiple samples, make a cocktail of all reagents except template and then aliquot to separate reactions to ensure homogeneity among reactions with respect to common constituents.

23. Subsequent primer extension results have seemed to improve as salts are purified away from the PCR-product template. In particular, residual guanidinium salts have seemed to decrease the efficiency and cleanliness of sequencing reactions; thus we use filters rather than minicolumns for PCR product purification.

24. Absolute concentration of PCR-product template is less important for the quality of the final data than is having equal concentration across the panel. Approximately 5 to 10 ng/μL of PCR-product template is generally sufficient, but a higher concentration can be helpful when feasible.

25. When extending multiple samples, make a cocktail of all reagents except template and then aliquot to separate reactions to ensure homogeneity among reactions with respect to common constituents.

26. We make an 80-mL solution and filter it through a Whatman Puradisc 25 AS 0.45-μm syringe filter to remove particles that nucleate high-intensity spots on gels. The filtered solution is stirred gently and polymerization is initiated by the addition of 400 μL 10% APS and 40 μL TEMED.

27. Load samples toward the center portion of the gel, as lanes in the center will electrophorese straighter than those toward the edges. In addition, electrophoresing dummy dye in lanes adjacent to those of primer extension products will help mitigate edge effects caused by solute gradients between loaded and empty lanes.

Acknowledgments

We are grateful to Randy Morse for the plasmid containing the estrogen-inducible activator and Steve Hanes for the plasmid containing the minimal *GAL1* promoter with *lexO* sites. This work was supported by Public Health Service grant CA095525 from the National Cancer Institute to M.P.K. and in part by a Texas Higher Education Coordinating Board ARP award to M.P.K.

References

1. Kladde, M. P., Xu, M., and Simpson, R. T. (1996) Direct study of DNA-protein interactions in repressed and active chromatin in living cells. *EMBO J.* **15,** 6290–6300.

2. Xu, M., Simpson, R. T., and Kladde, M. P. (1998) Gal4p-mediated chromatin remodeling depends on binding site position in nucleosomes but does not require DNA replication. *Mol. Cell. Biol.* **18,** 1201–1212.

3. Jessen, W. J., Dhasarathy, A., Hoose, S. A., Carvin, C. D., Risinger, A. L., and Kladde, M. P. (2004) Mapping chromatin structure in vivo using DNA methyltransferases. *Methods* **33,** 68–80.

4. Frommer, M., MacDonald, L. E., Millar, D. S., et al. (1992) A genomic sequencing protocol that yields a positive display of 5-methylcytosine residues in individual DNA strands. *Proc. Natl. Acad. Sci. USA* **89,** 1827–1831.

5. Clark, S. J., Harrison, J., Paul, C. L., and Frommer, M. (1994) High sensitivity mapping of methylated cytosines. *Nucleic Acids Res.* **22,** 2990–2997.

6. Hayatsu, H. (1976) Bisulfite modification of nucleic acids and their constituents. *Prog. Nucl. Acid Res.* **16,** 75–124.

7. Xu, M., Kladde, M. P., Van Etten, J. L., and Simpson, R. T. (1998) Cloning, characterization and expression of the gene coding for cytosine-5-DNA methyltransferase recognizing GpC sites. *Nucleic Acids Res.* **26,** 3961–3966.

8. Renbaum, P., Abrahamove, D., Fainsod, A., Wilson, G., Rottem, S., and Razin, A. (1990) Cloning, characterization, and expression in *Escherichia coli* of the gene coding for the CpG DNA from *Spiroplasma* sp. strain MQ-1 (M.SssI). *Nucleic Acids Res.* **18,** 1145–1152.

9. Caserta, M., Zacharias, W., Nwankwo, D., Wilson, G. G., and Wells, R. D. (1987) Cloning, sequencing, in vivo promoter mapping, and expression in *Escherichia coli* of the gene for the HhaI methyltransferase. *J. Biol. Chem.* **262,** 4770–4777.

10. Carvin, C. D., Dhasarathy, A., Friesenhahn, L. B., Jessen, W. J., and Kladde, M. P. (2003) Targeted cytosine methylation for in vivo detection of protein-DNA interactions. *Proc. Natl. Acad. Sci. USA* **100,** 7743–7748.

11. Warnecke, P. M., Stirzaker, C., Song, J., Grunau, C., Melki, J. R., and Clark, S. J. (2002) Identification and resolution of artifacts in bisulfite sequencing. *Methods* **27,** 101–107.

12. Sheen, J.-Y. and Seed, B. (1988) Electrolyte gradient gels for DNA sequencing. *Biotechniques* **6,** 942–944.

13. Kladde, M. P. and Simpson, R. T. (1998) Rapid detection of functional expression of C-5-DNA methyltransferases in yeast. *Nucleic Acids Res.* **26,** 1354–1355.

14. Kladde, M. P. and Simpson, R. T. (1994) Positioned nucleosomes inhibit Dam methylation in vivo. *Proc. Natl. Acad. Sci. USA* **91,** 1361–1365.

18

Protein Binding Microarrays (PBMs) for Rapid, High-Throughput Characterization of the Sequence Specificities of DNA Binding Proteins

Michael F. Berger and Martha L. Bulyk

Summary

DNA binding proteins play a number of key roles in cells, in processes including transcriptional regulation, recombination, genome rearrangements, and DNA replication, repair, and modification. Of particular interest are the interactions between transcription factors and their DNA binding sites, as they are an integral part of the transcriptional regulatory networks that control gene expression. Despite their importance, the DNA binding specificities of most DNA binding proteins remain unknown, as earlier technologies aimed at characterizing DNA-protein interactions have been time consuming and not highly scalable. We have developed a new DNA microarray-based technology, termed protein binding microarrays (PBMs), that allows rapid, high-throughput characterization of the in vitro DNA binding site sequence specificities of transcription factors in a single day. The resulting DNA binding site data can be used in a number of ways, including for the prediction of the genes regulated by a given transcription factor, annotation of transcription factor function, and functional annotation of the predicted target genes.

Key Words: DNA microarrays; transcription factors; DNA binding proteins; protein-DNA binding; DNA regulatory motifs; transcription factor binding sites.

1. Introduction

DNA binding proteins are important in various cellular processes including transcriptional regulation, recombination, genome rearrangements, and DNA replication, repair, and modification. The interactions between transcription factors and their DNA binding sites are of particular interest because they regulate gene expression required for progression through the cell cycle, through differentiation, and in response to environmental stimuli. However, only a small

From: *Methods in Molecular Biology, vol. 338: Gene Mapping, Discovery, and Expression:*
Methods and Protocols
Edited by: M. Bina © Humana Press Inc., Totowa, NJ

handful of sequence-specific transcription factors have been characterized well enough such that all the sequences that they can (and just as importantly cannot) bind are known. This sparseness of binding site sequence data is highly problematic because these sparse datasets are frequently used to search for genomic occurrences of these sites, with many false-positive and false-negative binding sites being predicted. Earlier technologies aimed at characterizing DNA-protein interactions have been time-consuming and not highly scalable, and microarray readout of chromatin immunoprecipitations (ChIP-chip, or genome-wide location analysis) requires that the given DNA binding protein be bound to its target sites when the cells are fixed *(1)*.

Recent advances in genomics and proteomics have set the stage for rapid, high-throughput characterization of DNA binding proteins. Overexpression and purification of DNA binding proteins of interest is a familiar technique that has been used to allow characterization of these proteins using various traditional biochemical techniques. Now, most researchers also have access to DNA microarraying facilities, if not at their own institution, then through another institution that provides microarraying services for a fee. Likewise, DNA microarray scanners and glass slides for printing of the microarrays are readily available.

We recently developed an in vitro DNA microarray technology, which we term protein binding microarrays (PBMs), for characterization of the sequence specificities of DNA-protein interactions. This technology allows the in vitro binding specificities of individual DNA binding proteins to be determined in a single day, by assaying the sequence-specific binding of a given DNA binding protein directly to double-stranded DNA microarrays spotted with a large number of potential DNA binding sites. Specifically, a DNA binding protein of interest is expressed with an epitope tag, purified, and then bound directly to triplicate double-stranded DNA microarrays. The protein-bound microarrays are then washed to remove any nonspecifically bound protein and labeled with a fluorophore-conjugated antibody specific for the epitope tag. In order to normalize the PBM data by relative DNA concentration, separate triplicate microarrays from the same print run are stained with the dye SYBR Green I, which is specific for double-stranded DNA (**Figs. 1** and **2**). The sequences corresponding to the significantly bound spots (**Fig. 3A**) are analyzed with a motif prediction tool in order to identify the DNA binding site motif for the given DNA binding protein (**Fig. 3B**).

This PBM technology will likely aid in the annotation of many regulatory proteins whose DNA binding specificities have not been characterized and in the construction of gene regulatory networks. For example, binding site data derived from PBMs on transcription factors from the yeast *Saccharomyces cerevisiae*, using whole-genome *S. cerevisiae* intergenic microarrays, corresponded well with binding site specificities determined from ChIP-chip. Furthermore,

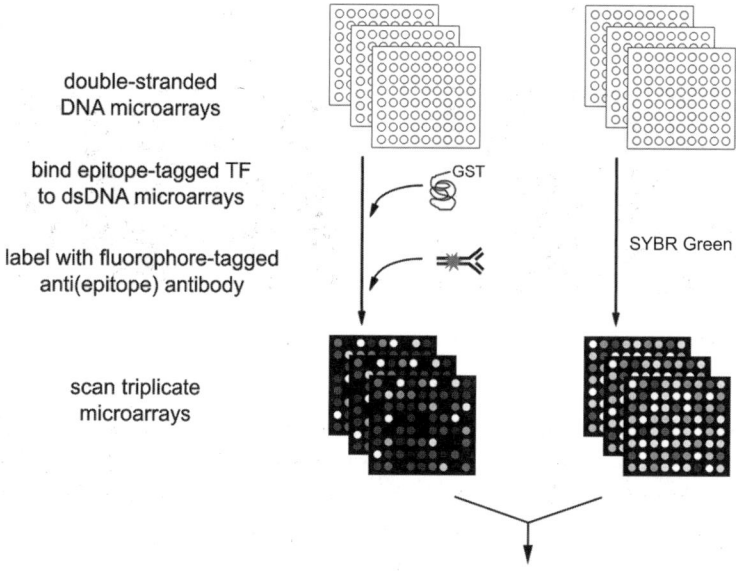

Fig. 1. Schema of protein binding microarray experiments. (Reproduced from **ref. 2** with permission from Nature Publishing Group.)

comparative sequence analysis of the PBM-derived binding sites indicated that many of the sites identified as bound in PBMs, including some not identified as bound in the ChIP-chip data, are highly conserved in other *sensu stricto* yeast genomes and thus are likely to be functional in vivo binding sites that may be utilized in a condition-specific manner *(2)*.

2. Materials

2.1. Preparation of Double-Stranded DNAs (dsDNAs)

1. Polymerase chain reaction (PCR) products corresponding to amplified noncoding regions of a genome of interest (*see* **Note 1**).
2. MultiScreen® PCR Filter Plates (Millipore, Billerica, MA) (*see* **Note 2**).

2.2. Printing and Processing of the dsDNA Microarrays

1. Corning® GAPS II or UltraGAPS 25 × 75-mm aminosilane-coated glass slides (Fisher Scientific) (*see* **Note 3**).
2. Black or orange light-protective plastic slide boxes (Fisher Scientific).
3. Vacuum desiccator (Fisher Scientific) and Drierite desiccant (Fisher Scientific).
4. Microarraying facility equipped with an OmniGrid® 100 microarrayer (Genomic Solutions, Ann Arbor, MI) with Stealth 3 pins (Telechem International, Sunnyvale, CA).

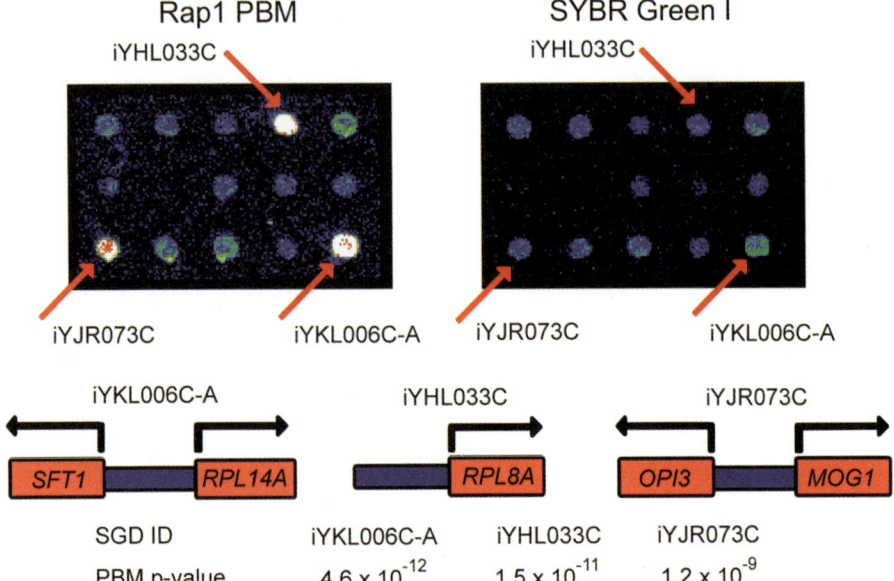

Fig. 2. Magnification of identical portions of a yeast intergenic microarray used in a PBM experiment (left) or stained with SYBR Green I (right). Fluorescence intensities are shown in false color, with white indicating saturated signal intensity, red indicating high signal intensity, yellow and green indicating moderate signal intensity, and blue indicating low signal intensity. The three labeled spots correspond to the intergenic regions depicted below, along with the *p* values derived from triplicate PBM and SYBR Green I microarray data. (Reproduced from **ref. 2** with permission from Nature Publishing Group.)

5. Handheld steamer (Conair, Stamford, CT).
6. Standard lab oven (VWR International, West Chester, PA) (*see* **Note 4**).
7. Stratalinker® 1800 UV Crosslinker (Stratagene, La Jolla, CA).

2.3. Staining the dsDNA Microarrays

1. 2X SSC: 300 m*M* NaCl, 30 m*M* sodium citrate (*see* **Note 5**). Make fresh before use. Prepare using sterile water and a stock solution of 20X SSC, pH 7.0 (*3*), sterilize by autoclaving, and store at room temperature.
2. SYBR Green I staining solution: 1:5000 dilution of SYBR Green I (Molecular Probes, Eugene, OR) in 2X SSC, 0.1% Triton X-100 (Fisher Scientific). Be sure that the SYBR Green I stock solution has been thawed completely at room temperature, in the dark to prevent photobleaching, and then vortexed before using. Make fresh before use.
3. Staining wash buffer: 2X SSC, 0.1% Triton X-100 (Fisher Scientific).
4. Forceps (Fisher Scientific).

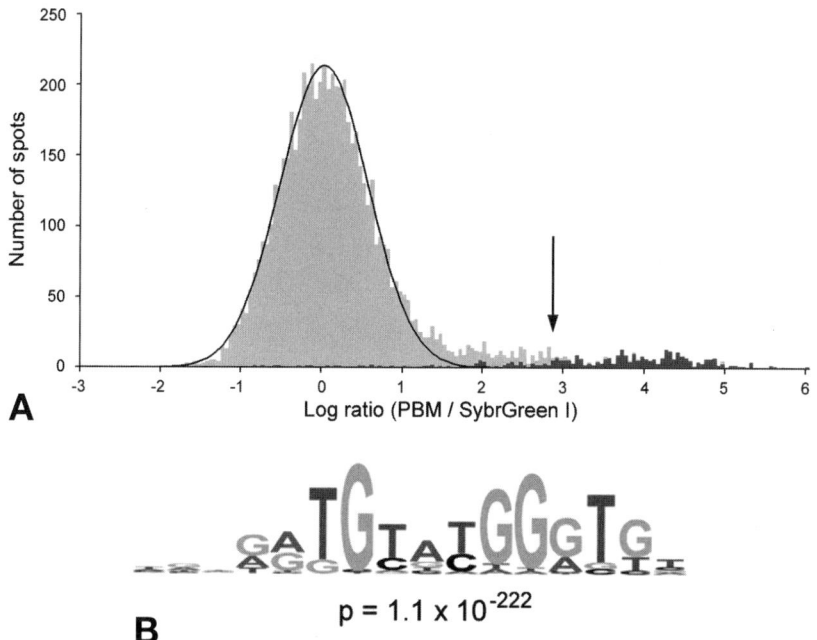

A

B

$p = 1.1 \times 10^{-222}$

Fig. 3. Identification of the DNA binding site motif from the significantly bound spots. (**A**) Distribution of ratios of PBM data, normalized by SYBR Green I data, for the yeast transcription factor Rap1 bound to yeast intergenic microarrays. The arrow indicates those spots passing a *p* value cutoff of 0.001 after correction for multiple hypothesis testing. Indicated in dark gray are spots with an exact match to a sequence belonging to the PBM-derived binding site motif. (**B**) Sequence logo *(16)* of the PBM-derived motif for the yeast transcription factor Rap1. (Reproduced from **ref.** *2* with permission from Nature Publishing Group.)

5. Polypropylene Coplin staining jars (Fisher Scientific) for 3 × 1-inch slides.
6. Adjustable speed platform shaker (Fisher Scientific).
7. Table-top centrifuge and rotor suitable for centrifuging microscope slides. We use an IEC Centra CL3R equipped with a 224 Microplate Rotor (Thermo Electron, Milford, MA).

2.4. Protein Binding Microarray Experiments

1. Purified DNA binding protein, epitope-tagged with glutathione *S*-transferase (GST), stored at −80°C in PBS (*see* **Note 6**).
2. Phosphate-buffered saline (PBS): 137 mM NaCl, 2.7 mM KCl, 4.3 mM Na$_2$HPO$_4$, 1.4 mM KH$_2$PO$_4$. Adjust pH to 7.4 using 1 M HCl, sterilize by autoclaving, and then store at room temperature.
3. Zinc acetate (ZnAc; Sigma, St. Louis, MO): 25 mM. Store in aliquots at −20°C, and then add to various buffers as required (*see* **Note 7**).

4. Prewetting buffer (PBS/0.01% TX-100): 0.01% Triton X-100 in PBS. Make fresh before use. Prepare using a stock solution of 10% Triton X-100 (Sigma), filter-sterilized using 0.22-μm filter unit (Fisher Scientific), and stored at room temperature.
5. Nonfat dried milk, (2% and 4% solutions; Sigma) in PBS. Allow to dissolve on a platform shaker, shaking very gently at 25 rpm, either overnight or for a few hours. Sterilize with a syringe (Fisher Scientific) equipped with a sterile 0.22-μm filter (Millipore) for the 2% milk solution, or with a 0.45-μm filter (Millipore) for the 4% milk solution (*see* **Note 8**).
6. Wash buffer 1 (PBS/0.1% Tween): 0.1% Tween-20 in PBS. Make fresh before use. Prepare using a stock solution of 20% Tween-20 (Sigma), filter-sterilized using a 0.22-μm filter unit (Fisher Scientific), and stored at room temperature.
7. Wash buffer 2 (PBS/ZnAc/0.5% Tween): 0.5% Tween-20 in PBS containing 50 μ*M* ZnAc. Make fresh before use. Prepare using a stock solution of 20% Tween-20 (Sigma), filter-sterilized using a 0.22-μm filter unit (Fisher Scientific), and stored at room temperature.
8. Wash buffer 3 (PBS/ZnAc/0.05% Tween): 0.05% Tween-20 in PBS containing 50 μ*M* ZnAc. Make fresh before use. Prepare using a stock solution of 20% Tween-20 (Sigma), filter-sterilized using a 0.22-μm filter unit (Fisher Scientific), and stored at room temperature.
9. Wash buffer 4 (PBS/ZnAc/0.01% TX-100): 0.01% Triton X-100 in PBS containing 50 μ*M* ZnAc. Make fresh before use. Prepare using a stock solution of 10% Triton X-100 (Sigma), filter-sterilized using a 0.22-μm filter unit (Fisher Scientific), and stored at room temperature.
10. Salmon testes DNA (Sigma).
11. Bovine serum albumin (New England Biolabs, Beverly, MA).
12. Alexa Fluor® 488-conjugated anti-GST polyclonal antibody (Molecular Probes) (*see* **Note 6**).
13. LifterSlips™ cover slips (Erie Scientific, Portsmouth, NH).
14. Kimwipes (Fisher Scientific).
15. Hydration chamber (*see* **Note 9**).
16. Polypropylene Coplin staining jars (Fisher Scientific) for 3 × 1-inch slides.
17. Adjustable speed platform shaker (Fisher Scientific).
18. Table-top centrifuge and rotor suitable for centrifuging microscope slides. We use an IEC Centra CL3R equipped with a 224 Microplate Rotor (Thermo Electron, Milford, MA).
19. Diamond-tipped glass scribe (Fisher Scientific).

2.5. Scanning of the Microarrays

1. ScanArray 5000 microarray scanner equipped with argon ion laser (488-nm excitation) and 522-nm emission filter (Perkin Elmer, Boston, MA).
2. Computer running Windows NT or Windows 2000 and equipped with at least 128 MB RAM, to run the above microarray scanner using ScanArray microarray analysis software.

3. Kimwipes (Fisher Scientific).
4. Canned air (Fisher Scientific).

2.6. Quantification of the Microarray Signal Intensities

1. GenePix microarray analysis software installed on an IBM-compatible computer with at least 1-GHz Pentium CPU, Windows 2000 or XP operating system, 512 MB RAM, and at least 12 MB disk space.
2. Masliner software. Masliner is a Perl script, so it can be run either locally or remotely by DOS box command lines on Win32 systems, or on a Linux computer. (http://arep.med.harvard.edu/masliner/pgmlicense.html.)
3. Excel software (Microsoft).

2.7. Postquantification Data Analysis

1. R statistics package (www.r-project.org) installed on a Linux, Macintosh, Unix, or Windows computer. The minimal system requirements for Windows are: Windows 95/98/ME/NT4/2000/XP Server, 16 MB RAM, and 50 MB disk space.
2. Mathematica software package (Wolfram Research, Champaign, IL) installed on a Windows, Macintosh, or Linux computer. For Windows, the minimal system requirements are 128 MB RAM and 500 MB disk space, and it is compatible with Windows 98/Me/NT 4.0/2000/XP.
3. BioProspector software *(4)* (http://motif.stanford.edu/distributions/). BioProspector is a C program. On our Linux machines, BioProspector takes up 113 kb.
4. ScanACE and MotifStats software *(5)* (http://atlas.med.harvard.edu/download/extra.html). These are both C++ programs. On our Linux machines, ScanACE takes up 1.6 MB and MotifStats takes up 115 KB.

3. Methods

3.1. Preparation of Double-Stranded DNAs (dsDNAs)

1. Amplify genomic regions by PCR. For the whole-genome yeast intergenic microarrays *(2)*, we performed 25 cycles at 94°C for 30 s, 60°C for 30 s, and 72°C for 90 s.
2. Filter PCR products using a 96-well MultiScreen® PCR filter plate according to the manufacturer's protocols. After application of a vacuum for 10 mins, plates may be air-dried in a clean chemical hood (*see* **Note 2**).

3.2. Printing and Processing of the dsDNA Microarrays

1. Suspend dsDNA in approximately 15 μL 3X SSC spotting buffer at a concentration between 100 and 500 ng/μL. Spot DNA onto GAPS II or UltraGAPS slides using an OmniGrid® 100 microarrayer. Approximately 0.7 nL will be deposited at each spot.
2. Lay slides face up and flat, and blow a moisture stream across their surface for approximately 5 s using a hand-held steam humidifier. Alternatively, one can incubate the microarrays in a humid chamber at 37°C for 48 h.

3. Bake microarrays in a standard VWR lab oven at 80°C for 2 h (*see* **Note 4**).
4. Apply 300 mJ to each slide using a Stratalinker® UV Crosslinker in order to cross-link the dsDNA to the substrate surface.

3.3. Staining the dsDNA Microarrays

1. Prepare SYBR Green I staining solution (*see* **Note 10**), and mix before using to stain microarrays.
2. Stain the microarrays using SYBR Green I staining solution for 12 min by shaking in a Coplin jar at room temperature at approx 100 to 125 rpm on a platform shaker.
3. Wash the microarrays in 2X SSC, 0.1% Triton X-100 staining wash buffer for 5 min at room temperature (*see* **Note 11**).
4. Wash the microarrays in 2X SSC for 2 min at room temperature.
5. Immediately spin slides dry in a table-top centrifuge by centrifuging for 5 min (500 rpm using an IEC Centra CL3R).

3.4. Protein Binding Microarray Experiments

1. Using a glass scribe, etch small grooves on the face (i.e., DNA side) of each micro-array, at a distance of a few millimeters beyond the borders of the printed area. This will help to confine all solutions to the center portion of the microarray through-out the protein binding microarray experiment.
2. Prewet the microarrays in PBS/0.01% TX-100 for at least 5 min by shaking in a Coplin jar at room temperature at approx 125 rpm on a platform shaker.
3. While the microarrays are being prewet, thaw the previously purified DNA binding protein of interest on ice.
4. Working with one microarray at a time, quickly remove the microarray from the Coplin jar, gently shake off any excess buffer, and wipe the back (i.e., the non-DNA side) and sides of the microarray with a Kimwipe (*see* **Note 12**).
5. Apply (*see* **Notes 13** and **14**) 250 μL of 2% milk solution (preblocking buffer) to the microarrays, cover with a LifterSlips™ cover slip, and allow to incubate in a hydra-tion chamber for 1 h at room temperature.
6. As soon as the microarray preblocking has been set up, preincubate the DNA bind-ing protein of interest with nonspecific competitors, for 1 h at room temperature. Specifically, dilute the thawed DNA binding protein to a 20 nM final concentra-tion in a 100-μL protein binding reaction mixture consisting of PBS, 50 μ*M* ZnAc, 2% (w/v) nonfat dried milk (Sigma), 51.3 ng/μL salmon testes DNA (Sigma), and 0.2 μg/μL BSA. The final 2% milk concentration is achieved by using a twofold dilution of the 4% milk solution that was prepared. A 100-μL reaction volume will be adequate for a printed microarray that encompasses approx two-thirds of the slide surface.
7. While the microarrays and the DNA binding protein are preblocking separately, thaw the Alexa Fluor® 488-conjugated anti-GST polyclonal antibody on ice, covered with either an ice bucket lid or aluminum foil, in order to prevent photobleaching (*see* **Note 10**).

8. Once the 1-h preblocking step is completed, wash the microarrays once with PBS/ 0.1% Tween for 5 min (*see* **Note 8**), followed by once with PBS/ZnAc/0.01% TX100 for 2 min.

9. Wipe the back and sides of the microarrays with a Kimwipe, apply the protein binding reaction mixture to the microarrays, cover with a LifterSlips™ cover slip, and allow to incubate in a hydration chamber for 1 h at room temperature.

10. As soon as the microarray protein binding reaction has been set up, dilute the Alexa Fluor® 488-conjugated anti-GST polyclonal antibody to a concentration of 0.05 mg/mL in 2% milk in 150 µL of PBS containing 50 µ*M* ZnAc, and allow to preincubate for 1 h at room temperature in the dark.

11. Once the 1-h protein binding step is completed, wash the microarrays once in PBS/ ZnAc/0.5% Tween for 3 min, followed by once in PBS/ZnAc/0.01% TX-100 for 2 min.

12. Wipe the back and sides of the microarrays with a Kimwipe, apply the preincubated antibody mixture to the microarrays, cover with a LifterSlips™ cover slip, and allow to incubate in a hydration chamber for 1 h at room temperature, covered with either an ice bucket or aluminum foil, to protect from light.

13. Once the 1-h antibody staining is completed, wash the microarrays three times with PBS/ZnAc/0.05% Tween, with each wash going for 3 min, followed by once with PBS/ZnAc for 2 min.

14. Immediately spin slides dry in a table-top centrifuge by centrifuging for 6 min (500 rpm using an IEC Centra CL3R).

3.5. Scanning of the Microarrays

1. Wipe the backs (i.e., the non-DNA sides) of the microarrays with a slightly dampened Kimwipe, to remove any streaks or spots caused by dried buffer.

2. Very quickly blow any lint off the spun-dry microarrays using canned air.

3. Scan the microarrays at a range of different laser power intensities or photomultiplier tube (PMT) gain settings per microarray, using an appropriate laser and filter set (488 nm excitation/522 nm emission for SYBR Green I and Alexa Fluor™ 488). We typically use approx three to six different settings (*see* **Note 15**), so that signal intensities are captured for even very low signal intensity spots, while ensuring that subsaturation signal intensities are captured for each of the spots on the microarray *(2,6)*.

3.6. Quantification of the Microarray Signal Intensities

1. Quantify microarray TIF images with GenePix Pro software (Axon Instruments). Analyze as a single-color image, and keep the feature size fixed throughout the alignment procedure.

2. Using Excel, GenePix, or other software, calculate the background-subtracted median intensities using the median local background.

3. Use masliner (MicroArray Spot LINEar Regression) software *(7)* to calculate the relative signal intensities over the full series of laser power (or PMT gain) setting scans in a semi-automated fashion. Masliner combines the linear ranges of multiple

A   **B**   **C**

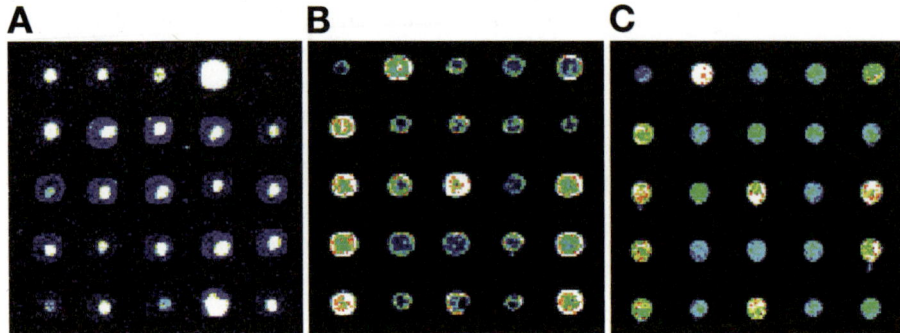

Fig. 4. Examples of DNA microarray spot quality. Identical portions of yeast inter-genic microarrays printed onto Corning® GAPS II slides, processed in different ways (see below) before UV crosslinking, and then stained with SYBR Green I. Images have been false-colored as in **Fig. 2**. (**A,B**) Examples of microarrays with poor spot quality. In both of these cases, the DNA is distributed nonuniformly, with either (**A**) high concentrations near the centers of spots, or (**B**) high concentrations along spot perimeters. Both of these microarrays resulted from two separate print runs, from which micro-arrays were UV crosslinked without first rehydrating and baking. (**C**) An example of a good-quality microarray. This microarray was rehydrated and then baked before being UV crosslinked.

scans from different scanner sensitivity settings onto an extended linear scale *(2,6, 7)*. The dynamic range of the final PBM and SYBR Green I-stained microarrays frequently have post-masliner fluorescence intensities that span 5 to 6 orders of magnitude *(2)*.

3.7. Postquantification Data Analysis

3.7.1. Microarray Data Quality Control (see **Note 16** and **Fig. 4**)

1. For microarray data on a given DNA binding protein, for each of the triplicate pro-tein binding microarrays, remove data corresponding to any flagged spots (i.e., spots that had dust flecks, and so forth).
2. Normalize the data from each of three triplicate microarrays according to total sig-nal intensity, so that the average spot intensity is the same for all three microarrays.
3. Within each individual microarray, separate the data into sectors, according to their local region on the slide. For example, for the whole-genome yeast intergenic arrays *(2)*, we sectored the spots into the 32 subgrids of the printed microarray.
4. Normalize the data again so that the mean spot intensity is the same over all the sectors. This serves to normalize for any region-specific inhomogeneities in the background and also binding and labeling reactions.
5. Remove any spots whose standard deviation (SD) divided by median value is greater than 2, i.e., spots with highly variable pixel signal intensities.

6. Average the background-subtracted, normalized signal intensities for all spots with reliable data in at least two of the three replicate microarrays, and calculate the SD/mean value. Remove any spots for which the SD/mean value is greater than 1.
7. Treat the SYBR Green I microarray data exactly the same way, except here remove any spots with fewer than 50% pixels with signal intensities greater than 2 SDs beyond the median background signal intensity, as these spots presumably do not have enough DNA present to allow accurate quantification of signal intensities *(2)* (*see* **Note 17**).

3.7.2. Identification of the 'Bound' Spots

1. Calculate the $\log_2$ ratio of the mean PBM signal intensity divided by the mean SYBR Green I signal intensity, and create a scatter plot of the log ratio vs the SYBR Green I signal intensities of the spots.
2. Although we expect that the log ratio should be independent of DNA concentration, we have found that higher DNA concentrations, as determined by higher SYBR Green I signal intensities, appear to bind proportionately less protein. To restore the independence of log ratio and SYBR Green I intensity, fit the scatter plot with a locally weighted least-squares regression using the LOWESS function (smoothing parameter = 0.5) *(8)* of the *R* statistics package.
3. Subtract the value of the regression at each spot from its log ratio, yielding a modified log ratio that is independent of DNA concentration.
4. Plot the distribution of all log ratios as a histogram (bin size = 0.05), which for a quite sequence-specific DNA binding protein is expected to resemble a Gaussian distribution with a heavy tail.
5. Determine the mode of the distribution by searching for the window of nine bins with the highest number of spots and taking the middle bin.
6. Reflect all values less than the mode and fit these values to a Gaussian function using the Mathematica software package. This provides the mean and SD of the distribution of nonspecifically bound spots.
7. Adjust the log ratios so that the peak of the distribution of nonspecifically bound spots is centered on zero.
8. Calculate a *p* value for each spot based on *z*, the number of standard deviations that the spot's log ratio departs from the mean of the Gaussian distribution, using the normal error integral *(9)* (*see* **Note 18**). The *p* value can be calculated easily in Microsoft Excel using the standard normal cumulative distribution function: normsdist(-*z*). This *p* value for each spot represents the probability that the spot is contained within the distribution of nonspecifically bound spots. Thus, spots with very small *p* values in the heavy upper tail of the real distribution are likely to be bound sequence-specifically by the given DNA binding protein.
9. To correct for multiple hypothesis testing, adjust all individual *p* values to a modified significance level using the modified Bonferroni method *(10,11)*. For significance testing of the PBM data, we recommend using an initial $\alpha = 0.001$, which corresponds to α' equal to approximately 1.5×10^{-7} for the highest ranking test case

when evaluating approx 6400 unique spots, which is the case for typical yeast intergenic microarray data *(2)*. Spots meeting or exceeding α' are considered 'bound' at a statistically significant threshold (*see* **Fig. 3A**).

3.7.3. Discovery of the DNA Binding Site Motif

1. To search for a DNA motif that is overrepresented in PBM data and thus is the likely DNA binding site motif of the given DNA binding protein, select the sequences from all the spots that had a Bonferroni-corrected *p* value less than or equal to 0.001.
2. For this set of input sequences, use BioProspector *(4)* (*see* **Note 19**) to perform separate motif searches at each width between 6 and 18 nucleotides to identify the highest scoring motifs at each width.
3. Identify all matches to the motif within all sequences spotted on the microarray, and then calculate the group specificity score *(5)* of each discovered motif. These tasks can be accomplished with the pair of programs ScanACE and MotifStats *(5)*, or with the software package MultiFinder *(12)*.
4. Choose the single motif with the lowest group specificity score *(5)* to be the most significant, using the set of all sequences spotted on the microarray as the background. We use this scoring metric because it indicates the degree to which the property of containing the sequence motif is specific to the input set of intergenic regions, as determined from the most significantly bound spots on the microarrays. A lower, and thus better, group specificity score indicates that the motif is more specific to the input set of spots (i.e., the spots beyond a 0.001 *p* value threshold in the PBM data, or the randomly selected spots in the computational negative controls [*see* **step 5** below]).
5. To assess the statistical significance of the DNA sequence motifs resulting from analysis of the PBM experiments, perform a set of computational negative control motif searches. Specifically, perform identical motif searches on at least 10 separate sets of randomly selected spots from the same microarrays used to perform the PBM experiments, with each of the 10 random sets containing the same number of sequences as the original input set for the given PBM dataset.
6. Motifs with group specificity scores that are more significant compared with the group specificity scores of the corresponding computational negative control sets are considered to likely correspond to the DNA binding site motif for the given DNA binding protein (*see* **Fig. 3B**). Examples of the ranges of group specificity scores for computational negative controls and for actual PBM data for yeast transcription factors can be found in **ref. 2**.

4. Notes

1. We have recently used whole-genome yeast intergenic microarrays in PBM experiments to identify the DNA binding site specificities of yeast transcription factors *(2)*. Microarrays spotted with coding regions are also expected to aid in identifying the sequence-specific binding properties of DNA binding proteins, even though it is currently thought that most in vivo functional regulatory sites will be located

in noncoding regions. Since PBM experiments are an in vitro technology, as long as there is sufficient sequence space represented on the DNA microarrays, one can expect to be able to derive a good approximation of the DNA binding site motif. Along these lines, one need not utilize microarrays spotted with amplicons representing genomic regions from the same genome as the DNA binding protein of interest; one can rather use microarrays spotted with a different genome's sequence. Similarly, microarrays spotted with synthetic dsDNAs can also be used in PBMs. Likewise, the dsDNAs need not be made by PCR amplification, but rather can be made by other means, such as primer extension. We have successfully performed PBMs using microarrays spotted with PCR products whose lengths ranged from approx 60 to approx 1500 bp *(2)* and also using microarrays spotted with synthetic dsDNAs ranging from approx 35 to approx 50 bp *(2,6)*.

2. Alternatively, PCR products may be precipitated with 1 M ammonium acetate and 2 vol of isopropanol, washed with 70% ethanol, dried overnight, and resuspended in 3X SSC printing buffer. The extra filtration provided by the MultiScreen® plates increases the purity of the double-stranded DNA.

3. Any remaining unused blank GAPS II or UltraGAPS slides from an opened package should be stored in a vacuum desiccator containing Drierite desiccant, as should all postprocessed, printed microarrays.

4. Baking the microarrays at 80°C for 2 h in a clean oven before UV-crosslinking may improve intraspot uniformity. However, baking may also result in a decreased shelf life of the microarrays.

5. Unless otherwise noted, buffers are prepared using distilled, deionized water (ddH$_2$O).

6. Purified DNA binding proteins should be aliquoted before storing at −80°C, to avoid unnecessary freeze/thaw. Antibody should be stored according to the manufacturer's recommendations. For long-term storage of Alexa Fluor® 488-conjugated anti-GST polyclonal antibody (Molecular Probes), we recommend aliquoting and storing at −20°C, per the manufacturer's recommendations. However, we have observed no noticeable decrease in signal intensity after storing this antibody at 4°C for 1 yr.

7. ZnAc is necessary only when performing PBM experiments on zinc finger or other zinc-coordinating proteins.

8. Filtering of the 2% and 4% milk solutions serves two purposes: (1) sterilization; and (2) removal of fine particulates that may contribute to noise in the PBM data. We have found that the 4% milk solution readily clogs a 0.22-μm filter unit, so we recommend using a 0.45-μm filter unit for syringe-filtering of the 4% milk solution. The 2% milk solution can be syringe-filtered with a 0.22-μm filter. Alternatively, 0.45-μm filters can be used for sterilizing both the 2% and 4% milk.

9. Microarrays are incubated in a hydration chamber to prevent excessive evaporation of the reaction mixture under the cover slip. An empty pipet tip box works nicely. Lift out the tip rack, fill the bottom of the pipet tip box with about half an inch of sterile water, and replace the tip rack. Wipe off the inside of the lid and the tip rack with ethanol using a Kimwipe before every use and between reaction steps in the PBM experiments.

10. Note that SYBR Green I and Alexa Fluor® 488-conjugated anti-GST polyclonal antibodies are light sensitive, and so measures should be taken to avoid their photobleaching in the course of the SYBR Green I staining and PBM experiments. We recommend turning off all overhead and benchtop lighting when handling these reagents and also when handling the microarrays once they have been stained with either SYBR Green I or the fluorophore-conjugated antibody.

11. All wash steps in Coplin jars are performed by shaking at room temperature at approx 125 rpm on a platform shaker. Shaking at speeds faster than approx 125 rpm may cause the Coplin jar to tip over while shaking.

12. Drying the back and sides of the microarray with a Kimwipe helps to prevent solution from leaking out from the edges of the cover slip. Because of the hydrophobicity of the active surface of the Corning® GAPS II and UltraGAPS glass slides, drying the front of the glass slide outside the perimeter of the DNA spots, using the etched grooves as a guide, can help to confine the protein or antibody solution, once dispensed onto the microarray, to the area of the microarray that contains the DNA spots. This must be done quickly so that the area containing the DNA spots does not dry.

13. Briefly centrifuge all reaction mixtures before applying to microarrays, to remove bubbles. When pipeting the reaction mixtures onto the microarrays, avoid pipeting any fine bubbles that may remain at the very top surface of the reaction mixtures. If nevertheless a few air bubbles become apparent once a reaction mixture has been applied to a microarray, very carefully attempt to remove the bubbles by pipetting, while avoiding removal of the reaction mixture. If some air bubbles still remain, they may be brought to the edge of the glass slide, and thus outside the spotted area, by gently rocking the cover slip as it is laid down on the microarray.

14. In applying reaction mixtures onto the microarrays, certain techniques can aid in spreading the mixture over the surface of the microarray and in increasing the homogeneity of the reaction mixture when it is applied to a prewet or washed microarray. The reaction mixture can be dispensed one droplet at a time, covering the entire surface where the DNA was spotted. The microarray can also be rocked back and forth to spread the reaction mixture uniformly across the spotted area. The use of LifterSlips™ cover slips helps to ensure a uniform distribution of the reaction mixture over the surface of the microarray.

15. For our ScanArray 5000 microarray scanner, we have found the PMT gain to be optimal between 70 and 80%. We typically fix the PMT gain setting and vary the laser power in increments of 10 to 15% (in terms of total laser power) such that there are no spots with saturated signal intensities in the lowest intensity scan.

16. It is important to ensure that each spot has enough DNA present to allow accurate quantification of its signal intensity, which is consequently used to estimate the degree of sequence-specific binding of a given DNA binding protein to that spot. If the DNA concentration at a particular spot is too low or if the DNA is spread nonuniformly throughout the pixels of a particular spot, accurate measurements are more difficult. For this reason, it is important to remove such error-prone spots from consideration. Since some spots may be noisy (i.e., spots with highly variable pixel

signal intensities) even after the use of this filter, we also remove noisy spots from consideration.

17. We found empirically that the following three additional filtering criteria helped to eliminate 'false-positive' calls (i.e., spots with no identifiable binding sites being erroneously identified as bound): (1) DNA length greater than 1500 bp; (2) low SYBR Green I raw signal intensity; and (3) low DNA density (SYBR Green I/length). These three additional filters together removed 2.7% of spots from consideration in our PBM experiments using yeast whole-genome intergenic microarrays *(2)*. Here we are not providing the actual values for the second and third filtering criteria because these values will vary somewhat among individual microarray scanners. We recommend that the user consider all three of these criteria as suggested guidelines to employ and adjust as may be appropriate.

18. This function is related to the probability of observing a data point greater than z standard deviations above the mean of a normal distribution. Strictly speaking, we are not calculating a true z-score, since here we do not calculate the p value relative to all the data, but rather just to the reflected left half of the distribution.

19. Other motif finders, such as AlignACE *(5,13)*, MEME *(14)*, and MDscan *(15)*, can also be used to identify the DNA binding site motif. We chose BioProspector over other available motif finding programs because it proved to be the most inclusive in accepting the largest number of input sequences in construction of yeast transcription factor binding site motifs *(2)*.

Acknowledgments

We thank Tom Volkert for technical assistance. This work was supported in part by National Institutes of Health grants from the National Human Genome Research Institute to M.L.B. (R01 HG002966 and R01 HG003420). M.F.B. was supported in part by a Graduate Research Fellowship from the National Science Foundation.

References

1. Bulyk, M. (2003) Computational prediction of transcription-factor binding site locations. *Genome Biol.* **5,** 201.
2. Mukherjee, S., Berger, M. F., Jona, G., et al. (2004) Rapid analysis of the DNA-binding specificities of transcription factors with DNA microarrays. *Nat. Genet.* **36,** 1331–1339.
3. Sambrook, J., Fritsch, E., and Maniatis, T. (1989) *Molecular Cloning: A Laboratory Manual,* 2nd ed., Cold Spring Harbor Laboratory Press, Cold Spring Harbor, NY.
4. Liu, X., Brutlag, D., and Liu, J. (2001) BioProspector: discovering conserved DNA motifs in upstream regulatory regions of co-expressed genes. *Pac. Symp. Biocomput.* 127–138.
5. Hughes, J. D., Estep, P. W., Tavazoie, S., and Church, G. M. (2000) Computational identification of cis-regulatory elements associated with groups of functionally related genes in *Saccharomyces cerevisiae. J. Mol. Biol.* **296,** 1205–1214.

6. Bulyk, M. L., Huang, X., Choo, Y., and Church, G. M. (2001) Exploring the DNA-binding specificities of zinc fingers with DNA microarrays. *Proc. Natl. Acad. Sci. USA* **98,** 7158–7163.

7. Dudley, A., Aach, J., Steffen, M., and Church, G. (2002) Measuring absolute expression with microarrays with a calibrated reference sample and an extended signal intensity range. *Proc. Natl. Acad. Sci. USA* **99,** 7554–7559.

8. Cleveland, W. and Devlin, S. (1988) Locally weighted regression: An approach to regression analysis by local fitting. *J. Am. Stat. Assoc.* **83,** 596–610.

9. Taylor, J. (1997) *An Introduction to Error Analysis*, 2nd ed., University Science Books, Sausalito, CA.

10. Bulyk, M., Johnson, P., and Church, G. (2002) Nucleotides of transcription factor binding sites exert interdependent effects on the binding affinities of transcription factors. *Nucleic Acids Res.* **30,** 1255–1261.

11. Sokal, R. and Rohlf, R. (1995) *Biometry: The Principles and Practice of Statistics in Biological Research*, 3rd ed., W. H. Freeman and Company, New York.

12. Huber, B. R. and Bulyk, M. L. (in review) Meta-analysis discovery of tissue-specific DNA sequence motifs from mammalian gene expression data.

13. Roth, F. P., Hughes, J. D., Estep, P. W., and Church, G. M. (1998) Finding DNA regulatory motifs within unaligned noncoding sequences clustered by whole-genome mRNA quantitation. *Nat. Biotechnol.* **16,** 939–945.

14. Bailey, T. and Elkan, C. (1995) The value of prior knowledge in discovering motifs with MEME. *Proc. Int. Conf. Intell. Syst. Mol. Biol.* **3,** 21–29.

15. Liu, X., Brutlag, D., and Liu, J. (2002) An algorithm for finding protein–DNA binding sites with applications to chromatin-immunoprecipitation microarray experiments. *Nat. Biotechnol.* **20,** 835–839.

16. Schneider, T. D. and Stephens, R. M. (1990) Sequence logos: a new way to display consensus sequences. *Nucleic Acids Res.* **18,** 6097–6100.

19

Quantitative Profiling
of Protein-DNA Binding on Microarrays

Jiannis Ragoussis, Simon Field, and Irina A. Udalova

Summary

Recent studies on genome-wide localization of transcription factor (TF) binding to DNA have shown that a large proportion of identified sequences do not contain consensus motifs predicted by databases of transcription factor binding sites, such as TRANSFAC. The main limitation of these databases is that they are based on a literature search of published examples of binding; consequently the data are not from a systematic survey and may be subject to sampling biases if investigators focused on particular motifs. Thus, there is an urgent need for systematic profiling of vertebrate transcription factor binding to DNA. We have developed a high-throughput platform for the quantitative analysis of protein-DNA interactions based on microarray technology.

Key Words: Transcription factor; DNA binding; double-stranded DNA microarray.

1. Introduction

Sequence-specific DNA binding transcription factors are the key elements in interpreting and transmitting the genetic DNA sequence information to the RNA polymerase II transcriptional machinery *(1)*. Their functionality depends on the cell-specific expression, nuclear translocation and modification in response to signaling events, and binding to specific DNA sequences as well as the chromatin remodeling state of a target promoter and interaction with other regulatory proteins. Several useful databases of transcription factor (TF) binding motifs exist, together with analytical tools that model protein-DNA interactions (reviewed in **ref. 2**). However, the current databases are limited by the lack of quantitative surveys of TF binding affinities to multiple sequence variants. For example, TRANSFAC NF-κB binding profile (based on published dichotomous data) was a poor predictor of quantitative binding to variant DNA motifs *(3)*. Moreover, the data in existing databases may be subject to sampling biases if investigators

From: *Methods in Molecular Biology, vol. 338: Gene Mapping, Discovery, and Expression: Methods and Protocols*
Edited by: M. Bina © Humana Press Inc., Totowa, NJ

focused on particular motifs and often are not freely available but must be purchased through subscription.

The established technique of systematic in vitro binding site selection (SELEX *[4]*) provides little information about low- and medium-affinity sites, which are required to generate accurate quantitative models of binding. The SELEX-SAGE method addresses the issue of low- and medium-affinity sites *(5)*, but it is laborious and involves several steps that are not scaleable for the analysis of hundreds of transcription factors. An alternative approach involves microarrays of double-stranded DNA molecules that allow the quantification of protein binding to DNA *(6)*. In fact, a systematic profiling of in vitro TF binding to yeast intergenic DNA sequences using this approach has shown significantly stronger correlation with genome localization data *(7)*. We have described a number of key improvements to the microarray technology that improve its specificity, reproducibility, and sensitivity and make it suitable for assaying many TF families *(8)*.

The quantitative data generated by this assay can be used for analyzing the binding specificities of any TF using various statistical approaches and for large-scale analysis of potential TF binding sites predicted from genomic DNA.

2. Materials

2.1. Design and Cloning of Protein Expression Constructs

1. RNeasy kit (Qiagen, cat. no. 74104).
2. ProSTAR® Ultra HF RT-PCR System (Stratagene, cat. no. 600166).
3. QIAquick Gel Extraction Kit (Qiagen, cat. no. 28704).
4. SuRE/Cut Buffer Set for Restriction Enzymes (Roche, cat. no. 1082035).
5. *Bam*HI (Roche, cat. no. 0220612).
6. *Xho*I (Roche, cat. no. 0703770).
7. QIAquick PCR Purification Kit (Qiagen, cat. no. 28104).
8. pET32(+)a bacterial expression vector (Novagen, cat. no. 69015-3).
9. Rosetta™ 2(DE3) cells (Novagen, cat. no. 71400).
10. 50X TAE electrophoresis buffer (Sigma, cat. no. T2192). 1X TAE buffer is prepared by diluting 50XAE buffer with deionized ultrapure water. 1X TAE: 0.04 M Tris-acetate, pH 8.3, 1 mM EDTA.
11. 100X TAE buffer (Sigma, cat. no. T9285). 1X TE buffer is prepared by diluting 100X TAE buffer with deionized ultrapure water. 1X TE: 0.01 M Tris-HCl, pH 8.0, 1 mM EDTA.

2.2. Ni-NTA Protein Purification

1. Imidazole buffers (from Novagen Ni-NTA Buffer Kit, cat. no. 70899-3):
   a. 1X Ni-NTA Bind buffer: 300 mM NaCl, 50 mM sodium phosphate buffer, 10 mM imidazole, pH 8.0.
   b. 1X Ni-NTA Wash buffer: 300 mM NaCl, 50 mM sodium phosphate buffer, 20 mM imidazole, pH 8.0.

    c. 0.5X Ni-NTA Elute buffer: 150 m*M* NaCl, 125 m*M* imidazole, 25 m*M* sodium phosphate buffer, pH 8.0.
2. Ni-NTA His Bind resin (Novagen, cat. no. 70666-3). Resin is stored in ethanol.
3. Calbiochem protease inhibitor cocktail set II, cat. no. 539132.
4. Sonicator (sonicating tip): Sonoplus HD 2070 (Bandelin); UW 2070 probe; MS73 tip.

2.3. DNA-Duplex Protein Purification

1. 2X Binding buffer: 10 m*M* Tris-HCl (pH 7.5), 1 m*M* EDTA, 2 *M* NaCl.
2. Protein–DNA Bind buffer: 50 m*M* Tris-HCl (pH 8.0), 0.1 m*M* EDTA, 50 m*M* NaCl, 10% glycerol, 0.01% NP-40.
3. Protein-DNA Wash buffer: 50 m*M* Tris-HCl (pH 8.0), 0.1 m*M* EDTA, 125 m*M* NaCl, 10% glycerol, 0.01% NP-40.
4. Protein-DNA Elute buffer: 50 m*M* Tris-HCl (pH 8.0), 0.1 m*M* EDTA, 500 m*M* NaCl, 10% glycerol, 0.01% NP-40.
5. Oligonucleotides (100 pmol/μL) from MWG Biotech or Metabion. Oligonucleotide sequences used in OCT-1 protein purification: 5'-AGCTGGGGGATTCCAGCTTCA AGTCAATCGGTCC-3'; 3'-TCGACCCCCTAAGGTCGAAGTTCAGTTAGCCA GG-5'-biotin.
6. Streptavidin-agarose *Streptomyces avidinii* (Sigma, cat. no. S1638).
7. Microcon YM-30 spin columns (Millipore, cat. no. 42410).

2.4. SDS-PAGE

1. MOPS SDS Running Buffer (Invitrogen, cat. no. NP0001).
2. 2X Sample buffer, Laemmli (Sigma, cat. no. S-3401).
3. NuPAGE 10% Bis-Tris Gel (Invitrogen, cat. no. NP0301BOX).
4. BenchMark™ Protein Ladder (Invitrogen, cat. no. 10747-012).
5. SimplyBlue™ SafeStain (Invitrogen, cat. no. LC6060).

2.5. Protein Quantification

1. Bovine serum albumin (BSA) (NEB, cat. no. B9001S).
2. Bradford reagent (Sigma, cat. no. B6916).
3. Spectrophotometer: Spectra Max 190 (Molecular Devices).

2.6. DNA-Duplex Preparation

1. 10X NEB3 buffer: 1 *M* NaCl, 500 m*M* Tris-HCl (pH 7.9), 100 m*M* MgCl$_2$, 10 m*M* dithiothreitol.
2. 10X NEB2 buffer: 500 m*M* NaCl, 100 m*M* Tris-HCl (pH 7.9), 100 m*M* MgCl$_2$, 10 m*M* dithiothreitol.
3. DNA Polymerase I, Large (Klenow) Fragment (NEB, cat. no. M0210S).
4. 10 m*M* Deoxy nucleotide triphosphates (dNTPs) set (Sigma, cat. no. DNTP10). Mix equal volumes of each dNTP to prepare dNTP mix (final concentration, 2.5 m*M* of each dNTP).

5. Oligonucleotides (100 pmol/µL) from MWG Biotech or Metabion. Oligonucleotide sequences used in microarray spotting: common primer *(9)*: 5'-H_3N-GGACCGAT TGACTTGA-3'. Specific 34-mer: 5'-AGCTNNNNNNNNNNNAGCTTCAAGTCA ATCGGTCC-3', where NNNNNNNNNN is a variant of the RYAKGNHAWY consensus sequence for OCT-1 binding.
6. Peltier Thermal Cycler: PTC-225 (MJ Research).

2.7. Slide Spotting, Processing, and DNA Staining

1. Microarray slides: codelink activated slides (GE Healthcare, cat. no. 300011).
2. 1X spotting solution: microarray Spotting buffer for aldehyde slides (Genetix, cat. no. K2060) or approx 150 mM Na_2HPO_4.
3. Blocking solution: 0.1 M Tris-HCl (pH 9), 50 mM ethanolamine.
4. Postcoupling wash solution: 4x SSC, 0.1% SDS.
5. SYBR Green I (Molecular Probes, cat. no. S-7563) or SYBR Gold (Molecular Probes, cat. no. S-11494).

2.8. Protein-DNA Binding and Detection

1. 2% w/v Dried skimmed milk (Marvel) in 1X PBS. 1XPBS is prepared by dissolving PBS tablets (Sigma, cat. no. P4417) in deionized ultrapure water. 1X PBS: 0.01 M phosphate buffer, 0.0027 M KCl, 0.137 M NaCl, pH 7.4. 2 g of dried skimmed milk are added per 100 mL of dH_2O.
2. Tween-20 (Sigma, cat. no. P1379).
3. Triton X-100 (Sigma, cat. no. T8787).
4. Poly dI-dC·Poly dI-dC (poly dIdC); Amersham Biosciences, cat. no. 27-7880-02).
5. 2X EMSA binding buffer: 12 mM HEPES, pH 7.8, 80 to 100 mM KCl, 1 mM EDTA, 12% glycerol, 1 µg of poly dIdC.
6. 10X PBS (Sigma, cat. no. D1408).
7. 8% w/v Dried skimmed milk (Marvel) in dH_2O.
8. α-His (H-15) antibody (200 µg/mL; Santa Cruz Biotechnology, cat. no. sc-803). Use as supplied.
9. Cy5-conjugated AffiniPure Goat Anti-Rabbit IgG (H+L) antibody (750 µg/mL) (Jackson ImmunoResearch, cat. no. 111-175-003). Resuspend in 3 mL of 50% glycerol to give stated antibody concentration.
10. Blocking solution: 0.1 M Tris-HCl (pH 9), 50 mM ethanolamine.
11. Microarray Scanner: Genepix 4000B (Axon Instruments, Molecular Dynamics).

3. Methods

3.1. Design and Cloning of Protein Expression Constructs

DNA-binding domains are clearly defined for many transcription factors. Thus, expressing isolated DNA binding domains may circumvent the problems of full-length mammalian protein expression in bacterial cells, such as inefficient expression, incorrect folding, and so on.

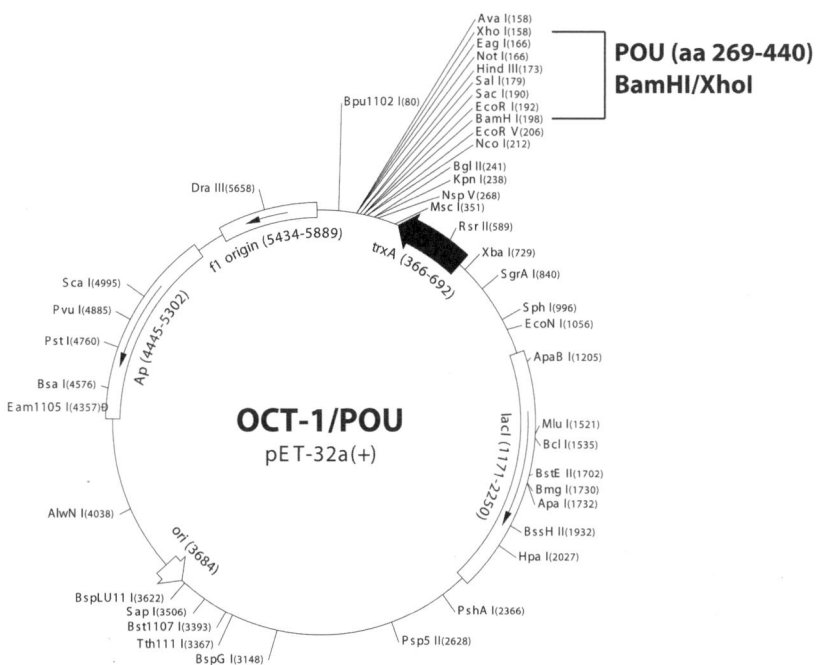

Fig. 1. Expressing construct for the POU domain of human OCT-1. The protein sequence corresponding to the DNA binding domain of human OCT-1 transcription factor (amino acids 269–440) was recovered by RT-PCR using the primers indicated in the text and total cDNA derived from Mono Mac 6 cells. It was cloned into *Bam*HI/*Xho*I sites of the pET32(+)a bacterial expression vector in frame with the N-terminus HIS-domain.

1. Examine the structure of the protein for various domains including the DNA binding one. Use the online Simple Modular Architecture Research Tool (SMART) for the domain analysis of protein sequences: http://smart.embl-heidelberg.de/.
2. Design primers for amplification of the DNA binding domain and/or full-length protein (*see* **Note 1**). Incorporate the restriction enzyme overhangs for unidirectional cloning (e.g., *Bam*Hi/*Xho*I; *see* **Fig. 1** for pET32(+)a):
   POU-*Bam*HI primer: 5'-aat**GGATCC**acaccaaagcgaatt
   POU-*Xho*I primer: 5'-aat**CTCGAG**tggtgggttgattct
3. Extract total RNA from the cells expressing the protein of interest using the RNeasy kit.
4. Perform two-step reverse transcriptase polymerase chain reaction (RT-PCR) using the ProSTAR® Ultra HF RT-PCR System with high-fidelity PfuTurbo® DNA polymerase. Use oligo-dT primer from the kit for the cDNA synthesis.
5. Analyze the molecular weight of the amplification product on 1% 1X TAE agarose gel.

6. Cut out the band corresponding to the right molecular weight product and extract the DNA fragment using the QIAquick Gel Extraction Kit. Elute in 45 µL of TE buffer.

7. Add 5 µL of the appropriate 10X buffer from the SuRE/Cut Buffer Set for Restriction Enzymes and incubate for 1 h at 37°C with the selected restriction enzymes. For example, if using the combination of *Bam*HI and *Xho*I enzymes, use Sure/Cut Buffer B.

8. Purify the *Bam*HI-*Xho*I fragment with the QIAquick PCR Purification Kit. Elute in 30 µL of TE buffer.

9. Clone the purified *Bam*HI-*Xho*I fragment into the *Bam*HI/*Xho*I sites of the pET32(+)a bacterial expression vector (**Fig. 1**).

10. Sequence to verify the nucleotide composition of the cloned fragment.

11. Transform the validated plasmid into the Rosetta™ 2(DE3) cells (Novagen, cat. no. 71400) to alleviate codon bias when expressing heterologous proteins in *E. coli* (*see* **Note 2**).

3.2. Purification Using Ni-NTA Resin

This method takes advantage of the His-tag presence in the recombinant protein (**Fig. 2**). It is simple and generally applicable but has the disadvantage that bacterial proteins containing multiple His residues will contaminate the protein of interest, and some of the protein isolated may not be functional.

1. To prepare Ni-NTA His Bind resin, take 65 µL per preparation, and spin at 6000*g* to pellet the resin.

2. Remove supernatant and add a large volume of Ni-NTA Bind buffer. Mix by inverting the tube. Spin at 6000*g* for 2 min in a benchtop centrifuge.

3. Repeat **step 2** twice.

4. Resuspend resin in the same volume of Ni-NTA Bind buffer as the aliquot originally taken. This will give approximately 100 µL resin from a 65-µL aliquot.

5. Harvest cells (50 mL of overnight culture), resuspend in 1 mL of Ni-NTA Bind buffer, and add 20 µL of protease inhibitors.

6. Sonicate at 40% of the maximum power six times with a 30-s constant burst, allowing the cell suspension to cool on ice between pulses.

7. Add 10 µL of 20% Triton X-100, and incubate at 4°C for 30 min, with shaking.

8. Spin at 16,000*g* for 20 min. Transfer supernatant to a fresh tube (save 2 µL for gel electrophoresis; *see* **Subheading 3.4.**).

9. Add 100 µL of Ni-NTA resin and incubate at 4°C for at least 15 min, with shaking.

10. Spin at approx 6000*g* for 5 min in a benchtop centrifuge

11. Remove supernatant (save 2 µL for the gel electrophoresis; *see* **Subheading 3.4.**) and wash once in 1 mL of Ni-NTA Bind buffer.

12. Aspirate supernatant and add 1 mL of Ni-NTA Wash buffer. Incubate on ice for 5 min with occasional mixing. Spin at approx 6000*g* for 2 min in a benchtop centrifuge to pellet.

13. Repeat **step 12** twice.

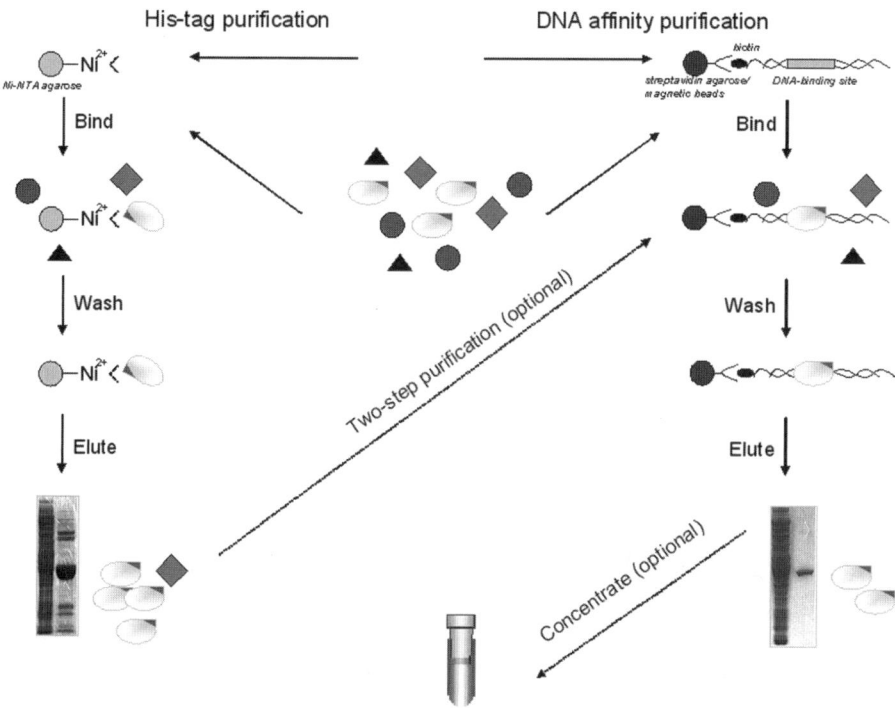

Fig. 2. Overview of protein purification stages. The OCT-1/POU domain was expressed in Rosetta™ 2(DE3) cells and purified by two independent techniques. Ni-NTA resin was used for purification of the OCT-1/POU-His-Tag protein by metal chelation chromatography (left). The His-Tag sequence binds to Ni^{2+} cations, which are immobilized on the Ni-NTA His-Bind Resin. Biotin-labeled oligoduplexes attached to streptavidin-agarose were used for purification of the OCT-1/POU-His-Tag protein by affinity chromatography (right). The DNA binding POU domain of OCT-1 specifically binds to the DNA motif encompassed in the oligoduplex. Optional steps were used to achieve the highest purity and yield of the protein and included: (1) two-step purification; (2) concentration of purified protein.

14. Elute in 200 μL of Ni-NTA Elute buffer at 4°C for at least 20 min. Pellet resin at 9000*g* and transfer supernatant to a fresh tube.
15. Concentrate and exchange into protein-DNA bind buffer (usually approx 120 μL) using Microcon YM-30 spin columns. (Note: if performing two-step purification, use this sample in purification on DNA oligoduplexes; follow **Subheading 3.3., step 10**).
16. Use 10 μL of the final concentrated sample in the gel electrophoresis.
17. Use 10 μL of the final concentrated sample in Bradford assay.
18. Add 1 μL of 10 mg/mL BSA to the remainder, aliquot, and store at −80°C.

3.3. Purification Using Biotinylated Oligoduplexes

This method purifies the protein based on its affinity to specific DNA sequences (**Fig. 2**). It has the advantage that only functional protein will be purified, but it requires reagents that are specially designed for each expressed protein. The DNA must be prepared in the form of a duplex consisting of two oligonucleotides, one of which is biotinylated and bound to streptavidin-coated agarose beads. *See* **Note 3** for experimental considerations.

1. Mix oligonucleotides (100 pmol/µL) in the following ratio: 45 µL of 100 µM biotinylated oligo (45 µM), 45 µL of 100 µM complimentary oligo (45 µM), and 10 µL of 10X NEB3 buffer.
2. To form the duplexes, anneal the mixture of two oligonucleotides in a PCR block: 1 cycle at 94°C for 1 min; and then 70 cycles at 94°C for 1 min and −1°C per cycle.
3. Take 100 µL of streptavidin-agarose (approx 45 µL sodium azide storage buffer, approx 55 µL agarose) and pellet by centrifugation at 6000g for 2 min. Discard supernatant.
4. Wash agarose twice in TE and once in 2X Binding buffer. Resuspend in 100 µL of 2X Binding buffer (to give approximately 155 µL of agarose).
5. To 55 µL of agarose in Binding buffer, add 22 µL biotinylated duplex + 33 µL H$_2$O. Final concentration of NaCl is 1 M.
6. Incubate at room temperature for 30 to 45 min with mixing.
7. Spin for 2 min at 6000g. Remove supernatant and wash twice with 1 mL of 1X Binding buffer. 1X Binding buffer is prepared by diluting 2X Binding buffer with water. Bring final volume to 200 µL.
8. Harvest cells (50 mL of overnight culture), resuspend in 1 mL of Protein-DNA Bind buffer, and add 20 µL of protease inhibitors.
9. Sonicate at 40% of the maximum power six times with a 10-s constant burst, allowing the cell suspension to cool on ice between pulses.
10. Add 10 µL of 20% Triton X-100 and incubate with shaking at 4°C for 30 min.
11. Spin at 16,000g for 20 min. (Viscous pellets may require longer.) Transfer supernatant to a fresh tube (save 2 µL for the gel electrophoresis; *see* **Subheading 3.4.**).
12. Add all 200 µL of DNA duplex attached to streptavidin-agarose and incubate at 4°C for at least 30 min (1 h for Oct-1) with shaking.
13. Spin at 9000g for 2 min in a benchtop centrifuge. If using a volume greater than 1.5 mL (in 15-mL tubes), spin at the maximum possible speed.
14. Remove supernatant. (Save 2 µL for the gel electrophoresis; *see* **Subheading 3.4.**). Add 1 mL of Protein-DNA Bind buffer. Spin at 9000g for 2 min.
15. Aspirate supernatant and add 1 mL of Protein-DNA Wash buffer. Incubate on ice with occasional gentle mixing for 5 min. Spin at 9000g for 2 min.
16. Repeat **step 15** twice.
17. Elute in 200 µL of Protein-DNA Elute buffer at 4°C for at least 30 min. Pellet resin at 9000g and transfer supernatant to a fresh tube.
18. Concentrate and exchange into Protein Bind buffer (usually approx 120 µL) using Microcon YM-30 spin columns.

19. Use 10 μL of the final concentrated sample in the gel electrophoresis.
20. Use 10 μL of the final concentrated sample in Bradford assay.
21. Add 1 μL of 10 mg/mL BSA to the remainder, aliquot and store at −80°C.

3.4. SDS-PAGE

This protocol assumes the use of the NuPAGE® gel system for high-performance gel electrophoresis. It consists of NuPAGE Bis-Tris Pre-Cast Gels and optimized buffers and provides sharp band resolution, long shelf-life, and high accuracy of results.

1. Prepare 1X gel running buffer by diluting 20X NuPAGE MOPS SDS Running Buffer in ultrapure water.
2. Add an equal volume of 2X Sample Buffer to each protein sample.
3. Place a NuPAGE 1.0 mm × 10-well 10% Bis-Tris Gel in an appropriate electrophoresis tank.
4. Add 1X MOPS buffer and flush out excess polyacrylamide from the wells of the gel with a syringe.
5. Add samples to each well either by syringe or pipet with duck-billed tips. Use 10 μL of BenchMark™ Protein Ladders a marker.
6. Run the gel for 1 h at 200 V (approx 100 mA).
7. Remove the gel from the tank and its plastic case.
8. Wash it in ultrapure water by shaking for 5 min. Discard the water.
9. Repeat **step 8** twice.
10. Stain the protein bands by adding approximately 50 mL of SimplyBlue™ SafeStain and shaking for at least 1 h.
11. Destain by shaking in ultrapure water for at least 2 h.
12. Replace with fresh water and shake for a further 15 min.
13. Discard the water and take a photograph of the gel or obtain an image by scanning with a flat-bed scanner (**Fig. 3**).

3.5. Protein Quantification

After buffer exchange and concentration, the protein concentration is estimated in a Bradford assay. The assay is based on the observation that the absorbance maximum for an acidic solution of Coomassie Brilliant Blue G-250 shifts from 465 to 595 nm when binding to protein occurs. Both hydrophobic and ionic interactions stabilize the anionic form of the dye, causing a visible color change. The dye reagent reacts primarily with arginine residues and less so with histidine, lysine, tyrosine, tryptophan, and phenylalanine residues. Quantification is therefore approximate owing to differences in amino acid composition of standards (typically BSA or IgG) and sample protein.

1. Prepare the following BSA standards in Protein-DNA bind buffer (all in mg/mL): 0.0, 0.1, 0.2, 0.4, 0.8, 1.2, 1.6, 2.0, 2.6, and 3.2.

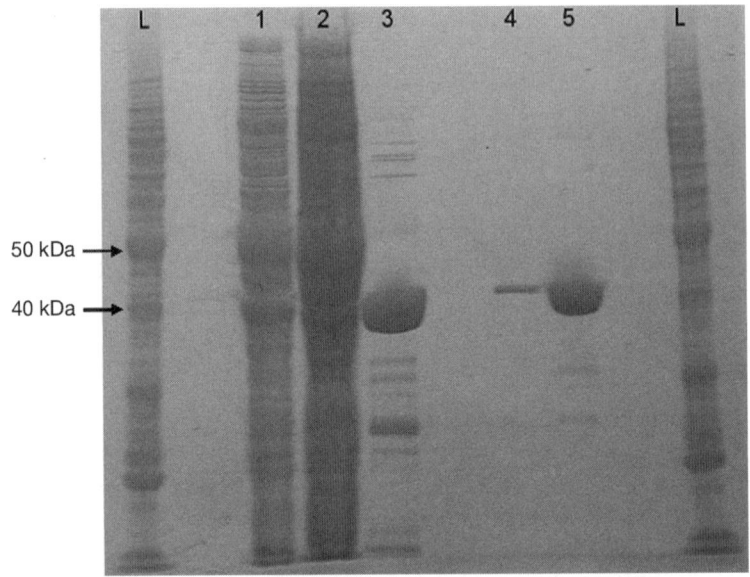

Fig. 3. Fractions of protein purification procedure analyzed on SDS-polyacrylamide gel stained with SimplyBlue™ SafeStain. The OCT-1/POU-His-Tag recombinant protein was isolated by two-step purification method to achieve the highest purity. Lane 1, total cell extract (*see* **Subheading 3.2., step 8**). Lane 2, supernatant after binding to Ni-NTA His resin (*see* **Subheading 3.2., step 11**). Lane 3, protein eluted from Ni-NTA His Resin (*see* **Subheading 3.2., step 16**). This protein fraction was used in subsequent purification on biotinylated DNA-duplexes. Lane 4, supernatant after binding to the OCT-1 DNA duplex-streptavidin-agarose- (*see* **Subheading 3.3., step 12**). Lane 5, protein eluted from the OCT-1 DNA duplex-streptavidin-agarose (*see* **Subheading 3.3., step 17**). Lane L, protein ladder.

2. Place 5 µL of each dilution of the BSA standard into the wells of a Greiner UV transparent multiwell plate in duplicate.

3. Use empty wells to place 5 µL of the concentrated purified protein sample in duplicate.

4. To each of the used wells add 250 µL of Bradford reagent. Cover the plate and incubate at RT for 5 min.

5. Measure absorbance at 595 nm by spectrophotometry. The Molecular Devices Spectra Max 190 is currently in use.

6. Construct a protein standard curve (**Fig. 4**).

7. Read off the protein from the average A_{595} of the unknown. These standards are in the linear range suitable for measuring Ni-NTA resin purified protein. It may be more useful to use additional concentrations in the range 0.1 to 10 mg/mL for the lower yields obtained with DNA-duplex-based or two-step purifications.

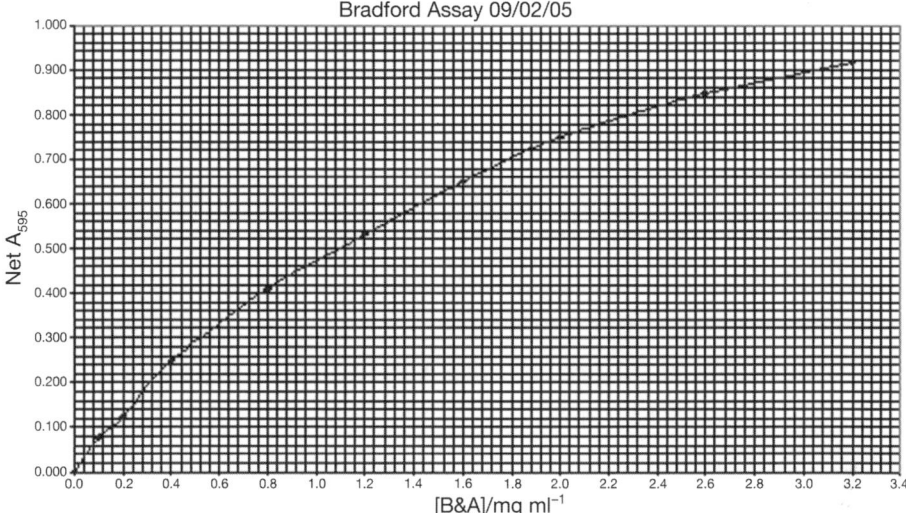

Fig. 4. Typical standard curve for estimating protein concentration.

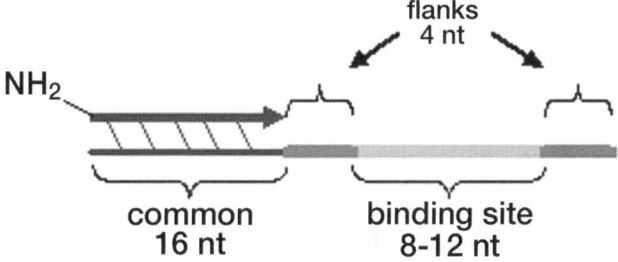

Fig. 5. Structure of the DNA duplex used for producing the microarrays. The 5' end of the extension primer (top strand) carries the amino modification. The binding site is flanked by 4-nt-long sequences.

3.6. Preparation of DNA Duplex for Spotting

A common amino-modified 16-mer is annealed to a 34-mer containing a complementary region and a transcription factor binding site (**Fig. 5**). The latter is then made double-stranded in a simple extension reaction.

1. In each well of a 96-well plate mix the following: 30 μL of 100 μ*M* specific primer, 15 μL of 100 μ*M* common primer, and 5 μL of 10X NEB3 buffer.
2. Anneal the mixture of two oligonucleotides in a PCR block: 1 cycle at 94°C for 1 min; and 70 cycles at 94°C for 1 min and −1°C per cycle.
3. Transfer a 20-μL aliquot to a 0.6-mL tube and add the following extension mix (*see* **Note 4**): 2 μL (10 U) of DNA Polymerase I, Large (Klenow) Fragment, 3 μL of 10X EcoPol or NEB2 buffer, 2 μL of 2.5 m*M* dNTP mix, and 3 μL of dH$_2$O.

4. Incubate at RT for 30 min and then at 37°C for 15 to 30 min.
5. Add 90 μL of 100% ethanol and 3 μL of 3 *M* sodium acetate (pH 5.2).
6. Place at −20°C overnight and then precipitate by centrifugation at 16,000*g* for 30 min in a benchtop centrifuge at 4°C.
7. Aspirate supernatant and add 120 μL of ice-cold 70% ethanol. Wash DNA by spinning at 16,000*g* for 30 min at 4°C.
8. Aspirate supernatant and allow samples to dry at RT. Resuspend in 30 μL 1X spotting solution. Transfer duplexes to a 384-well plate. Spin plate at 200*g*. Samples are now ready to be spotted.

3.7. Slide Spotting and SYBR Staining

The Amersham Biosciences Lucidea Array Spotter (*see* **Note 5**) is used to deposit DNA on CodeLink Activated Slides (**Fig. 6**). First the slides are washed and blocked against nonspecific binding and then processed for staining.

1. Condition the pen set by washing several times (with standard recommended buffers) before starting the spotting run.
2. Upon completion of spotting, slides should be left in the machine at 70% humidity for at least 30 min.
3. Place the slides in an open rack in a tightly sealed saturated NaCl humidification chamber overnight. Slides are then ready for use.
4. Wash with Blocking solution for 30 min. Note: SDS is not added.
5. Wash with postcoupling wash solution.
6. Pipet 200 μL of a 1:2500 dilution of SYBR Green I or SYBR Gold in 4X SSC onto slide spotting area. Cover with Parafilm and incubate in a humid chamber for 45 min.
7. Wash with 4X SSC/0.1% Triton X-100 for 5 min twice.
8. Wash with 4X SSC for 5 min twice.
9. Dry slides by centrifugation at 200*g* for 3 min.
10. Scan in Axon GenePix 4000B scanner (*see* **Subheading 3.9.**).

3.8. Protein-DNA Binding on Microarrays

1. Block slide with 2% w/v dried skimmed milk in 1X PBS for 1 h at RT.
2. Wash slide with 1X PBS/0.1% Tween-20 (*see* **Note 6**) for 3 min.
3. Wash slide with 1X PBS/0.01% Triton-X100 for 3 min.
4. Probe slide with 80 μL of protein binding reaction mix (40 μL of 2X EMSA buffer, 20 μL of 8% w/v dried skimmed milk in dH$_2$O, *x* μL of protein, and dH$_2$O up to 80 μL) for 1 h at RT.
5. Wash slide with PBS/1% Tween-20 for 3 min five times.
6. Wash slide with PBS/0.01% Triton-X100 for 3 min three times.
7. Probe slide with 100 μL primary Ab reaction mix (25 μL of 8% w/v dried skimmed milk in dH$_2$O, 10 μL of 10X PBS, 1 μL of α-His Antibody, 64 μL of dH$_2$O) for 1 h at RT or 30 min at 37°C.
8. Wash slide with PBS/0.05% Tween-20 for 3 min three times.

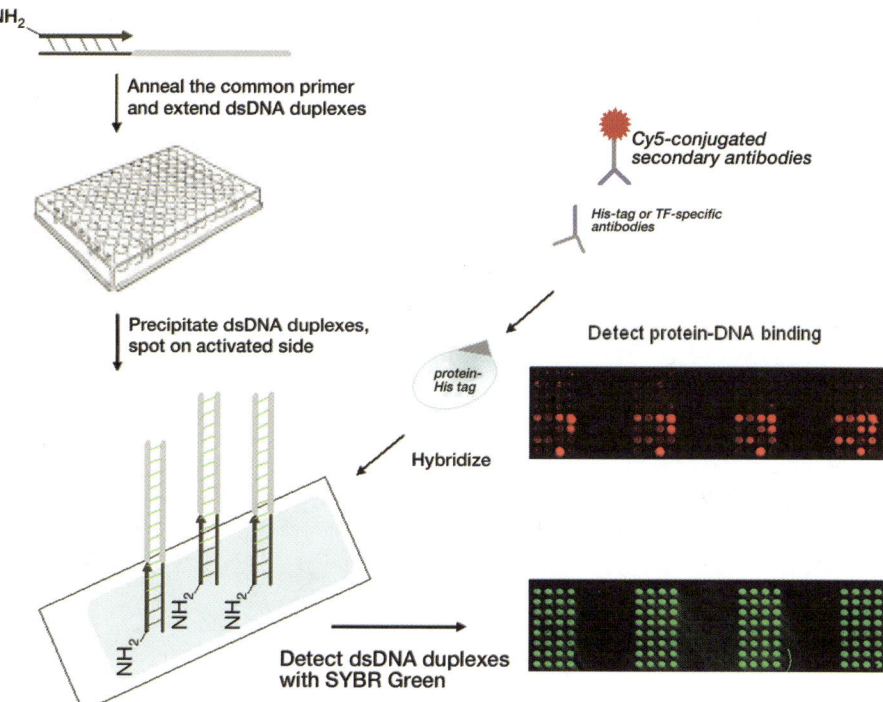

Fig. 6. Overview of microarray-assisted analysis of protein-DNA binding. All oligo-nucleotides were designed to carry common and binding site-specific parts. To the common part, a complementary oligonucleotide, modified at the 5' end with an amino group, was annealed, and the complementary DNA strand was extended over the site-specific part by polymerization. Duplexes were purified and spotted in quadruplicate onto polyacrylamide/ester-coated Codelink slides. Top panel shows a fluorescence DNA binding detection of recombinant OCT-1/POU-His-Tag protein, using the combination of rabbit anti-HIS and a Cy5-conjugated anti-rabbit IgG antibodies. Bottom panel shows SYBR Green staining of the microarray.

9. Wash slide with PBS/0.01% Triton-X100 for 3 min three times.
10. Probe slide with 100 μL secondary Ab reaction mix (25 μL 8% w/v dried skimmed milk in dH$_2$O, 10 μL of 10X PBS, 2 μL of Cy5-conjugated α-rabbit IgG Antibody, 63 μL of dH$_2$O) for 1 h at RT or 30 min at 37°C.
11. Repeat **step 8**.
12. Repeat **step 9**.
13. Wash slide with PBS for 3 min.
14. Rinse slide with dH$_2$O (optional).
15. Dry slide by centrifugation at 200g for 3 min and/or air spray.

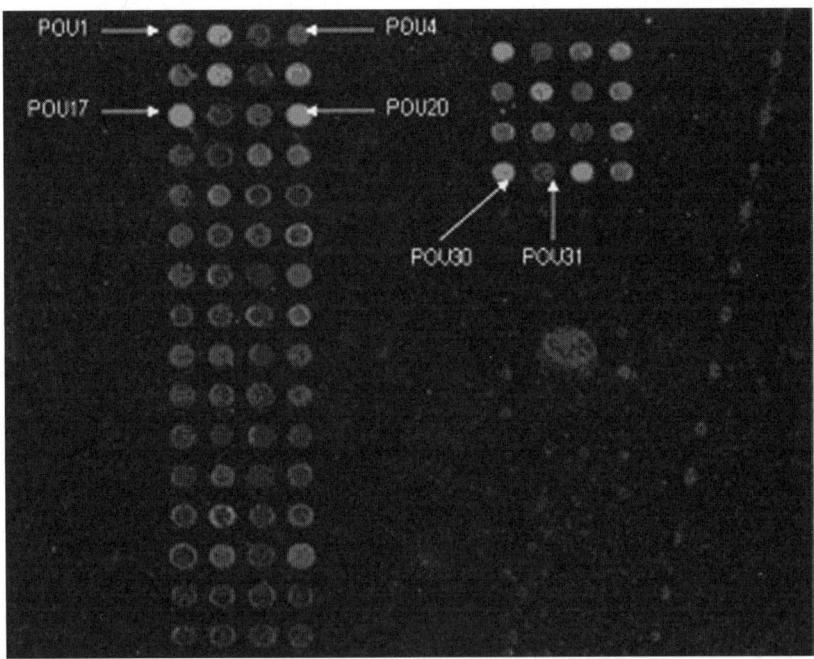

Fig. 7. Codelink slide probed overnight with 3 µg of purified Oct-1 (in 80 µL of Protein binding reaction mix) scanned as described in **Subheading 3.9.** Arrows indicate the spots containing duplexes binding OCT-1 at various affinities (*see* **Fig. 8**).

3.9. Slide Scanning

Slides are scanned in the Axon Instruments GenePix 4000B in conjunction with GenePix Pro 4.1 software (**Fig. 7**).

1. Set "hardware settings" as follows:
   a. Pixel size 10 µm.
   b. Lines to average 2.
   c. Focus position 0 µm.
   d. "Green channel" for SYBR Green/Gold detection: wavelength 532, power 100%, photomultiplier (PMT) gain 400.
   e. "Red channel" for Cy5-labeled Ab detection: wavelength 635, power 100%, PMT gain 700.
2. "Preview scan" the slide first, and then scan the spotted area at high resolution. This ensures that fluorophores receive the same amount of photobleaching prior to obtaining a final image.
3. Save the image as a Multi-image TIFF file.

3.10. Image Analysis Using GenePix

1. Set spot IDs and parameters. Spot IDs are contained within a GenePix Array List (GAL) text file; this also details feature (spot) parameters:
   "Block1 = a, b, c, d, e, f, g," where
   a = distance in µm between the first (top left) feature and the left edge of the slide.
   b = distance in µm between the first (top left) feature and the top edge of the slide.
   c = feature diameter (µm).
   d = number of columns.
   e = column spacing (µm).
   f = number of rows.
   g = row spacing (µm).
   (*See* **Note 7**.)
2. Open the slide image TIFF file in GenePix Pro 4.1, and then load the array list.
3. Manually adjust the circles to fit each feature.
4. Mark poor spots as bad or absent. Save the results output as a GenePix Results (GPR) file.

3.11. Data Analysis

1. Open the GPR file in Microsoft Excel or similar spreadsheet program. Most of the data provided are redundant; the key value is signal intensity minus local background for each feature:
   F635 Median–B635 (for Cy5-labeled Ab).
   F532 Median–B532 (for SYBR Green/Gold).
2. For estimating the DNA concentration by SYBR Green/Gold, choose a feature of average SYBR intensity (F532 Median–B532) and normalize data on this, i.e., divide all values by this. We have arbitrarily chosen block 4, column 3, row 4. Features flagged as bad should be excluded from further analysis, as are those that have an SYBR value of <0.5 or >2.0. Such features may be relatively common owing to inconsistencies in deposition of DNA by spotting equipment.
3. Exclude features flagged as bad before further analysis.
4. Divide raw intensity minus background (F635 Median–B635) by the corresponding SYBR value. This gives the relative binding affinity for each spot, normalized for DNA concentration (**Fig. 8**).
5. Take the mean value of the replicate features for each sequence.
6. Perform a final normalization on one of the medium-binding sequences, to which a value of 1000 units is assigned. POU17 is used to normalize data from Oct-1 protein binding (*see* **Note 8**).

4. Notes

1. Make sure that the amplified fragments will be cloned in-frame with the Start Codon if the Start Codon is provided by the expression vector.
2. The cells carry a plasmid with the tRNA genes that decode seven rare codons (AGA, AGG, AUA, CUA, GGA, CCC, and CGG) to improve the yield of mammalian proteins.

Oligos	RYAKGNHAWY	Exp 6				Average	Exp 5				Average	Exp 14				Average
pou1	GCATGAAATC	97.93	69.24	74.17	42.71	71.01	61.65	58.63	69.81	49.35	59.86	262.35	257.43	296.41	310.66	281.71
pou2	ACATGCAAAC	177.28	150.15	83.55	76.08	121.77	89.32	115.21	148.43	112.84	116.45	293.20	308.37	317.56	279.62	299.69
pou3	GCAGGCTAAC	7.30	3.87	12.13	5.82	7.28	4.32	5.43	8.58	7.19	6.38	15.64	20.40	21.70	17.84	18.90
pou4	ACAGGATATC	13.93	7.88	31.08	10.96	15.96	8.06	7.68	9.45	7.87	8.27	24.89	33.78	30.53	34.57	30.94
pou5	GTAGGAAAAC	6.45	4.03	11.57	4.29	6.58	3.30	4.27	5.57	4.42	4.39	14.69	17.71	23.24	17.02	18.17
pou6	GCATGATAAT	458.18	568.43	239.87	252.76	379.81	341.09	448.29	518.27	436.79	436.11	675.32	750.07	797.65	690.55	728.40
pou7	ATATGACAAC	11.30	3.85	13.00	5.36	8.38	3.80	4.30	5.31	7.98	5.35	18.76	22.04	23.56	20.91	21.32
pou8	GCAGGCAATT	194.14	223.93	156.35	81.98	164.10	163.16	118.97	155.44	143.93	145.38	470.16	436.31	496.25	533.86	484.14
pou9	ACATGCCATT	111.15	76.47	147.97	95.06	107.66	80.45	84.27	89.88	72.86	81.86	301.14	324.77	405.74	347.44	344.77
pou10	ACAGGGCAAC	7.78	4.30	10.82	5.79	7.17	7.40	8.12	7.61	5.67	7.20	20.56	27.34	32.06	28.39	27.09
pou11	GTATGCTATC	34.79	20.78	53.70	37.80	36.77	28.74	29.33	28.98	22.94	27.50	88.59	105.99	106.91	111.93	103.36
pou12	GTAGGACATT	129.38	82.76	169.81	104.30	121.56	91.00	93.03	99.00	44.35	81.85	306.91	342.79	339.21	376.79	341.43
pou13	ACAGGAAAAT	28.01	12.20	44.07	32.72	29.25	22.50	19.79	23.89	21.12	21.83	70.63	82.99	97.49	91.60	85.67
pou14	GCATGTCAAC	55.63	35.58	102.77	69.90	65.97	42.13	56.69	58.14	53.97	52.73	177.99	228.47	309.35	278.98	248.69
pou15	ATAGGCAATC	38.90	27.16	64.78	41.69	43.13	27.82	32.82	36.46	27.69	31.20	124.65	126.52	150.87	132.03	133.52
pou16	ATATGAAATT	343.29	170.69	492.02	297.07	325.77	206.81	245.02	297.52	238.19	246.89	585.02	638.34	743.59	733.36	675.08
pou17	GTATGCAAAT	1000.00	1000.00	1000.00	1000.00	1000.00	1000.00	1000.00	1000.00	1000.00	1000.00	1000.00	1000.00	1000.00	1000.00	1000.00

Fig. 8. Fragment of the Excel spreadsheet with binding data analysis. Four measurements of the relative binding affinity are presented for each sequence in each experiment. They are raw signal intensity normalized for (1) background; (2) DNA (by corresponding SybrGreen value); (3) one of the sequences with assigned value of 1000 units. POU17 (GTATGCAAAT) was used to normalize data for Oct-1 protein binding. The average relative binding affinity was derived based on the four measurements for each sequence.

3. Experimental considerations and design: Work on the assumption that 0.05 mg functional protein is the maximum that may be isolated from a 50-mL culture. This equates to 1 mg from 1 L.

For example, in the case of a 50-kDa protein: 0.05 mg of 50 kDa protein = 1.0 × 10^{-9} mol protein. Therefore the isolation requires at least 1.0 × 10^{-9} mol of oligoduplex. With 45 μ*M* concentration of oligoduplex 1.0 × 10^{-9} mol will be contained in:

$$\frac{1.0 \times 10^{-9} \text{ mol}}{4.5 \times 10^{-8} \text{ mol/mL}} = 0.022 \text{ mL} = 22 \text{ μL of oligoduplex.}$$

Each oligoduplex is labeled with one biotin. Thus, in 1.0 × 10^{-9} mol of oligoduplex (or 22 μL) there will be 1.0 × 10^{-9} mol × 244 = 2.44 × 10^{-7} g biotin = 0.24 μg biotin.

1 mL streptavidin-coated agarose binds 15 μg of biotin. Thus, the minimal amount of streptavidin-agarose that would retain all biotin-labeled oligoduplexes will be 1/15 × 0.24 = 0.016 mL = 16 μL. However, we routinely use at least 55 μL of streptavidin-agarose to ensure that all biotin-labeled duplexes are retained. For proteins that bind DNA at lower affinity, use a five times excess of oligoduplexes to the protein.

4. Note. To check whether extension has worked, run 1 μL of extended sample and 1 μL of annealed sample on a 4% w/v agarose gel (150 V, 20 min). Observe shift in migration.

5. The following Lucidea Array Spotter Control settings are used (In this example the duplexes are distributed across four 384-well plates):

    Slides per dip: 15.
    Plates: 4.
    Spotting mode: Normal.
    Replicates: 4.
    Repeats: 1.
    Complete last row: Yes.
    Row pitch: 250 μm.
    Column pitch: 250 μm.
    Slide thickness: 0.9 to 1.1 mm.
    Humidity control: 70%.

6. Wash with shaking in a slide box with approx 75 mL solution.

7. Example of Spot IDs and paramenters:

    ATF 1.
    31   5.
    Type – GenePix ArrayList V1.0.
    BlockCount = 28.
    BlockType = 0.
    "Block1 = 3875, 5000, 160, 4, 250, 16, 250."
    "Block2 = 6125, 5000, 160, 4, 250, 16, 250."
    "Block3 = 8375, 5000, 160, 4, 250, 16, 250."

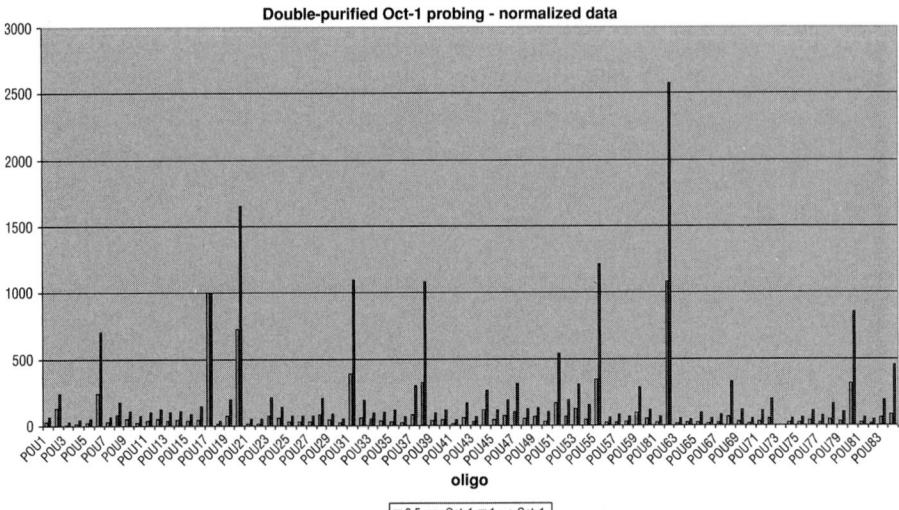

Fig. 9. Example of normalized binding data showing linearity in binding intensities for the protein concentration used. The graph shows the relative binding affinities of OCT-1 sequences estimated at two different protein concentrations (0.5 and 1 mg) of OCT-1/POU-His-Tag protein.

"Block4 = 10625, 5000, 160, 4, 250, 16, 250," and so on.

Block	Column	Row	ID	Sequence
1	1	1	1-Oct	GCATGAAATC
1	2	1	2-Oct	ACATGCAAAC
1	3	1	3-Oct	GCAGGCTAAC
1	4	1	4-Oct	ACAGGATATC
1	1	2	5-Oct	GTAGGAAAAC
1	2	2	6-Oct	GCATGATAAT
1	3	2	7-Oct	ATATGACAAC
1	4	2	8-Oct	GCAGGCAATT

and so on.

8. Several experiments with varying protein concentration may be required to ensure that an adequate signal intensity is obtained without saturating the higher binding sequences (**Fig. 9**).

References

1. Kadonaga, J. T. (2004) Regulation of RNA polymerase II transcription by sequence-specific DNA binding factors. *Cell* **116,** 247–257.
2. Benos, P. V., Lapedes, A. S., and Stormo, G. D. (2002) Is there a code for protein-DNA recognition? Probab(ilistical)ly. *Bioessays* **24,** 466–475.

3. Udalova, I. A., Mott, R., Field, D., and Kwiatkowski, D. (2002) Quantitative prediction of NF-kappa B DNA-protein interactions. *Proc. Natl. Acad. Sci. USA* **99,** 8167–8172.
4. Tuerk, C. and Gold, L. (1990) Systematic evolution of ligands by exponential enrichment: RNA ligands to bacteriophage T4 DNA polymerase. *Science* **249,** 505–510.
5. Roulet, E., Busso, S., Camargo, A. A., Simpson, A. J., Mermod, N., and Bucher, P. (2002) High-throughput SELEX SAGE method for quantitative modeling of transcription-factor binding sites. *Nat. Biotechnol.* **20,** 831–835.
6. Bulyk, M. L., Gentalen, E., Lockhart, D. J., and Church, G. M. (1999) Quantifying DNA-protein interactions by double-stranded DNA arrays. *Nat. Biotechnol.* **17,** 573–577.
7. Mukherjee, S., Berger, M. F., Jona, G., et al. (2004) Rapid analysis of the DNA-binding specificities of transcription factors with DNA microarrays. *Nat. Genet.* **36,** 1331–1339.
8. Linnell, J., Mott, R., Field, S., Kwiatkowski, D. P., Ragoussis, J., and Udalova, I. A. (2004) Quantitative high-throughput analysis of transcription factor binding specificities. *Nucleic Acids Res.* **32,** e44.
9. Bulyk, M. L., Huang, X., Choo, Y., and Church, G. M. (2001) Exploring the DNA-binding specificities of zinc fingers with DNA microarrays. *Proc. Natl. Acad. Sci. USA* **98,** 7158–7163.

20

Analysis of Protein-DNA Binding
by Streptavidin–Agarose Pulldown

Kenneth K. Wu

Summary

Binding of nuclear transactivators to sequence-specific regulatory elements on the promoter regions is of fundamental importance in gene expression and regulation. DNA-bound transactivators recruit transcription coactivators or repressors and an array of associated proteins that interact with the basal transcription factors, thereby activating the transcription machinery. Analysis of the large complex of proteins that bind to DNA is an important step in elucidating the mechanisms by which gene expressions are regulated. Commonly used techniques to determine DNA-protein binding such as the electrophoretic mobility shift assay (EMSA) have limited value for analyzing simultaneously a large number of proteins in the complex. We describe here a streptavidin-agarose pulldown assay that is capable of analyzing quantitatively binding of an array of proteins to DNA probes. The assay is easy to perform and does not require radiolabeled probes. It involves incubation of nuclear extract proteins with 5'biotinylated double-stranded DNA probes and streptavidin-agarose beads. The complex is pulled down, and proteins in the complex are dissociated and analyzed by Western blotting. This method has been shown to be useful in determining the regulation of binding of transactivators, p300/CBP, and associated proteins to the cyclooxygenase-2 (COX-2) promoter.

Key Words: Protein-DNA binding; electrophoretic mobility shift assay; transcription factors; streptavidin-agarose; biotinylated oligonucleotide; cyclooxygenase-2; promoter activity.

1. Introduction

Transcriptional activation of euchromatin genes depends on binding of transcriptional activators to sequence-specific DNA binding sites at the promoter regions of genes *(1)*. DNA bound transcriptional activators recruit coactivators,

From: *Methods in Molecular Biology, vol. 338: Gene Mapping, Discovery, and Expression:
Methods and Protocols*
Edited by: M. Bina © Humana Press Inc., Totowa, NJ

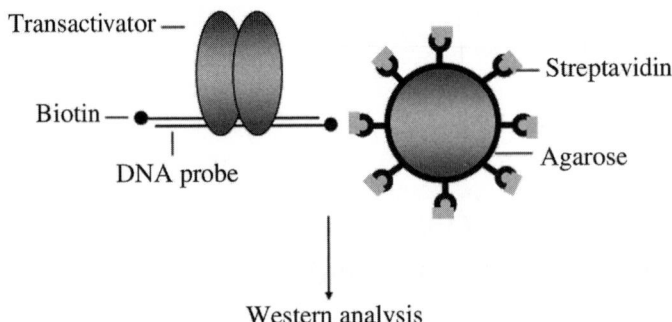

Fig. 1. Schematic illustration of the principle of streptavidin-agarose pulldown assay of protein-DNA binding. Transactivators (shown as footballs) bind to sequence-specific cis-acting elements of a biotinylated double-stranded oligonucleotide probe. Biotin binds streptavidin immobilized on agarose beads. The complex is centrifuged. Transactivators are dissociated and analyzed by Western blotting. The assay is suitable for studying binding of a single transactivator or multiple transactivators simultaneously. Furthermore, it is suitable for quantitative analysis of not only transactivators but also proteins that interact with transactivators.

corepressors, and associated proteins that interact with basal transcription factors in the transcription machinery and activate the RNA polymerase II to initiate transcription. Mitogenic factors and proinflammatory mediators induce gene expression by enhancing the DNA binding activity of responsive transactivators. Basal gene expression and gene expression induced by exogenous stimuli are regulated by binding of a combination of distinct transactivators to their specific binding sites (enhancer elements) at the promoter region that recruit coactivators or corepressors and associated proteins to form a complex. Thus, quantitative analysis of the complex that binds to specific DNA motifs is crucial for understanding diverse biological functions. Protein-DNA binding may be analyzed by a number of assays including DNase protection assay, electrophoretic mobility shift assay (EMSA), molecular beacon assay, chromatin immunoprecipitation (ChIP) assay, and a recently described streptavidin-agarose pulldown assay (SAPA) (2–8). The SAPA assay will be described in detail in this chapter.

The SAPA assay (**Fig. 1**) takes advantage of an extremely high binding affinity of biotin to avidin ($K_d = 10^{-15} M$), which has been widely applied to detecting proteins by enzyme immunoassay and immunohistochemical analysis of proteins (9,10). The SAPA assay was developed for detecting proteins that bind to oligonucleotides. The principle of the SAPA assay is to incubate nuclear extracts with biotinylated double-stranded DNA probes in the presence of streptavidin-conjugated agarose beads. Transactivators in the nuclear extracts bind to sequence-specific binding sites at the biotinylated probe, and the complex binds streptavidin–agarose via biotin. The mixture is isolated by centrifugation. Transactivators, coactivators, corepressors, and associated proteins in the complex are dissolved and analyzed by immunoblotting. This technique provides a semiquantitative analysis of an array of proteins that interact with promoter response elements and the transcription machinery. The length of the biotinylated DNA probe may be as short as 20 bp, harboring a single specific binding site, or a 500-bp cyclo-oxygenase-2 (COX-2) core promoter that harbors multiple binding sites. Several internal controls should be included in this pulldown assay. The most crucial is the inclusion of a probe that does not harbor any recognizable enhancer element. We have routinely used a 21-mer as a universal control *(8)*. Under special circumstances, a mutant probe should be included as a control. For example, a 24-mer COX-2 promoter fragment containing a C/EBP binding site (5'-ACCGGC<u>TTA CGCAA</u>TTTTTTTAAG-3') and a C/EBP mutant (5'-ACCGGCGCGATAGTTT TTTTAAG-3') are used to illustrate the specific binding of C/EBP isoforms to this regulatory element *(11)*. In our initial experiments, we compared the pulldown by streptavidin-conjugated agarose vs plain agarose beads and found that plain beads did not pull down biotinylated DNA-protein complexes. This control may be unnecessary for each experiment but should be included when a new batch of streptavidin-conjugated agarose beads is used.

This binding assay has been shown in our experiments to be suitable for analyzing transcriptional factor binding activities in several cell types including human foreskin fibroblasts, human endothelial cells (human umbilical vein endothelial cells, ECV 304 and EA.hy927 cells), and murine RAW 264.7 macrophages *(8,11–13)*. It is reasonable to assume that it is useful in evaluating protein-DNA interaction in any cell type.

This assay is suitable for analyzing multiple transcription factors that bind to a promoter. By using a 500-bp COX-2 promoter probe, we have shown that this assay is useful in identifying and quantifying simultaneously all C/EBP isoforms including the C/EBPβ truncated forms, all NF-κB isoforms (P65 RelA, P68 RelB, P75 C-Rel, P50 NF-κB1, and P52 NF-κB2), CREB/ATF isoforms, C-Jun and C-Fos isoforms, p300/CBP, PCAF, $TFII_B$, Med7, and Srb7 to the COX-2 promoter *(8)*. The assay appears to be versatile in analyzing all the proteins in the complex as long as suitable specific antibodies with reasonable affinity for the candidate proteins are available.

The assay is relatively simple and does not require radiolabeled probes. We have tested its feasibility for studying promoter regulation in a prototypic pro-inflammatory gene, COX-2. Our results show that it is useful in determining the temporal and spatial relationship between transactivator and p300 coactivator binding to the core COX-2 promoter region and COX-2 transcriptional regulation by proinflammatory mediators.

2. Materials

2.1. Nuclear Extract Isolation

1. Phosphate-buffered saline (PBS), pH 7.4 (Sigma, St. Louis, MO).
2. The following reagents are prepared and stored in stock concentrations:
   a. 0.5 M Sodium fluoride (NaF; Sigma), stored at 4°C.
   b. 100 mM Phenylmethylsulphonyl fluoride (PMSF; Sigma) solution in isopropanol, stored at –20°C.
   c. 0.1 M Dithiotreitol (DTT; Invitrogen), stored at –20°C.
   d. 1 mg/mL Leupeptin (Sigma), stored at –20°C.
   e. 1.25 M β-Glycerophosphate disodium salt (Sigma), stored at 4°C.
   f. 1 M Sodium vanadate (Sigma), stored at –20°C
   g. 1 M Potassium chloride (KCl; Aldrich, Milwaukee, WI) stored at room temperature (RT).
   h. 1 M HEPES (Sigma), stored at 4°C.
   i. 1 M Magnesium chloride hexahydrate (MgCl$_2$, Sigma), stored at RT.
   j. 2 M Sucrose (Sigma), stored at RT.
   k. 10% Igepal CA-630 (NP-40; Sigma), stored at RT.
   l. 5 M Sodium chloride (NaCl; Sigma), stored at RT.
   m. 0.5 M EDTA (Invitrogen), stored at RT.
3. Glycerol (Sigma), stored at RT.
4. PBS buffer containing inhibitors (PBSI): 0.5 mM PMSF, 25 mM β-glycerophosphate, 10 mM NaF, stored at 4°C.
5. Buffer A: 10 mM HEPES, pH 7.9, 1.5 mM MgCl$_2$, 10 mM KCl, 300 mM sucrose, 0.5% NP-40, stored at 4°C.
6. Buffer B: 20 mM HEPES, pH 7.9, 1.5 mM MgCl$_2$, 420 mM NaCl, 0.2 mM EDTA, 2.5% glycerol, stored at 4°C.
7. Buffer D: 20 mM HEPES, pH 7.9, 100 mM KCl, 0.2 mM EDTA, 8% glycerol, stored at 4°C.
8. Cell scraper (Corning Incorporated Life Sciences, Acton, MA).
9. Ultrasonic dismembrator model 500 (Fisher Scientific, Pittsburgh, PA).
10. Microcentrifuge (Eppendorf, Hamburg, Germany, cat. no. 5415D).
11. Temperature-controlled room (Biocold Environmental, Fenton, MO).

2.2. Protein Assay

1. BCA Protein Assay Reagent Kit (Pierce, Rockford, IL): Reagent A containing sodium carbonate, sodium bicarbonate, bicinchoninic acid, and sodium tartrate in 0.1 M sodium hydroxide, and Reagent B containing 4% cupric acid.
2. Albumin standard 2 mg/mL.
3. 96-Well plate (Corning).

4. Microplate spectrophotometer, Benchmark Plus (Bio-Rad, Hercules, CA).
5. Microplate Manager version 5.2 (Bio-Rad).

2.3. Binding Assay

Stepwise, the assay is divided into (1) the pulldown procedure; (2) sodium dodecyl sulfate-polyacrylamide gel electrophoresis (SDS-PAGE); and (3) Western blot analysis. Materials for each procedure are described separately.

2.3.1. Pulldown Procedure

1. Streptavidin immobilized on 4% beaded agarose activated by cyanogen bromide (Sigma).
2. Single-stranded 5'-biotinylated oligonucleotides (Integrated DNA Technologies, Coralville, IA) (*see* **Note 1**).
3. PBSI buffer.
4. Rocking platform labquake (Barnstead International, Dubuque, IA).
5. Microcentrifuge.

2.3.2. SDS-PAGE

1. 2X Laemmli sample buffer: 950 μL of Laemmli sample buffer mixed with 50 μL of 2-mercapthoethanol (Bio-Rad).
2. 4 to 15% Gradient polyacrylamide gel in Tris-HCl (Bio-Rad).
3. 10X Running buffer containing 250 m*M* Tris, 1.92 *M* glycine, 1% SDS (all from Sigma). 1X Running buffer is prepared by diluting 100 mL of 10X buffer with 900 mL deionized water.
4. 10X Transfer buffer containing 250 m*M* Tris-HCl, 1.92 *M* glycine (both from Sigma). Stored at RT. 1X Transfer buffer is prepared by mixing 100 mL of 10X transfer buffer with 200 mL of methanol and 700 mL of deionized water.
5. Methanol (AAPER Alcohol and Chemicals Company, Shelbyville, KY).
6. Prestained broad-range SDS-PAGE standards (Bio-Rad).
7. Mini-Trans Blot Electrophoretic Transfer Cell (Bio-Rad).
8. Minigel holder cassette (Bio-Rad).
9. Fiber pads (Bio-Rad).
10. Trans-Blot Transfer Medium Supported Nitrocellulose Membrane, 0.2 μm (Bio-Rad).
11. Filter paper (Bio-Rad).

2.3.3. Western Blot

1. Rabbit affinity-purified polyclonal antibodies specific for transactivators and coactivators (various sources). Nonimmune IgG control (various sources) (*see* **Note 2**).
2. IgG-horseradish peroxidase-conjugated polyclonal antibody (*see* **Note 3**).
3. Blotting grade blocker nonfat dry milk (Bio-Rad).
4. PBS with Tween 0.05% (PBST) (Sigma).
5. Rocking platform Labquake (Barnstead International).
6. Materials for detection of proteins by chemiluminescence:

a. Kodak X-Omatic Cassette (Eastman Kodak, Rochester, NY).
b. Kodak BioMax MS film (Eastman Kodak).
c. X-ray film processor Konica SRX-201-A (Source One Healthcare Technologies, Mentor, OH).
d. SuperSignal® West Pico Chemiluminescent Substrate kit (Pierce) (*see* **Note 4**).
e. Rocking platform Labquake (Barnstead International).
f. Plastic Sheet (Reports Cover, C-Line Products, Mount Prospect, IL) (*see* **Note 5**).

3. Methods

3.1. Nuclear Extract Isolation

1. All procedures are done on ice or in a controlled temperature room at 4°C.
2. Remove medium from cultured cells and wash them once with cold PBS. Add PBSI buffer (1 mL/10-cm dish) and harvest cells with a rubber scraper. Cells are collected into a 50-mL conical tube, which is centrifuged at 550g for 5 min (*see* **Note 6**).
3. Remove supernatant. Transfer the pellet to a 1.5-mL microcentrifuge tube, and centrifuge at 1500g for 30 s.
4. To buffers A, B, and D, add the following inhibitors: 0.5 mM PMSF, 1 mM Na$_3$VO$_4$, 0.5 mM DTT, 1 µg/mL leupeptin, 25 mM β-glycerophosphate, 10 mM NaF.
5. Remove supernatant and resuspend the pellet in 2 package cell volume of buffer A with inhibitors. Keep on ice for 10 min. Vortex briefly, and centrifuge at 2600g for 30 s.
6. Remove supernatant and resuspend the pellet in 2/3 package cell volume of buffer B with inhibitors.
7. Sonicate the mixture for 5 s. Centrifuge at 10,400g for 5 min (*see* **Note 7**).
8. Dilute the supernatant isovolumetrically with buffer D with inhibitors. Aliquot and store at −80°C (*see* **Note 8**).

3.2. Protein Assay

1. Prepare standard albumin concentrations at 50, 100, 250, 500, 750, 1000, 1500, and 2000 µg/mL.
2. Pipet 10 µL of standards and samples each in duplicate onto a 96-well plate.
3. Mix 50 parts of buffer A with 1 part of buffer B.
4. Add 200 µL of mixed buffer A and B to standards and samples.
5. Incubate plate at 37°C for 30 min.
6. Read the protein content at 562 nm on a microplate spectrophotometer.

3.3. Binding Assay

3.3.1. Pulldown Procedure

1. Dissolve single-stranded biotinylated oligonucleotides in deionized sterile water. The final concentration is 1 µg per 1 to 2 µL.
2. Add an equal quantity of sense and antisense biotinylated oligonucleotide solution to a microfuge tube, mix, and place it in a 100°C water bath for 1 h. Remove the tube and allow it to cool down at RT (*see* **Note 9**).

3. Gently mix the streptavidin-agarose bead suspension with a vortex and place the suspension in a 1-mL (tuberculin) syringe. Pull the syringe piston forward and backward gently two to three times to ensure an even suspension (*see* **Note 10**).

4. Add 2 drops of the streptavidin-agarose bead suspension to a mixture of 400 µg of nuclear extract proteins and 4 µg of double-strand biotinylated oligonucleotides in 500 µL of PBSI buffer (*see* **Note 11**).

5. Place the mixture on a rocking platform at RT and rock the mixture at a gentle speed for 2 h (*see* **Note 12**).

6. Centrifuge the mixture at 550 g for 1 min (*see* **Note 13**).

7. Discard the supernatant.

8. Wash the pellet with PBSI for three times.

9. Suspend the pellet in 40 µL of 2X Laemmli sample buffer.

10. Incubate the sample at 95°C for 5 min (*see* **Note 14**).

11. Centrifuge the sample at 7000*g* for 30 s in a microfuge and collect the supernatant (*see* **Note 15**).

3.3.2. SDS-PAGE

1. Remove 4 to 15% gradient "ready to use" gel from plastic, remove the comb, and cut the tape along the black line across the entire gel. Pull out the gel (*see* **Note 16**).

2. Set up the electrophoresis unit.

3. Place the gel onto the minigel holder cassette of the electrophoresis unit.

4. Load the 10-well gel with 5 µL of standards or 25 µL of the supernatant protein samples.

5. Add 1X running buffer to the electrophoresis cell.

6. Cover the electrophoresis cell with the cell lid.

7. Connect the electrophoresis unit to a power supply. Set the voltage at 150 V.

8. Run the gel at 150 V until the marker dye has run off the edge of the gel.

9. Remove gel and rinse with 1X transfer buffer precooled in ice (4°C).

10. Set up the electrophoretic transfer cell.

11. Soak fiber pad, filter paper, and nitrocellulose membrane in 1X transfer buffer before placing them in the gel cassette.

12. Place nitrocellulose membrane between gel and anode. Inspect to make sure that no air bubble is trapped between gel and membrane.

13. Connect the electrophoretic transfer unit to a power supply.

14. Place the unit on ice or at 4°C cold room.

15. Run the gel at 90 V for 1 h.

3.3.3. Western Blotting

1. Remove the nitrocellulose membrane from the electrophoretic transfer unit.

2. Immerse the membrane in PBS containing 5% nonfat milk and gently rock it on a rocking platform at RT for 1 h.

3. Remove the PBS milk solution and add primary antibodies at appropriate dilutions. Incubate at RT for 1 h.

4. Wash the membrane in PBST solution at RT for 5 min. Repeat three times.

5. Incubate membrane in horseradish peroxidase-conjugated secondary antibodies at RT for 1 h.
6. Wash the membrane in PBST at RT for 5 min. Repeat three times.
7. Rinse the membrane.
8. Mix an equal volume of stable peroxide and luminol/enhancer solutions provided in the SuperSignal® West Pico kit.
9. Add the above solutions to the membrane and incubate it at RT for 5 min on a rocking platform.
10. Remove the membrane, blot it quickly with paper towel, and place it between two plastic sheets that have been cut to fit the size of the X-ray film cassette.
11. In a dark room, place the membrane in an X-ray film cassette containing a Kodak X-ray film. Expose the film to X-ray for a suitable time period, usually 5 to 10 s.
12. View the protein bands on X-ray film and analyze their intensities by densitometry.

4. Notes

1. We design oligonucleotides based on the promoter sequences containing the enhancer element(s) to be tested. The length of the oligonucleotide may be a short 20-mer containing a single binding site or as long as 500 bp containing multiple enhancer elements. The sequence is sent to Integrated DNA Technologies (IDT). 5'-biotinylated sense and antisense single-stranded oligonucleotides are synthesized and the sequence is verified by IDT. The oligonucleotides are lyophilized and delivered to our laboratory in powder form. The turnaround time is generally reasonably short.
2. We select affinity-purified rabbit polyclonal antibodies (IgG) specific for transcription factors to be investigated and affinity-purified rabbit nonimmune IgG as control whenever available. The concentrations (dilutions) of specific primary antibodies used in our experiments are generally prepared according to the manufacturers' recommendation.
3. We use goat horseradish peroxidase-conjugated IgG that recognizes rabbit IgG as the secondary antibody when the primary antibody is prepared from rabbit serum.
4. The chemiluminescent kit comes with two separate bottles of reagents: stable peroxide and luminol/enhancer.
5. For convenience, we use plastic sheets from C-Line Products. However, there are many alternatives. For example, Saran™ Wrap will do the work.
6. A 50-mL conical tube is used to ensure a maximal yield of cells.
7. In our experience, a 5-s sonication is sufficient to break the cells. Sonication for a longer period may disrupt the nucleus.
8. Nuclear extracts so prepared do not have detectable cytosolic proteins.
9. As 4 μg of biotinylated double-stranded oligonucleotides are used for each assay, we mix 4 μg of sense with 4 μg of antisense oligonucleotides to generate 4 μg of 5'-biotinylated double-stranded oligonucleotide in each experiment.
10. The streptavidin-agarose suspension is thick and difficult to mix well. The method described here provides a good mix. After gently stirring with a vortex, the suspension is transferred into a 1-mL tuberculin syringe and mixed; then 2 drops of

the mixed suspension are added to the mixture containing nuclear extract proteins and biotinylated oligonucleotides.

11. Each drop contains about 20 μL of streptavidin-agarose suspension.
12. The incubation time may be shortened to 1 h but should not be shorter than 1 h.
13. It is important to centrifuge the complex at a low speed. A higher speed of centrifugation may disrupt the complex.
14. This step causes dissociation of transcription factors from the complex.
15. A short spin in a microfuge is sufficient to remove the complex.
16. We have found the Bio-Rad minigel to be convenient. However, minigels from other manufacturers may be used.

Acknowledgments

The author wishes to thank Dr. Ying Zhu for his contribution to the development of this assay and Drs. Katarzyna Cieslik and Wu-Guo Deng for applying the assay to the study of pharmacologic and pathophysiologic control of transactivator and cotransactivator binding to cyclooxygenase-2 and inducible nitric oxide synthase promoters. This work was supported by grants from the NIH (R01 HL-50675 and P50-NS-23327).

References

1. Ptashne, M. and Gann, A. (1997) Transcriptional activation by recruitment. *Nature* **386,** 569–577.
2. Faber, S., O'Brien, R. M., Imai, E., Granner, D. K., and Chalkley, R. (1993) Dynamic aspects of DNA/protein interactions in the transcriptional initiation complex and the hormone-responsive domains of the phosphoenolpyruvate carboxykinase promoter in vivo. *J. Biol. Chem.* **268,** 24976–24985.
3. Garner, M. M. and Revzin, A. (1981) A gel electrophoresis method for quantifying the binding of proteins to specific DNA regions: application to components of the *Escherichia coli* lactose operon regulatory system. *Nucleic Acids Res.* **9,** 3047–3060.
4. Demczuk, S., Harbers, M., and Vennstrom, B. (1993) Identification and analysis of all components of a gel retardation assay by combination with immunoblotting. *Proc. Natl. Acad. Sci. USA* **90,** 2574–2578.
5. Heyduk, T. and Heyduk, E. (2002) Molecular beacons for detecting DNA binding proteins. *Nat. Biotechnol.* **20,** 171–176.
6. Storek, M. J., Ernst, A., and Verdine, G. L. (2002) High-resolution footprinting of sequence-specific protein-DNA contacts. *Nat. Biotechnol.* **20,** 183–186.
7. Luo, R. X., Postigo, A. A., and Dean, D. C. (1998) Rb interacts with histone deacetylase to repress transcription. *Cell* **92,** 463–473.
8. Deng, W. G., Zhu, Y., Montero, A., and Wu, K. K. (2003) Quantitative analysis of binding of transcription factor complex to biotinylated DNA probe by a streptavidin-agarose pulldown assay. *Anal. Biochem.* **323,** 12–18.

9. Bayer, E. A. and Wilchek, M. (1980) The use of the avidin-biotin complex as a tool in molecular biology. *Methods Biochem. Anal.* **26,** 1–45.

10. Wilchek, M. and Bayer, E. A. (1989) Avidin-biotin technology ten years on: has it lived up to its expectations? *Trends Biochem. Sci.* **14,** 408–412.

11. Zhu, Y., Saunders, M. A., Yeh, H., Deng, W. G., and Wu, K. K. (2002) Dynamic regulation of cyclooxygenase-2 promoter activity by isoforms of CCAAT/enhancer-binding proteins. *J. Biol. Chem.* **277,** 6923–6928.

12. Deng, W. G., Zhu, Y., and Wu, K. K. (2004) Role of p300 and PCAF in regulating cyclooxygenase-2 promoter activation by inflammatory mediators. *Blood* **103,** 2135–2142.

13. Deng, W. G. and Wu, K. K. (2003) Regulation of inducible nitric oxide synthase expression by p300 and p50 acetylation. *J. Immunol.* **171,** 6581–6588.

21

Isolation and Mass Spectrometry of Specific DNA Binding Proteins

Mariana Yaneva and Paul Tempst

Summary

A subset of the proteome that binds to specific DNA sequences is at the center of genome function, integrity, and dynamics. We present a detailed protocol that allows the isolation of any specific DNA binding protein and its subsequent identification by mass spectrometry. The procedure involves prefractionation of crude nuclear extract by phosphocellulose (P11) chromatography, followed by a series of positive/negative selections on wild-type and site-mutated ligand DNA in a magnetic microparticulate format. DNA-affinity capture requires concatamerized and biotinylated ligand, selective salt conditions, and improved competitor DNAs. The amount of protein(s) captured on DNA-magnetic beads is generally sufficient for successful MALDI-TOF and TOF/TOF MS-based protein identification. As an example, we describe the procedures used to isolate and identify four specific transcription factors from 2×10^9 promyelocytic NB4 cells.

Key Words: DNA binding; affinity capture; magnetic beads; transcription factors; mass spectrometry.

1. Introduction

Mass spectrometry accounts for essentially all *de novo* protein identification. For the most part, applications fall into one of two categories, either systems biology or standard identification of proteins isolated by traditional means such as multicolumn chromatography. The latter purification schemes can be very costly and time consuming, yet they typically yield only one or a few purified proteins, which can be identified within a day or so by a proteomics facility. Molecular and cellular biologists could take better advantage of the much improved protein identifying capabilities and throughput if robust, facile methods were available to scale down and accelerate the purification process, a bridge,

From: *Methods in Molecular Biology, vol. 338: Gene Mapping, Discovery, and Expression: Methods and Protocols*
Edited by: M. Bina © Humana Press Inc., Totowa, NJ

as it were, between traditional life sciences research and the analytical power of proteomics. This has been particularly the case in the field of nucleic acid biochemistry, including the study of DNA binding proteins that function in transcriptional regulation, replication, and maintenance of chromosomal integrity.

We describe an efficient, accelerated method for affinity capture of transcription factors on specific DNA-magnetic particles, to yield final preparations in a form and amounts that are compatible with standard MALDI-TOF MS-based protein identification. A major obstacle to developing this approach into a widely applicable protocol is the inadvertent, extensive binding of nonspecific proteins to the bait. This problem can be addressed at two levels. First, a single batch-fractionation on P11 serves to increase the relative concentration of the transcription factors of interest, removes several of the abundant nonspecific DNA binding proteins, and may resolve multiple factors in different fractions that can subsequently be used for parallel affinity capture. Second, a number of counteracting measures can be implemented. Protein fractions are incubated with magnetic beads carrying concatamerized DNA binding sites, and in the presence of short oligo(dI:dC) competitor, resulting in higher specific-ligand density and displacement of proteins that preferentially bind to DNA nicks and ends. The beads are easily collected from larger volumes using high-powered magnets. Preclearance with mutant DNA-beads (negative selection) greatly reduces the background of nonspecific proteins in the final preparation. The number of negative selections depends on the abundance and binding kinetics of the respective transcription factors and is determined experimentally. In the case of low-affinity binders, negative selection is preceded by a single positive selection, which results in much reduced protein mass and volume. Affinity capture of high-affinity binding transcription factors, on the other hand, can be performed in high salt, increasing the stringency of the procedure.

A targeted proteomic analysis of this nature should always be preceded by a thorough molecular biological analysis of the system under study. Proteomics is not a stand-alone science, and the field would benefit greatly if more projects were initiated by the traditional, hypothesis-driven research groups. The best way to popularize it among biologists and clinical scientists is by bringing a large portion of proteomic research activities into their own laboratories, leaving only the final readout (i.e., protein identifications) to specialized facilities. The approach and protocol that are presented here should be considered in that context.

2. Materials

1. Cell culture medium for growth of NB4 cells: RPMI-1670 (Life Technologies) supplemented with 10% fetal calf serum (Sigma), nonessential amino acids, and penicillin and streptomycin at 5 µg/mL each.
2. Cell lysis buffer: 10 mM Tris-HCl, pH 8.0, 10 mM NaCl, 8 mM MgCl$_2$.

3. Glass/glass (Dounce) homogenizer with tight pestle B (Fisher Scientific).
4. Buffers for extraction of nuclear proteins.
   a. Low salt buffer: 20 mM Tris-HCl, pH 7.5, 20 mM KCl, 1.5 mM MgCl$_2$, 0.2 mM EDTA, 25% glycerol.
   b. High salt buffer: 20 mM Tris-HCl, pH 7.5, 1.2 M KCl, 1.5 mM MgCl$_2$, 0.2 mM EDTA, 25% glycerol.
5. Protease inhibitors cocktail (Amersham Pharmacia Biotech): one tablet is dissolved in 2 mL of water and used as 25X stock solution. Protein inhibitors cocktail (Sigma) used as 100X solution is equally effective. Phenylmethylsulfonyl fluoride (PMSF) is dissolved in *iso*-propanol as 0.2 M stock solution and stored at 4°C.
6. Phosphocellulose P11 (Whatman).
7. Buffer for chromatography on P11 column (buffer D): 20 mM HEPES-KOH, pH 7.9, 0.2 mM EDTA, 0.5 mM dithiothreitol (DTT), 0.01% NP-40, 0.2 mM PMSF, and 10% glycerol. The NaCl concentration in this solution varies from 75 mM to 1 M.
8. DNA binding buffer: 20 mM HEPES-KOH, pH 7.9, 0.1 M KCl, 0.2 mM EDTA, 0.5 mM DTT, 0.01% NP-40, 10% glycerol.
9. Poly(dI:dC) (Amersham Pharmacia): 10 mg/mL stock solution is prepared in H$_2$O and stored at –20°C.
10. Large-scale preparation of genomic *E. coli* DNA is performed using CsCl gradient centrifugation as described previously (*1*).
11. Custom synthesized oligodeoxynucleotides (e.g., by IDT, Coraville, IA):
    a. Single-stranded oligonucleotides (dI:dC), 30 bp in length.
    b. Regular and 5'-biotinylated single-stranded oligonucleodtides with 6-carbon linker. The conditions for preparation of double-stranded oligonucleotides are described in **Subheading 3.4.**
12. Magnetic beads coated with streptavidin, M280 (Dynal Biotech).
13. Cobalt magnetic discs, 6-mm-diameter (cat. no. CR30352075, Edmund Scientific).
14. Deoxyribonucleotides (Roche Molecular Biochemicals).
15. "Vent" DNA polymerase (New England BioLabs).
16. Buffer for PCR: 10 mM KCl, 10 mM (NH$_4$)$_2$SO$_4$, 3.5 mM MgSO$_4$, 0.1% Triton X-100, and 20 mM Tris-HCl, pH 8.8 (at 25°C).
17. 0.1% Coomassie Brilliant Blue R250 (Bio-Rad) made in 40% methanol and 10% acetic acid. Store the solution at room temperature.
18. Antibodies for Western blotting: antibodies to PU.1, Fos, Jun, and RARα (Santa Cruz Biotechnology) and rabbit anti-GABPα polyclonal antibodies (custom-raised by Pocono Rabbit Farm, Canadis, PA) against KHL-coupled, C-terminal synthetic peptides (Yaneva et al., unpublished data).

3. Methods

3.1. Preparation of Nuclear Extract (NE)

1. Human promyelocytic leukemia NB4 cells are cultured under 5% CO$_2$ at 37°C to a density of 0.5 to 1.5 × 10^6 cells/mL. Cell cultures are passaged twice a week to maintain the cell density between 0.5 and 1.5 × 10^6 cells/mL.

2. Nuclear extract from these cells is prepared essentially as previously described *(2)*. All procedures are performed at 4°C. Collect the cells by centrifugation for 10 min at 800*g*, wash twice with cold PBS, and suspend in cold hypotonic lysis buffer at 15 to 20 × 10^7 cells/mL.
3. Homogenize the cells in Dounce homogenizer (pestle B) with 15 up and down strokes, and incubate on ice for 15 min *(see **Note 1**)*.
4. Collect the nuclei by centrifugation at 3000*g* for 15 min, and subsequently suspend the pellet by homogenization in a minimal volume of low salt extraction buffer, containing protease inhibitors cocktail at concentration recommended by the manufacturer.
5. Place the thick suspension on a magnetic stirrer in a cold room, and add one-half vol of high salt extraction buffer slowly, dropwise, to reach 0.42 *M* KCl final concentration *(see **Note 2**)*.
6. Stir the suspension for an additional 30 min, and centrifuge for 30 min at 15,000*g*.
7. Dialyze the clear supernatant for 5 h at 4°C against 50 vol of buffer D for chromatography on a P11 column *(see **Note 3**)*.

3.2. Chromatography on a P11 Phosphocellulose Column

To avoid binding of nonspecific DNA binding proteins to "specific" DNA-beads *(see **Note 4**)*, nuclear protein extract must be first fractionated. We found that chromatography on phosphocellulose column is one of the best procedures for this purpose *(see **Note 5**)*.

1. Prepare the P11 column exactly as instructed by the manufacturer *(see **Note 6**)*, and equilibrate with buffer D, containing 75 m*M* NaCl, for 36 to 48 h at 0.4 mL/min.
2. As a rule, approx 3 to 5 mg NE is loaded onto 1 mL P11 resin.
3. Elute the bound proteins either with a linear gradient of 0.075 to 0.85 *M* NaCl or stepwise with 0.1, 0.3, 0.5, and 0.85 *M* NaCl, in the same buffer. Gradient or stepwise, the entire elution should be done with 20 column vol and at a flow rate of 0.4 mL/min.
4. After chromatography, analyze every other fraction for the presence of the desired DNA binding activity by EMSA.
5. Pool the active fractions, and dialyze overnight at 4°C against 50 vol of buffer D containing 0.1 *M* NaCl, if necessary *(see below)*.
6. Bind the proteins directly to DNA-magnetic beads, constructed with concatamerized DNA and equilibrated with the same buffer.

3.3. Concatamerization of DNA Bait

1. Multimers of DNA binding sites are generated by a self-priming PCR method *(4)* using two direct repeats of complementary single-stranded oligonucleotides with either wild-type or mutant versions of specific binding sites, as required.
2. Only the forward oligonucleotides are biotinylated at the 5' end, using a 6-carbon spacer arm.

Cycles: 6   7

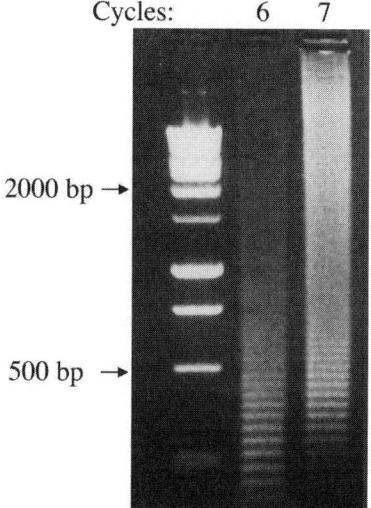

2000 bp →

500 bp →

Fig. 1. Concatamerization of specific DNA oligonucleotides by self-priming PCR. PCR products with the concatamers of AP.1-specific oligonucleotide were analyzed by electrophoresis in 1% agarose gel and stained with ethidium bromide. Lane 1, molecular weight markers; lanes 2 and 3, PCR products after six and seven PCR cycles, respectively.

3. PCR (50-µL vol): 460 ng of each oligonucleotide, 8 µ*M* dNTPs, and 2 U of "Vent" polymerase in PCR buffer. Cycling conditions must be optimized for each pair of oligonucleotides (*see* **Note 7**).
4. Purify the PCR products using the QiaQuick kit (Qiagen).
5. Analyze concatamerized DNA by electrophoresis in 1% agarose gel/Tris-borate-EDTA buffer. Each PCR yields about 3 to 5 µg of DNA, ranging between 200 bp and 5 to 10 kb in length (**Fig. 1**).

3.4. Preparation of DNA-Magnetic Beads

1. To prepare double-stranded oligonucleotides, mix complementary single-stranded oligonucleotides in a 1:1 molar ratio (50 µ*M* each) in 10 m*M* Tris-HCl, pH 8.0, 10 m*M* MgCl$_2$, 100 mM KCl, heat at 88°C for 3 min, and then gradually cool down for annealing: 10 min at 65°C, 1 min at 55°C, 10 min at 37°C, 5 min at room temperature. Annealed oligos are stored at –20°C.
2. Attach the 5'-biotinylated double-stranded oligonucleotides or concatamerized DNA, (DNA)$_n$, to M280 magnetic beads coated with streptavidin using the KilobaseBINDER kit (Dynal Biotech), according to the manufacturer's instructions.
3. For quantitations, end-label small aliquots of double-stranded oligonucleotides or (DNA)$_n$ with T4 polynucleotide kinase (New England Biolabs) and [γ-^{32}P]ATP (Perkin Elmer Life Sciences), and then use as a tracer to monitor final attachment.

The amount of DNA attached to the beads varies from 3 to 9 µg per mg beads for the concatamers DNA (*see* **Note 8**) and about 200 pmol per mg beads for the double-stranded oligonucleotides.

3.5. Protein Binding to DNA-Magnetic Beads

1. The precise composition of the DNA binding buffer, KCl concentration, divalent cation concentration, and other special requirements for DNA binding of a specific protein must be established in EMSA prior to the capture (*see* **Notes 9** and **10**). All procedures are carried out at 4°C.
2. The beads with attached concatameric DNA are first washed in DNA binding buffer containing double-stranded oligo(dI:dC) and poly(dI:dC) at 0.1 mg/mL each.
3. If necessary, the protein fractions after P11 chromatography should be dialyzed against 50 vol of the binding buffer containing the optimal salt concentration.
4. At this point, the capture should proceed in a different way depending on the salt concentration at which the protein was eluted from the P11 column (*see* **Note 11** and **Fig. 2**):
   a. Proteins eluted with <500 m*M* KCl should be first concentrated by capture on beads with wild-type DNA (positive selection), then eluted, and incubated with beads with mutant DNA (negative selection). After elution from these beads, the proteins are finally bound to wild-type DNA-beads (positive selection).
   b. Proteins eluted with >500 m*M* KCl usually bind to DNA with higher affinity and thus require only negative selection (one to three rounds). Again, the number of the negative selections depends on the abundance of the respective activity.
5. The number of binding to mutant DNA-beads will depend on the abundance of the activity in this fraction, the less abundant proteins require more initial extract and proportionally more rounds of negative selections than the more abundant ones.
6. All capturing schemes must end with positive selection.
7. The general empirical rules for DNA-affinity capture of site-specific proteins are outlined in **Note 12**.
8. The actual binding to the DNA-beads, wild-type or mutant, should be done as follows:
   a. Mix the active in DNA binding proteins (eluted from the P11 column or from beads) with 1 mg concatamerized DNA-magnetic beads in the presence of the competitor DNA oligo(dI:dC) and poly(dI:dC) at 0.1 mg/mL each (*see* **Note 13**).
   b. Incubate for at least 3 h at 4°C with rotation on a LabQuake shaker (Labindustries).
9. For collection of the beads from large volumes of nuclear extracts or column fractions, place a high-powered cobalt magnetic discs underneath 15-mL or 50-mL Falcon tubes during centrifugation in a GS-6KR rotor (Beckman) for 5 min at 700*g*.
10. For smaller volumes, in Eppendorf tubes, use benchtop magnets (Dynal Biotech).
11. Wash the beads with binding buffer (≥1 mL per mg beads) and three times with the same buffer containing *E. coli* double-stranded and single-stranded DNA (*see* **Note 14**), at 0.1 mg/mL each, as a nonspecific competitor.

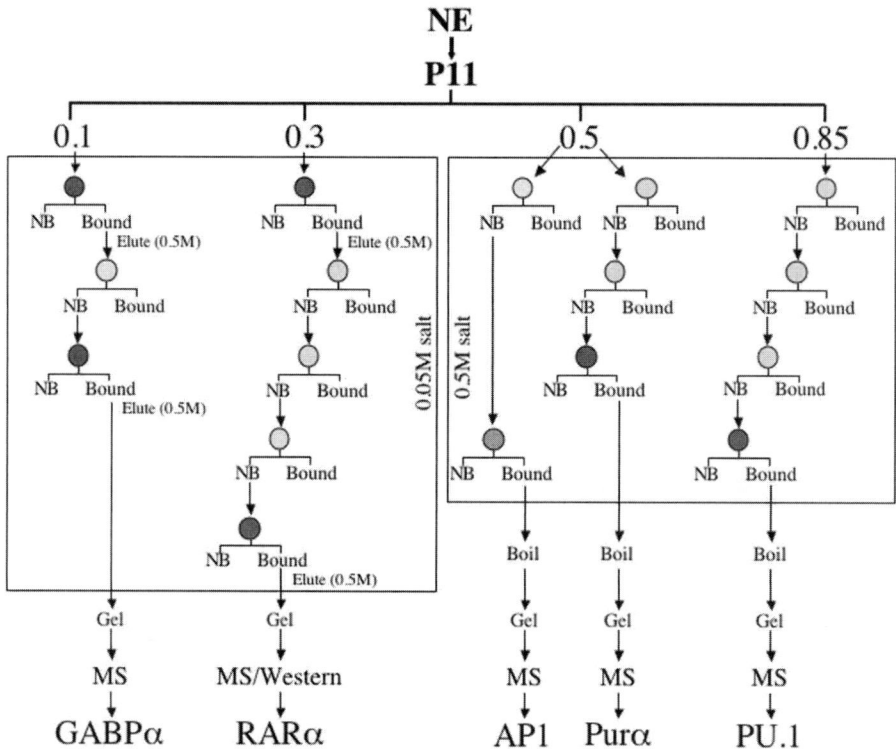

Fig. 2. DNA-specific, affinity microcapture diagram of nucleic factors. Sequence of positive (wt; filled circles) and negative (mut; open circles) selections performed on 0.1, 0.3, 0.5, and 0.85 *M* salt elution P11 fractions is shown. The (DNA)$_n$-coated magnetic beads were prepared as described in **Subheading 3.4.** The boxes cluster all the steps that were carried out at a similar salt concentration of either 0.05 or 0.5 *M*, as indicated. Arrows mark the steps where, and what, competitor DNA was added to the incubation mixtures; competitor is carried along with unbound fractions but is freshly added to any bead eluates before the next selection cycle. Elution, when done, was with 0.5 *M* KCl; tightly bound proteins were recovered from the beads by boiling in gel loading buffer. The proteins indicated on the bottom of the diagram (with the exception of RARα) were identified by MALDI-TOF MS. NE, nuclear extract; NB, not bound. (Reprinted from Analytical Chemistry, with permission.)

3.6. Elution of the Proteins from the DNA-Magnetic Beads

1. Elute the proteins from the DNA-beads with 50 µL binding buffer, containing 0.5 *M* NaCl, for 15 min on ice.
2. Separate the eluate from the beads on a magnet for 2 min on ice.

3. Denature the eluted proteins with 10 µL 6X Laemmli sample buffer and boiling for 3 min.
4. Subject proteins to analysis by SDS gel electrophoresis.
5. Proteins that bind to DNA with high affinity do not always elute from the beads under the above conditions. To analyze these proteins, suspend the beads after elution directly in 50 µL Laemmli sample buffer, heat for 3 min at 95°C for complete denaturation, and subject to analysis by SDS gel electrophoresis and Western blotting.
6. For mass spectrometric identification, separate the eluted proteins on 4 to 15% SDS gel and stain the gel with 0.1% Coomassie Blue R250 (*see* **Note 15**).

3.7. Analysis of the Eluted Proteins

3.7.1. Electrophoretic Mobility Shift Assay (EMSA)

1. EMSA is performed essentially as described previously *(4)*, with minor modifications. In brief, carry out prebinding of nuclear extract (5–10 mg/mL), or respective protein fraction, to the poly(dI-dC) at 25°C for 10 min in buffer containing 4% glycerol, 1 mM MgCl$_2$, 0.5 mM EDTA, 0.5 mM DTT, 25 mM NaCl, 10 mM Tris-HCl (pH 7.5), and 0.05 mg/mL poly(dI-dC)·poly(dI-dC).
2. Add the probe (3.5 fmol, approx 2×10^4 cpm) to the above reaction, mix, and incubate at 25°C for 20 min.
3. Add 1 µL of 10X gel loading buffer, containing 250 mM Tris-HCl (pH 7.5), 0.2% bromophenol blue, 0.2% xylene cyanol, and 40% glycerol, to the reaction.
4. Load onto a 6% native gel (which should be prerun for 90 min at 100 V) in 0.5X nondenaturing Tris-Borate-EDTA buffer. Perform the electrophoresis at 25°C and 100 V for about 3.5 h.
5. Transfer the gel onto Whatman paper, vacuum dry, and expose to Hyperfilm (Amersham Pharmacia) for the desired period at −80°C and with an intensifier screen.

3.7.2. Western Blotting

Whenever a specific antibody is available, use Western blotting to detect the presence of the captured protein in the eluates. Only one-tenth (or less) of the eluted material should be used for Western blotting anlysis.

1. Protein solutions are separated first by electrophoresis in 4 to 15% gradient polyacrylamide sodium dodecylsulfate gels, and then transferred onto a PVDF membrane in Tris-glycine-methanol buffer.
2. After the membranes are blocked for at least 1 h at room temperature with PBS, 0.05% Tween-20, 5% dry milk, the respective antibodies are added to the membranes at concentrations of 1 µg/mL in PBS, 0.05% Tween-20, for 2 h at room temperature.
3. Anti-mouse or anti-rabbit-horseradish peroxidase-conjugated antibodies (Santa Cruz) and the Enhanced chemiluminescence kit (Pierce) are used for visualization of the immune complexes.

3.7.3. Identification of the Captured Proteins by Mass Spectrometry

1. Excise the bands from the stained/destained SDS gel, corresponding to the proteins eluted from the DNA-magnetic beads, as well as the proteins that remain on the beads after elution (*see* **Note 16**).
2. Digest these gel slices with trypsin.
3. Fractionate the mixtures on a Poros 50 R2 RP microtip.
4. Analyze the resulting peptide pools by matrix-assisted laser-desorption/ionization reflectron time-of-flight (MALDI-reTOF) MS using a BRUKER UltraFlex TOF/TOF instrument (Bruker Daltonics).
5. Take selected experimental masses (m/z) to search a nonredundant protein database (NR; National Center for Biotechnology Information, Bethesda, MD), utilizing the PeptideSearch (Matthias Mann, Southern Denmark University, Odense, Denmark) algorithm.
6. A molecular weight range twice the predicted weight should be covered, with a mass accuracy restriction better than 40 ppm, and should allow a maximum of one missed cleavage site per peptide.
7. Perform mass spectrometric sequencing of selected peptides by MALDI-TOF/TOF (MS/MS) analysis on the same prepared samples, using the UltraFlex instrument in "LIFT" mode.
8. Take fragment ion spectra to search the NR database using the MASCOT MS/MS Ion Search program (Matrix Science, London, UK) *(11)*. Any identification thus obtained should be verified by comparing the computer-generated fragment ion series of the predicted tryptic peptide with the experimental MS/MS data.

In general, positive identifications are made on the basis of a Mascot MS/MS score ≥ 64 ($p < 0.05$) for a single peptide, or a score of ≥ 35 for each of two or more peptides (combined score ≥ 80). Alternatively, a Mascot MS/MS score ≥ 35 for a single peptide should be combined with a peptide mass fingerprinting (PMF) result that yielded $\geq 15\%$ sequence coverage.

4. Notes

1. Incubation of NB4 cells in the hypotonic buffer causes the cells to swell. The homogenization disrupts cell membranes mechanically while the nuclei remain intact. This process can be monitored in a phase-contrast microscope. After homogenization, no large cell should be seen, only small dense dots representing the nuclei.
2. Slow addition of high salt buffer to the nuclear suspension prevents dissociation of the chromatin through local excess of salt leading to lysis of the nuclei and increased viscosity. Blood cells are particularly sensitive to exposure to high salt, and frequently the extract becomes viscous. This problem can be solved by short sonication (2 min at 50% duty cycle and output control 3 on a Branson Sonifier 450) without any effect on the protein quality.
3. During this dialysis step, a white protein precipitate is formed, which is removed after the dialysis by centrifugation for 20 min at 15,000*g*. In our experience, this

precipitation is not specific. An overall protein loss occurs owing to the drop in salt from 0.42 to 0.15 *M*.

4. One of the major reasons for fractionating nuclear proteins prior to DNA-affinity capture is to reduce the binding of nonspecific proteins. We conducted experiments on direct binding of crude nuclear extract to specific oligo DNA-beads (AP.1, NFκB, Sp.1) under the condition established in EMSA *(3)*. The profiles of the proteins eluted from these DNA-beads looked nearly identical, regardless of the difference in DNA sequences attached, suggestive of a high degree of nonspecific binding. Several major bands from the gel containing such proteins were excised and identified by MS as mostly nonspecific RNA or DNA binding proteins, some with a preference for binding to free DNA ends (PARP, DNA-dependent protein kinase, Ku protein, splicing factors, DNA helicase, and so on) Thus, under conditions in which specific binding is observed in EMSA, most of the DNA binding sites on the beads are occupied nonspecifically.

5. Ion exchangers (DEAE-cellulose, BioRex70) or heparin-sepharose, phenyl-sepharose, and gel-filtration columns (Sephacryl S300) have been used as a prefractionation step for affinity purification of nucleic acid binding proteins *(7,8)*. However, we chose to fractionate nuclear extracts on P11 phosphocellulose as it provides a unique combination of ion exchange and affinity properties. P11 chromatography is a simple, popular procedure that has been used as a first step in many proven transcription factor purification schemes; it allows enrichment of multiple factors, cofactors, and corepressors in separate fractions *(2,9,10)*. In this way, parallel affinity captures of several different factors, each from a separate P11-fraction but all derived from a single batch of nuclear extract, can be performed.

6. For best results, the instructions for preparation of the P11 resin from Whatman that come with the product must be followed strictly. The P11 column chromatography can be run with FPLC equipment (Amersham Pharmacia Biotech); the flow rate must not exceed 0.4 mL/min to avoid pressure buildup.

7. It should be noted that PCR-based concatamerization reactions are sequence specific and must be carefully optimized first in pilot experiments. For example, for GABP-binding concatamers, the PCR conditions were: 95°C for 2 min followed by 14 cycles at 95°C for 1 min, 55°C for 1 min, and 72°C for 3 min; for PURα: 95°C for 2 min followed by 9 cycles at 95°C for 1 min, 55°C for 1 min, and 72°C for 3 min; for AP.1: 92°C for 2 min followed by 7 cycles at 95°C for 1 min, 55°C for 1 min, and 72°C for 3 min (**Fig. 1**); for PU.1: 92°C for 2 min followed by 14 cycles at 95°C for 1 min, 55°C for 1 min, and 72°C for 3 min; for RARα: 95°C for 2 min followed by 9 cycles of 95°C for 1 min, 55°C for 1 min, and 72°C for 5 min.

8. The efficiency of concatamer binding to the M280 beads depends on the length of the DNA. According to the manufacturer's specifcations, binding of 1000 bp DNA is on the order of 2 to 5 µg/mg beads; lower and higher molecular weight $(DNA)_n$ bind at up to approx 12 µg/mg.

9. Effect of salt: we observed a striking correlation between the salt concentration at which a given protein eluted from the P11 column and the stability of the DNA-protein complex in the presence of salt as measured by EMSA. Proteins eluting at

≤300 mM NaCl formed less stable complexes with DNA than proteins that eluted at ≥500 mM. For example, formation of the GABP-DNA complex was adversely affected by the presence of 500 mM compared with 50 mM KCl, whereas high salt had no such destabilizing effect on the PU.1-DNA *(3)*. In some cases (e.g., AP.1 and PURα), the presence of high salt caused a slight change in the mobility of the DNA-protein complexes, but the DNA binding specificity was preserved, as confirmed in subsequent capturing experiments.

10. The effect of Mg^{2+} on DNA binding must also be optimized *a priori* in an EMSA. Low Mg^{2+} concentrations would reduce the probability for degradation of DNA concatamers, attached to the beads, by endogenous nucleases. We noticed that five chosen proteins bound equally well to their cognate DNA in the presence or absence of Mg^{2+} *(3)*. Consequently, all further DNA captures could be performed in the absence of magnesium.

11. As nonspecific protein binding appears to be the major obstacle for successful transcription factor affinity capture, the problem could be reduced to the adequate removal of such interfering proteins. Conceptually, a suitable medium could be an immobilized DNA sequence that binds many or all of the nonspecific proteins but not the specific one. We envisioned that this could be readily accomplished by using DNA ligands consisting of minimally modified binding sites, for instance, by one or more point mutations, just enough to abolish specific factor binding in vitro. The mutated DNA sequences should be derived from pilot competition EMSA experiments performed in the presence of an excess of unlabeled, mutant DNA probes. Point mutations that fully abolish competition will be then selected. Preclearing of protein solutions with beads loaded with such DNA is referred to as *negative selection*.

12. Based on our previously reported results *(3)*, we propose the following general, empiric rules for future DNA affinity capture and identification of transcription factors by MS, as illustrated in **Fig. 2**.

    a. After initial fractionation of nuclear extract on a P11 column, it is critical to determine whether the protein of interest will form a stable complex with its target DNA in the presence of at least 500 mM salt. This is generally the case for all proteins that elute from P11 at or higher than 500 mM salt (e.g., AP.1, PURα, PU.1, and others).

    b. Binding affinities may also be known from prior molecular characterization of the promoter of interest.

    c. Affinity and salt tolerance, often directly linked, are the primary determinants of the order in which positive and negative selections are then carried out: "negative/positive" for high-affinity, high-salt binders; "positive/negative/positive" for low-affinity, low-salt binding transcription factors.

    d. The addition of a positive step prior to the negative selections serves a dual purpose: (1) the protein mass is drastically reduced, and (2) protein eluates are very concentrated, allowing a further microcapture format. This is not possible for high-affinity binders, as approximately 1 M salt is required for elution, in most cases resulting in some degree of denaturation, which complicates both assaying and subsequent affinity captures. Consequently, the first positive selection must be omitted. This can be compensated, in part, by the option to include 0.5 *M* salt throughout the entire selection process, thereby also increasing capturing stringency, albeit by a different mechanism. Both options appear to be mutually exclusive.

    e. Regardless of the selection sequence, the number of negative selections on mutant DNA-magnetic beads always depends on the abundance of the transcription factors in the respective fraction. Low-abundant species require more nuclear extract and proportionally more rounds of negative selection than the higher abundant ones, for example, one round for GABP and AP.1, two for PURα, three for PU.1, but more than three for RARα (**Fig. 2**). All capturing schemes end with a positive selection, including extensive washing of the beads with *E. coli* competitor ds- and ssDNA.

    f. Final elution is done with 0.5 *M* NaCl for low-affinity binders, or by boiling in Laemmli gel-loading buffer for the others, followed by gel electrophoresis and identification by the mass spectrometric method of choice.

    g. The purification scheme generates proteins in amounts and form that are readily compatible with MALDI TOF-based peptide mass fingerprinting.

13. Poly(dI:dC) competitor at a concentration of 0.1 mg/mL, already high, is insufficient to prevent the unwanted nonspecific DNA-protein interactions in binding to DNA-magnetic beads. We established experimentally that nonspecific protein binding is significantly reduced in the presence of double-stranded oligodI:dC and does not visibly increase when concatamerized DNA instead of oligonucleotide is used as affinity ligand on the beads *(3)*. From the premise that several major, non-specifically binding proteins (e.g., DNA-PK, Ku autoantigen, PARP) have high affinities for DNA breaks or ends, we reasoned that the resultant background could be reduced by using (1) beads with DNA affinity ligands of a lower ends-to-binding site ratio and, conversely, (2) competitor DNA of a higher ends-to-weight ratio. The latter option was implemented by including a 30-bp long, synthetic double-stranded oligo(dI:dC) in the binding reaction, in addition to the standard polymeric competitor *(6)*. An 0.1 mg/mL oligo(dI:dC) concentration was empirically found to be the highest that did not visibly interfere with protein-DNA binding in EMSA. As for the first option, long (≥1000 bp) 5'-biotinylated double-stranded concatamers of specific oligonucleotides that are produced in a self-priming PCR effectively increase ligand density on the beads without added DNA ends.

14. We prepare single-stranded *E. coli* DNA by heating double-stranded genomic DNA in boiling water for 20 min and quick chilling on ice.

15. Silver staining of the sodium dodecylsulfate gels prior to mass spectrometric analysis is not advisable as silver binds to traces of competitor DNA left in the final preparations.

16. Clean work while handling the samples and processing the gel slices (wearing gloves, using clean glassware and tools) is imperative for successful identification since the mass spectrometry can detect the presence of protein contaminants, particularly keratins.

References

1. Ausubel, F. M., Brent, R., Kingston, R. E., et al. (eds.). Large-scale preparation of bacterial genomic DNA, in *Current Protocols in Molecular Biology.* Wiley Interscience, New York, pp. 2.4.3.
2. Dignam, J. D., Martin, P. L., Shastry, B. S., and Roeder, R. G. (1983) Eukaryotic gene transcription with purified components. *Methods Enzymol.* **101,** 582–598.
3. Yaneva, M. and Tempst, P. (2003) Affinity capture of specific DNA-binding proteins for mass spectrometric identification. *Anal. Chem.* **75,** 6437–6448.
4. Ma, Y., Qin, S., and Tempst, P. (1998) Differentiation-stimulated activity binds an ETS-like, essential regulatory element in the human promyelocytic *defensin-1* promoter. *J. Biol. Chem.* **273,** 8727–8740.
5. Hemat, F. and McEntee, K. (1994) A rapid and efficient PCR-based method for synthesizing high-molecular-weight multimers of oligonucleotides. *Biochem. Biophys. Res. Commun.* **205,** 475–481.
6. West, R., Yaneva, M., and Lieber, M. (1998) Productive and nonproductive complexes of Ku and DNA-dependent protein kinase at DNA termini. *Mol. Cell. Biol.* **18,** 5908–5920.
7. Kadonaga, J. T. (1991) Purification of sequence-specific binding proteins by DNA affinity chromatography. *Methods Enzymol.* **208,** 10–23.
8. Gadgil, H., Jurado, L. A., and Jarrett, H. W. (2001) DNA affinity chromatography of transcription factors. *Anal. Biochem.* **290,** 147–178.
9. Ge, H. and Roeder, R. G. (1994) Purification, cloning, and characterization of a human coactivator, PC4, that mediates transcriptional activation of class II genes. *Cell* **78,** 513–523.
10. Roeder, R. G. (1996) The role of general initiation factors in transcription by RNA polymerase II. *Trends Biochem. Sci.* **21,** 327–335.
11. Perkins, D. N., Pappin, D. J. C., Creasy, D. M., and Cottrell, J. S. (1999) Probability-based protein identification by searching sequence databases using mass spectrometry data. *Electrophoresis* **20,** 3551–3567.

22

Isolation of Transcription Factor Complexes by In Vivo Biotinylation Tagging and Direct Binding to Streptavidin Beads

Patrick Rodriguez, Harald Braun, Katarzyna E. Kolodziej, Ernie de Boer, Jennifer Campbell, Edgar Bonte, Frank Grosveld, Sjaak Philipsen, and John Strouboulis

Summary

Efficient tagging methodologies are an integral aspect of protein complex characterization by proteomic approaches. Owing to the very high affinity of biotin for avidin and streptavidin, biotinylation tagging offers an attractive approach for the efficient purification of protein complexes. The very high affinity of the biotin/(strept)avidin system also offers the potential for the single-step capture of lower abundance protein complexes, such as transcription factor complexes. The identification of short peptide tags that are efficiently biotinylated by the bacterial BirA biotin ligase led to an approach for the single-step purification of transcription factor complexes by specific in vivo biotinylation tagging. A short sequence tag fused N-terminally to the transcription factor of interest is very efficiently biotinylated by BirA coexpressed in the same cells, as was demonstrated by the tagging of the essential hematopoietic transcription factor GATA-1. The direct binding to streptavidin of biotinylated GATA-1 in nuclear extracts resulted in the single-step capture of the tagged factor and associated proteins, which were eluted and identified by mass spectrometry. This led to the characterization of several distinct GATA-1 complexes with other transcription factors and chromatin remodeling cofactors, which are involved in activation and repression of gene targets. Thus, BirA-mediated tagging is an efficient approach for the direct capture and characterization of transcription factor complexes.

Key Words: Biotinylation tagging; BirA; transcription factors; chromatin; mass spectrometry; size fractionation; GATA-1.

From: *Methods in Molecular Biology, vol. 338: Gene Mapping, Discovery, and Expression: Methods and Protocols*
Edited by: M. Bina © Humana Press Inc., Totowa, NJ

1. Introduction

Completion of the sequencing of an ever increasing number of genomes has led to a shift of focus toward the characterization of the protein complement of cells, i.e., the proteome. A key aspect of proteomic analysis is the development of simple methodologies for the efficient isolation of protein complexes for peptide analysis and identification by powerful mass spectrometric approaches. This is particularly challenging for the analysis of nuclear proteins involved in transcriptional regulation such as transcription factors and their chromatin-associated cofactors because of their relatively lower abundance, the different parallel functions that they execute (e.g., activation and repression involving different partners), and the often transient nature of their interactions. Transcription factor purification approaches involving several prepurification steps are laborious and costly and most likely result in the isolation of only the most abundant of the protein complexes formed by the factor. We describe here the application of in vivo biotinylation tagging as a simple approach for the efficient direct purification of transcription factor complexes from crude nuclear extracts *(1)*.

Biotin is a naturally occurring cofactor essential for certain metabolic enzymes such as carboxylases. Specific protein-biotin ligases are responsible for covalently attaching biotin to these enzymes. The key to using biotinylation lies in the fact that biotinylated substrates can be bound very tightly by the naturally occurring proteins avidin and streptavidin ($K_d = 10^{-15}$), a fact that has been widely exploited in many affinity-based biochemical applications. In addition, in vivo biotinylation tagging offers a number of advantages for protein purification purposes. First, there are few naturally biotinylated proteins (mostly cytoplasmic and mitochondrial), ensuring that nonspecific background binding remains low. Second, the very high affinity of (strept)avidin for biotin allows high stringencies to be employed during purification without fear of losing binding of the tagged protein.

The biotinylation tagging approach described here is based on previous work on the screening of a combinatorial synthetic peptide library for efficient biotinylation by the bacterial BirA biotin ligase *(2)*. This led to the identification of a number of short sequence tags that can be very efficiently biotinylated in vitro *(2,3)*. Such tags were subsequently utilized for the efficient in vivo biotinylation of tagged proteins in bacterial cells through the coexpression of BirA *(4,5)*. We have applied this approach in mammalian cells and demonstrated its efficiency in specifically biotinylating nuclear proteins in cultured cells and transgenic mice through the coexpression of the BirA biotin ligase together with the tagged protein (**Fig. 1A**) *(1)*. Our work is focused primarily on the biotinylation tagging of hematopoietic transcription factors in erythroid cells. Most of our work to date has been carried out with GATA-1, a critical transcription factor for

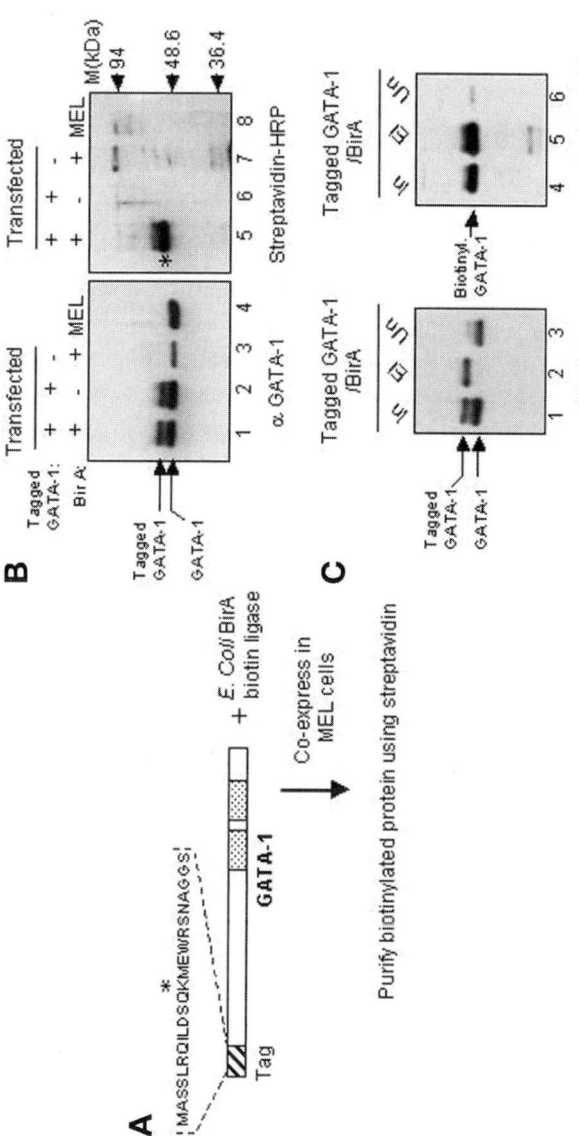

Fig. 1. (A) Scheme for the specific biotinylation of tagged GATA-1 by BirA biotin ligase in mouse erythroleukemic (MEL) cells. The sequence of the 23-amino-acid peptide tag fused to the N-terminus of GATA-1 is shown. The asterisk indicates the lysine residue that becomes specifically biotinylated by BirA. Speckled boxes indicate the positions of the two GATA-1 zinc fingers. (B) Biotinylation of tagged GATA-1 in MEL cells. Left: Western blot with an anti-GATA-1 antibody to detect endogenous and tagged GATA-1 proteins. Right: Western blot of the same extracts with streptavidin-HRP conjugate to detect biotinylated GATA-1. Biotinylated GATA-1 (asterisk) is clearly visible in the right panel only in the lane of the double transfected cells. (C) Efficiency of GATA-1 biotinylation and binding to streptavidin beads. Left: Western blot using anti-GATA-1 antibody to detect binding of tagged GATA-1 to streptavidin beads Input and unbound material are shown in lanes 1 and 3. Right: the same filter stripped and reprobed with streptavidin-HRP to detect the binding of biotinylated GATA-1 to streptavidin beads (lane 5). Lane 6 shows that very little tagged GATA-1 remains unbound by streptavidin. In, input (nuclear extract); El, eluted material; Un, unbound material. (Reproduced with permission from **ref. 1**. Copyright [2003] National Academy of Sciences, USA.)

erythroid cell differentiation. We have been able to very efficiently biotinylate GATA-1 in cultured mouse erythroleukemic cells (**Fig. 1B** and **C**) *(1)*, leading to the isolation and characterization of GATA-1 protein complexes by direct binding of nuclear extracts to streptavidin beads. Using this approach we identified a number of GATA-1 complexes containing other essential hematopoietic transcription factors (FOG-1, Gfi-1b, and TAL-1) and chromatin remodeling and modification complexes (MeCP1 and ACF/WCRF). These complexes were implicated in the transcriptional activation and repression of different subsets of target genes *(6)*. Thus, biotinylation tagging has proved to be a very efficient method for the single-step purification and characterization of transcription factor complexes. It should also be noted that we have no evidence so far that biotinylation tagging adversely affects the physiological properties of the tagged protein *(1)*.

In this chapter we describe protocols for the binding of nuclear extracts expressing a specific biotin-tagged protein to streptavidin beads and the preparation of the eluted material for analysis by mass spectrometry. We do not provide protocols for the stable transfection of cultured cells, as these will vary depending on the cell line/type used in each case. We routinely prepare large-scale nuclear extracts from a few liters of cultured cells, we test for the presence of the biotin-tagged protein in high molecular weight complex(es) by gel filtration using a Superose 6 column, and we then carry out the binding of the tagged factor to streptavidin paramagnetic beads. We normally check the efficiency of the biotin tagging and the binding to streptavidin beads by testing the nuclear extract (input), the bound material (eluate), and the flowthrough (unbound) by Western blotting using first an antibody against the tagged protein followed by streptavidin-horseradish peroxidase (HRP) conjugate on the same blot. In this way the fraction of the tagged protein that becomes biotinylated in vivo and subsequently captured by the streptavidin beads can be determined. As shown in **Fig. 1**, for GATA-1 the biotinylation efficiency and capture by the beads is nearly 100%. The proteins eluted from the beads are fractionated by sodium dodecyl sulfate-polyacrylamide gel electrophoresis (SDS-PAGE) and the gel is stained and photographed (**Fig. 4**). The gel lane with the fractionated proteins is excised and cut into small pieces (or gel plugs) along its entire length. The gel plugs are then processed for protein identification by mass spectrometry. The following sections describe in detail all these techniques. We also provide an overview of the background binding in experiments using nuclear extracts and the specific enrichment observed when purifying biotin-tagged transcription factors (**Fig. 4**). Lastly, we provide protocols for the size fractionation of nuclear extracts using a preparative Superose 6 column and for the cleavage of proteins bound to streptavidin beads using TEV protease (**Fig. 3**). Both approaches are presented with the aim of reducing the background in protein purification by biotinylation tagging. Given the increase in the potential applications of biotinylation tagging for protein purification, the description of such protocols may prove a useful resource.

2. Materials

2.1. Cell Culture

1. Dulbecco's modified eagle's medium (DMEM, Cambrex BioScience, Belgium), supplemented with 10% fetal bovine serum (FBS, Hyclone, Belgium).
2. Penicillin (used at 100 U/mL final concentration) and streptomycin (used at 100 µg/ mL final concentration), stored at −20°C (100X stock from Cambrex BioScience, Belgium).
3. 100% Dimethylsulfoxide (DMSO, Merck, Germany), used at 2% final concentration.
4. Neomycin, stock prepared as 100 mg/mL in phosphate-buffered saline (PBS; *see* **Subheading 2.2., item 1**), filter-sterilized, aliquoted, stored at −20°C, and used at 400 µg/mL final concentration.
5. Puromycin, 1000X stock aliquoted and stored as above and added to 1 µg/mL final concentration.

2.2. Nuclear Extract Preparation

1. Dulbecco's PBS (Cambrex Bio Science, Belgium).
2. Protease inhibitors: Complete (Roche, Germany). Use 1 tablet for 50 mL of solution.
3. Cell resuspension buffer: 2.2 M sucrose in 10 mM HEPES-KOH, pH 7.9, 25 mM KCl, 0.15 mM spermine, 0.5 mM spermidine, 1 mM EDTA (with protease inhibitors added as above).
4. Standard household blender with rotating blades for homogenizing cells.
5. Nuclear lysis buffer: 10 mM HEPES, pH 7.9, 100 mM KCl, 3 mM MgCl$_2$, 0.1 mM EDTA, 20% glycerol (with protease inhibitors added as above).
6. Coomassie Plus Assay reagent (Pierce, Rockford, IL).
7. Protein standards: dilutions of 0, 100, 200, 300, 400, and 500 µg/mL of chicken egg albumin (Sigma-Aldrich, St. Louis, MO) in ddH$_2$O prepared from a 20 mg/mL stock solution; 10 µL of each standard diluted in 1 mL (final volume) of ddH$_2$O is used for obtaining a standard curve.
8. Spectrophotometer: Ultraspec II (LKB Biochrom), cuvets (Sterna, Germany).
9. Conductivity meter: Philips PW 9526.
10. Standards for determining salt concentration: 100, 200 300, and 400 mM KCl diluted in ddH$_2$O from a 1 M KCl stock solution; 10 µL of each standard are diluted in 1 mL (final volume) of ddH$_2$O for obtaining a standard curve.

2.3. Size Fractionation by Superose 6 Gel Filtration

1. Superose 6 analytical grade column: HR 10/30 with a total bed volume of 24 mL. Preparative grade column: XK 50/600 with a bed height of 30.3 cm, total bed volume of 589 mL (both from Amersham Biosciences, UK). Range of separation of 5000 to 5,000,000 Daltons. Ethanol (20%) is used as preservative. The column is connected to an AKTA FLPC system (Amersham Biosciences).

2. Gel filtration calibration kit: dextran blue and high molecular weight standards (Amersham Biosciences, UK).
3. Column running buffer: 20 mM HEPES, 0.5 mM EGTA, 1 mM MgCl$_2$, 200 mM KCl, 10% glycerol. All buffers used for gel filtration column should be filtered prior to use.
4. 100% Trichloroacetic acid (TCA; Sigma-Aldrich).

2.4. SDS-Polyacrylamide Gel Electrophoresis

1. Pre-cast NuPAGE 4 to 12% Bis-Tris gel (Invitrogen, UK).
2. Gel electrophoresis buffer: 1X MOPS buffer diluted from 20X stock solution and NuPAGE antioxidant (both from Invitrogen).
3. Sample loading buffer: final concentration 62.5 mM Tris-HCl, pH 6.8, 25% glycerol (v/v from 100% stock), 2% SDS (v/v from 20% stock in ddH$_2$O), 0.01% bromophenol blue (w/v), 5 % β-mercaptoethanol (v/v, 100% stock). Can be prepared as a 4X stock solution.
4. Broad-range prestained SDS-PAGE molecular weight standards (Bio-Rad, Hercules, CA).
5. SimplyBlue Safestain gel staining solution (Invitrogen).

2.5. Western Blotting

1. ProTran nitrocellulose membrane (Schleicher & Schuell, Germany).
2. Gel-blotting paper (Schleicher & Schuell).
3. Blotting buffer: 25 mM Tris-HCl (made directly from the solid), 192 mM glycine (made directly from the solid), 20% methanol.
4. Tris-buffered saline (TBS): 10 mM Tris-HCl, pH 7.4, 150 mM NaCl. Can be prepared as a 10X stock and stored at room temperature.
5. Blocking buffer: 5% bovine serum albumin (BSA; Roche, Germany) in 1X TBS, prepared fresh.
6. Washing buffer: 1X TBS adjusted to 0.5 M NaCl (using a 5 M NaCl stock solution), 0.3% Triton X-100 (Sigma-Aldrich).
7. Primary and secondary antibody dilution: in blocking buffer with 0.2% NP-40.
8. Secondary antibodies: antirabbit (1:50,000 dilution), antimouse (1:15000 dilution) from Amersham Biosciences, antirat (1:3000 dilution) and antigoat (1:4000 dilution) from DakoCytomation (Denmark). All antibodies are purchased as horseradish peroxidase (HRP) conjugates.
9. Enhanced Chemiluminescence (ECL) kit (Amersham Biosciences).
10. Bio-Max MR film (Kodak, Rochester, NY).

2.6. Streptavidin Binding

1. Streptavidin paramagnetic beads (Dynabeads M280, Dynal, Sweden).
2. Chicken egg albumin (Sigma-Aldrich).
3. Binding buffer: 1X TBS, 0.3% NP-40 (Nonidet-40; Sigma-Aldrich).

4. HENG buffer: 10 mM HEPES-KOH, pH 9, 1.5 mM MgCl$_2$, 0.25 mM EDTA, 20% glycerol, 1 mM phenylmethylsulfonyl fluoride (PMSF; prepared as a 100X stock in ethanol and stored at –20°C).
5. Wash buffer: HENG buffer with 250 mM KCl and 0.3% NP-40.
6. Elution buffer: 1X sample loading buffer.
7. Magnets: Dynal MPC-1 and MPC-S magnets, for large and small volumes, respectively (Dynal, Sweden).

2.7. TEV Protease Cleavage

1. Nuclear extract dilution buffer: 20 mM Tris-HCl, pH 7.5, 0.45% NP-40.
2. Tobacco Etch Virus (TEV) protease (AcTEV Protease, Invitrogen, Scotland).

2.8. Sample Preparation for Mass Spectrometry

1. Trypsin, sequencing grade (Roche); 10X stock made by dissolving lyophilized powder in 1 mM HCl to 100 ng/µL final concentration.
2. 50 mM Ammonium bicarbonate (Sigma-Aldrich). Dissolved in ddH$_2$O and filter sterilized.
3. 100% Acetonitrile (Sigma-Aldrich).
4. 100% Formic acid (Sigma-Aldrich).
5. Gel slice destaining solution: 25 mM ammonium bicarbonate in 50% acetonitrile prepared by mixing equal volumes of the stock solutions given above.
6. 50 mM Iodoacetamide (Sigma-Aldrich) prepared in 50 mM ammonium bicarbonate.
7. 6.5 mM Dithiothreitol (DTT; Sigma-Aldrich) prepared in 50 mM ammonium bicarbonate.

3. Methods

3.1. Cell Culture

1. Mouse erythroleukemia (MEL) cells are Friend virus transformed erythroid progenitors arrested at the proerythroblast stage of differentiation (*7*). MEL cells are cultured in DMEM supplemented with 10% FBS at 37°C and 5% CO$_2$. Cells are grown to a maximum of 2×10^6 cells/mL and routinely split to a density of 5×10^4 cells/mL. Cells appear to be semiadherent and rounded, with a smooth surface (*see* **Note 1**).
2. For induction, cells are diluted to 5×10^5 cells/mL and cultured in DMEM with 2% DMSO (v/v) for at least 3 d. Cells become smaller but remain round and when pelleted appear pink/red owing to hemoglobinization.

3.2. Nuclear Extract Preparation

1. Cells are harvested in 1 L centrifuge bottles by centrifugation at 640g for 40 min at 4°C in a Beckman J4 centrifuge and washed once with 100 mL of ice-cold PBS. Resuspended cells are transferred into 50-mL Falcon tubes and repelleted in an Eppendorf 5810R benchtop centrifuge at 2540g for 10 min at 4°C. The supernatant is discarded.

2. The cell pellet is gently resuspended by pipeting up and down in 200 mL of cell resuspension buffer with protease inhibitors added. Cells are equilibrated to the new osmotic conditions for 20 min on ice.

3. Cells are lysed in a blender using a single 30-s pulse at setting 3 (*see* **Note 2**). Excessive foaming should be avoided.

4. Lysis efficiency is checked under the microscope by staining a 10-μL aliquot with an equal volume of Unna stain (Methylgreen-Pyronin). Nuclei appear blue, whereas intact cells appear with a blue nucleus surrounded by a nonstained cytoplasm. Optimal lysis should result in more than 90% of the nuclei appearing free of cytoplasm.

5. Nuclei are pelleted by ultracentrifugation at 141,000g using the SW28 rotor for 2 h at 4°C. A clean white pellet corresponding to the nuclei should be visible at the bottom of the tube. The top layer (cellular debris/cytoplasm) is discarded.

6. Nuclei are resuspended in 15 mL of nuclear lysis buffer, and proteins are extracted by the dropwise addition of a 3.3 M KCl solution with gentle agitation on ice, until the final concentration is approx 350 to 400 mM (*see* **Note 3**). Nuclear lysis and protein extraction are allowed to proceed by incubating on ice for 20 min. Two phases should be visible: one is clear and represents the soluble nuclear extract fraction, whereas the other phase appears viscous and represents the insoluble fraction of mostly chromatin fragments.

7. Insoluble material is removed by ultracentrifugation at 300,000g using the SW50.1 rotor for 1 h at 4°C. The supernatant (soluble nuclear extract, approx 17–18 mL) is aliquoted in 1 to 5-mL aliquots.

8. A small aliquot is used to measure protein concentration using the Bradford method. First, a standard curve is obtained by adding 200 μL of Bradford reagent to 800 μL of each of the chicken egg albumin standards, incubating for 5 min at room temperature, and measuring the absorbance at 595 nm in a spectrophotometer. The absorbance of dilutions of the aliquot of the nuclear extract are measured in the same way. The concentration of the nuclear extract is determined using the standard curve, taking the dilution of the sample into account.

9. The salt concentration of the solution is determined by measuring the conductance of the sample with a conductivity meter. First, a calibration curve of conductance is obtained using the KCl dilutions as standards. The conductivity of a diluted aliquot of the nuclear extract is then measured and its approximate salt concentration is estimated using the calibration curve.

10. Nuclear extracts are snap frozen in liquid nitrogen and stored at −70°C (*see* **Note 4**).

3.3. Analytical Superose 6 Gel Filtration (see Note 5)

1. An aliquot of nuclear extract is thawed on ice and centrifuged for 5 min at full speed using a microcentrifuge at 4°C. The volume of extract loaded on the column should not exceed 1% of the column volume.

2. The pump and the column are equilibrated with column running buffer. The running program is set up as follows: 100 μL/min flow rate; sample volume loop 200 μL; fraction volume 500 μL; elution length 1 column volume of running buffer;

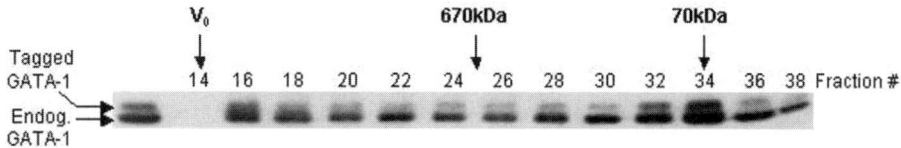

Fig. 2. Superose 6 fractionation profiles of nuclear extracts from MEL cells express-ing biotin-tagged GATA-1 detected by Western blotting. The tagged GATA-1 protein migrates with a slower mobility than that of the endogenous GATA-1 owing to the extra tag sequences fused to it. The elution of the molecular weight markers is indicated at the top. V_0, void volume.

alarm pressure set at 0.5 MPa (*see* **Notes 6** and **7**). After each run the column is washed with 2 column volumes of running buffer.

3. The calibration standards are run through the column to establish the elution vol-ume of protein complexes and of free protein monomers (*see* **Notes 8–10**).

4. Collected fractions are concentrated by trichloroacetic acid precipitation (TCA), as follows: 125 µL of cold 100% TCA are added to each 500-µL fraction, mixed well, and incubated on ice for 30 min.

5. Proteins are pelleted by centrifugation at full speed in a microcentrifuge at 4°C for 20 min.

6. The supernatants are discarded and the pellets washed with at least 500 µL 1% ice-cold TCA (in ddH$_2$O) and recentrifuged as above.

7. The supernatants are discarded again, and the pellets are washed with ice-cold acetone and recentrifuged as above.

8. The pellets are air-dried (on ice) and resuspended in 50 µL of sample loading buffer (*see* **Note 11**). The fractionation patterns of specific proteins are determined by SDS-PAGE and Western blotting, as described below in **Subheadings 3.8.** and **3.9.** An example of a GATA-1 Superose 6 fractionation profile of nuclear extracts from MEL cells expressing biotin-tagged GATA-1 is shown in **Fig. 2**. From this, it is clear that both endogenous and tagged GATA-1 elute in high molecular weight (>670 kDa) fractions. The fractionation profile of tagged GATA-1 follows that of the endogenous GATA-1 protein.

3.4. Preparative Gel Filtration by Superose 6

1. **Steps 1** and **3** are as in **Subheading 3.3.**

2. Running program: 4 mL/min flow rate; sample volume loop 5 mL; fraction volume 10 mL; elution length 1.5 column volumes; alarm pressure set at 0.65 Mpa (*see* **Note 12**).

3. Fractions are collected in 15-mL Falcon tubes and used for precipitation (*see* **step 4**) or for binding to streptavidin beads (*see* **Subheading 3.6.**).

4. Proteins are precipitated with 20% TCA (2.5 mL of 100% TCA are added to every 10-mL fraction) and kept on ice for 1 h. The tubes are centrifuged in an Eppendorf 5810R benchtop centrifuge at maximum speed (2540g) for 20 min at 4°C. The

pellet is subsequently washed with ice-cold 1% TCA (in ddH$_2$O). At this step the pellet can be carefully resuspended and transferred into microfuge tubes. The samples are repelleted by spinning as above, or in a microfuge for 20 min, full speed at 4°C. The pellets are washed with ice-cold acetone, centrifuged again as above, and air-dried on ice. The protein pellets are finally resuspended in 50 µL of sample loading buffer and denatured by boiling before loading on an SDS-PAGE gel (*see* **Subheading 3.8.**).

3.5. Binding to Streptavidin Beads

1. Nuclear extract (5–10 mg) is thawed on ice and diluted to 150 mM KCl final concentration by the dropwise addition of ice-cold HENG with gentle shaking (*see* **Note 13**). NP-40 is adjusted to a 0.3% final concentration.
2. We routinely use 200 µL of resuspended streptavidin beads per 5 mg of nuclear extract. The beads are blocked with 200 ng/mL chicken egg albumin (CEA) in a 1 mL final volume (made up with HENG buffer), for 1 h at room temperature on a rotating platform.
3. The beads are immobilized using a magnetic rack, and the blocking solution is removed. The beads are then resuspended in the diluted nuclear extract and incubated on a rotating platform 4°C for 2 h to overnight.
4. The beads are immobilized on ice using the magnetic rack. The supernatant, corresponding to the unbound fraction or flowthrough is collected and saved.
5. The beads are washed in 1 mL of washing buffer as follows: 2 quick rinses followed by three washes, 10 min each, at room temperature on a rotating platform.
6. After the last wash, the beads are resuspended in 50 µL of sample buffer. Bound proteins are eluted by boiling the beads for 5 min.
7. The eluted material is fractionated by SDS-PAGE and processed for analysis by mass spectrometry as described below in **Subheading 3.10.**

3.6. Binding of Preparative Superose 6 Fractions to Streptavidin Beads

1. After the fractionation profile of the protein(s) of interest has been determined by SDS-PAGE and Western blotting (as below), the peak fractions are pooled in a suitable sterile container. (We conveniently use a sterilized glass measuring cylinder.) All work is carried out on ice or in the cold room.
2. KCl and NP-40 concentrations are adjusted to 150 mM and 0.3%, respectively (as above) with gentle mixing.
3. The diluted fractions are divided equally into separate 50-mL Falcon tubes, so that tubes are not more than three-fourths full, and resuspended streptavidin M280 beads (equilibrated and blocked as in **Subheading 3.5., step 2** above) are added followed by overnight incubation at 4°C on a rotating platform. We use approximately 10 µL of streptavidin M280 beads for every 10-mL fraction.
4. The beads are immobilized using the Dynal MPC-1 magnet, and the supernatant corresponding to the unbound fraction is removed and saved. The immobilized beads from each tube are resuspended in washing buffer and pooled by transferring to a microfuge tube.

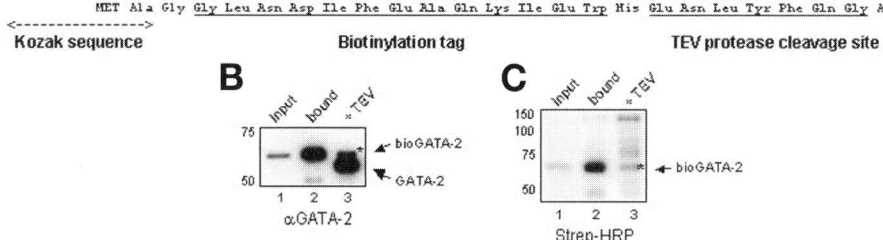

A

GGCCGCCATC ATG GCT GGT GGC CTG AAT GAC ATC TTT GAG GCC CAG AAG ATC GAG TGG CAT GAG AAC CTG TAC TTC CAG GGA GCC
          MET Ala Gly <u>Gly Leu Asn Asp Ile Phe Glu Ala Gln Lys Ile Glu Trp</u> His Glu Asn Leu Tyr Phe Gln Gly Ala
<----------------->
Kozak sequence   **Biotinylation tag**   **TEV protease cleavage site**

Fig. 3. TEV protease cleavage of bioGATA-2 bound to streptavidin beads. (**A**) Sequence and translation of the shorter biotinylation tag and the TEV protease cleavage site. (**B**) and (**C**) Lane 1, nuclear extract from MEL cells expressing tagged GATA-2 (there is no endogenous GATA-2 expressed in MEL cells). Lane 2, biotinylated GATA-2 (bioGATA-2) bound to the beads. Lane 3, TEV protease cleavage of bound GATA-2. (**A**) Detection with anti-GATA-2 antibody. (**B**) Detection of the blot shown in (A) with streptavidin-HRP. TEV cleavage of bioGATA-2 results in a downshift in size of the protein and loss of the biotin-tag (compare lanes 2 and 3 in A and B, respectively). Remaining uncleaved bioGATA-2 is indicated with an asterisk in (B).

5. The beads are washed four to five times at room temperature in 1 mL washing buffer for 5 to 10 min each.
6. The bound proteins are eluted by boiling the beads in sample buffer (50 μL of sample buffer per 20 μL of immobilized beads). The eluted material is fractionated by SDS-PAGE and processed for analysis by mass spectrometry as described below.

3.7. TEV Protease Cleavage

With the aim of reducing the nonspecific background observed in streptavidin binding experiments, we developed a modified version of the biotin tag for the N-terminal tagging of fusion proteins. This tag consists of a shorter (14) amino-acid sequence than the one previously used; it also very efficiently biotinylated by the BirA biotin ligase *(3)*. The biotin tag sequence is followed by a 7-amino-acid cleavage site for the highly specific tobacco etch virus (TEV) protease (**Fig. 3A**). In this way, the biotinylated protein and associated complexes can be specifically released from the streptavidin beads by cleaving off with the TEV protease (**Fig. 3B** and **C**).

1. Immediately after the washing step (**step 5, Subheading 3.5.**) the streptavidin paramagnetic beads are resuspended in 1X TBS/0.3% NP-40 buffer (95 μL of buffer for every 50 μL of resuspended beads used in the outset of the experiment).
2. TEV protease (5–10%, v/v) is added to the resuspended beads followed by incubation for 1 to 3 h at 16°C with shaking (*see* **Note 14**).

3. The supernatant can be collected (it should contain the cleaved protein) and, if necessary, can be concentrated by TCA precipitation, as in **Subheading 3.4., step 4** above.

4. The efficiency of protease cleavage can be monitored by testing an aliquot of the supernatant by SDS-PAGE and Western blotting (**Fig. 3B** and **C**). The material that remains bound to the beads after TEV protease cleavage can be eluted by boiling in sample buffer and tested by SDS-PAGE. Successful cleavage results in loss of the biotin tag (as visualized by streptavidin-HRP; **Fig. 3C**) accompanied by a downshift in the size of the tagged protein, thus resulting in faster migration by SDS-PAGE (as visualized by the tagged protein-specific antibody; **Fig. 3B**).

3.8. SDS-PAGE

We preferably use NuPAGE precast gels since they give clear and reproducible results in terms of resolution and sharpness of the protein bands. This is of particular importance if the gel is to be processed for mass spectrometry. There is also the added advantage of using gradient precast gels (e.g., 4–12%) for resolving proteins in a wide range of molecular weights. We use the Invitrogen electrophoresis system for running the NuPAGE gels.

1. Precast gels are removed from the plastic envelope and rinsed in ddH$_2$O. The sticker near the bottom of the gel and the comb are removed carefully.
2. 800 mL of 1X MOPS electrophoresis buffer are prepared.
3. The gel is placed in the electrophoresis system (Invitrogen). The outside chamber is filled with 600 mL of 1X MOPS.
4. 500 µL of antioxidant are added to the remaining 200 mL of the 1X MOPS buffer, mixed, and used to fill the inner chamber.
5. The wells of the gel are rinsed twice with the running buffer.
6. The samples in sample loading buffer are boiled for 5 min to denature the proteins and loaded directly onto the gel; 5–10 µg of nuclear extract are loaded per lane.
7. The gel is electrophoresed at 200 V constant voltage for 60 to 75 min.

3.9. Western Blotting

Samples that have been separated by SDS-PAGE are electrophoretically transferred onto nitrocellulose membrane by "wet blotting." For protein transfer we use the Trans-Blot electrophoretic transfer cell (Bio-Rad).

1. Four gel-blotting papers are cut to the size of the gel. A "sandwich" is set up consisting of a sponge, two pieces of blotting paper, the gel, the membrane, another two pieces of blotting paper, and a sponge (*see* **Note 15**). The membrane is prewetted in water followed by transfer buffer.
2. The "sandwich" is placed in the transfer tank containing transfer buffer prechilled at 4°C, such that the membrane is between the gel and the anode.
3. Blotting is carried out under constant amperage at 390 mA for 70 min in the cold room (*see* **Note 16**).

4. At the end of the transfer, the "sandwich" is disassembled, and the membrane is rinsed in 1X TBS/0.05% NP-40 (*see* **Note 17**).

5. The membrane is blocked at room temperature in freshly prepared blocking buffer for 1 h on a rocking platform.

6. The blocking buffer is discarded, replaced by the primary antibody (in this case anti-GATA-1 N6 antibody diluted 1:5000) in blocking buffer/0.2% NP-40, and incubated overnight at 0°C on a rotating wheel.

7. The primary antibody is removed and the membrane washed three times with 50 mL of washing buffer, 15 min each wash at room temperature (*see* **Note 18**).

8. A freshly prepared secondary antibody (in this case antirat diluted 1:3000 in blocking buffer) is added to the membrane and incubated for 1 h at room temperature on a rocking platform.

9. The secondary antibody is discarded and the membrane is washed as in **step 8**, followed by one wash for 5 min in 1X TBS/0.05% NP-40.

10. The membrane is lifted out of the wash buffer, excess liquid is removed by touching it to a clean tissue, and the membrane is placed in a clean tray such as a plastic weigh boat.

11. 3 mL of ECL solution (per filter) is prepared according to the manufacturer's instructions, immediately added to the membrane, and shaken gently for 1 min to ensure an even coverage of the membrane by the liquid.

12. Excess liquid is again removed by touching the membrane to a clean tissue. The membrane is wrapped in cling wrap and exposed to film in an autoradiography cassette in a dark room, *as soon as possible.*

3.10. Preparation of Samples for Mass Spectrometry (see Note 19)

The procedures described below are for the analysis by liquid chromatography-tandem mass spectrometry (LC-MS/MS) using a Q-Tof Ultima API mass spectrometer. The treatment of samples may vary depending on the type of analysis and the instrument used. It is best to consult with the mass spectrometry facility where the analysis is to be carried out for the processing of samples.

1. Following electrophoresis by SDS-PAGE the gel is stained overnight with Colloidal Blue, according to the manufacturer's instructions.

2. The gel is destained in several changes of ddH$_2$O until the background (i.e., the non-protein-containing part of the gel) is completely destained. This usually takes several hours (i.e., more than 12 h).

3. The destained gel is photographed to provide a record of the purification experiment.

4. 20 to 25 microfuge tubes are rinsed in 60% acetonitrile.

5. The entire lane is cut out lengthwise and divided into at least 20 gel slices. Each gel slice is placed in a separate tube.

6. Each gel slice is destained in 100 µL of destaining solution (25 m*M* ammonium bicarbonate in 50% acetonitrile) for 20 to 30 min. This step is repeated until the gel slice becomes completely destained (usually three to four times). Alternatively, gel slices can be destained overnight at 4°C.

7. Each gel slice is dehydrated in 100 μL of 100% acetonitrile for 5 to 10 min at room temperature. The plug become hard and white at this step.
8. The gel slices are reduced with freshly prepared 6.5 mM DTT solution for 45 to 60 min at 37°C.
9. The solution is discarded and proteins in the gel slices are alkylated by adding 100 μL of 54 mM iodoacetamide solution and incubating for 60 min at room temperature in the dark.
10. The solution is discarded and the gel slices are washed in 100 μL of gel slice destaining solution for 15 min at room temperature. This step is repeated once more.
11. The washing solution is discarded and the gel slices are dried in 100 μL of 100% acetonitrile for 10 min. The solution is again discarded and the gel slices are dried at room temperature.
12. Proteins are in-gel-digested in 15 μL of 10 ng/μL modified trypsin at (diluted from the 100X stock in 50 mM ammonium bicarbonate) for 30 min on ice (*see* **Note 20**). Then 15 μL of 50 mM ammonium bicarbonate are added to the samples followed by overnight incubation at 37°C.
13. Samples are equilibrated to room temperature. Then 30 μL of 2% acetonitrile in 0.1% formic acid are added to the samples and incubated at room temperature for 15 min. The samples are then vortexed briefly and sonicated for 1 min.
14. The supernatants are collected in separate tubes and the remaining gel slices are treated with 30 μL of 50% acetonitrile in 0.1% formic acid and incubated as above. Samples are again vortexed and sonicated as above, and the supernatants are collected and pooled with the corresponding supernatants from **step 13**.
15. The samples are vacuum dried in a vacuum centrifuge for 45 to 60 min until they are dry.
16. The eluted peptides are now ready for analysis by mass spectrometry.

Figure 4A shows an example of a preparative binding of nuclear extracts from MEL cells expressing biotinylated GATA-1 (lane 3) and control extracts from cells expressing the BirA biotin ligase only (lane 5). The control binding experiment shows that background consists of a few strongly stained bands against a backdrop of more faintly staining bands. Analysis by mass spectrometry identified the strongly staining bands as corresponding to naturally biotinylated proteins such as carboxylases *(1)*. The bulk of the remaining background proteins corresponded to abundant nuclear proteins such as splicing factors, proteins involved in ribosome biogenesis, and so on (**Fig. 4B**). The low background binding (<1% of the total) of transcription factors and chromatin remodeling and modification proteins is of note (**Fig. 4B**). By contrast, analysis of the GATA-1 purification gel slice shows a large increase in the binding of transcription factors and chromatin remodeling and modification proteins, thus indicating specific copurification with biotin-tagged GATA-1 (**Fig. 4C**) *(6)*. A number of these interactions have been validated by independent immunoprecipitation experiments using nuclear extracts from nontransfected MEL cells and

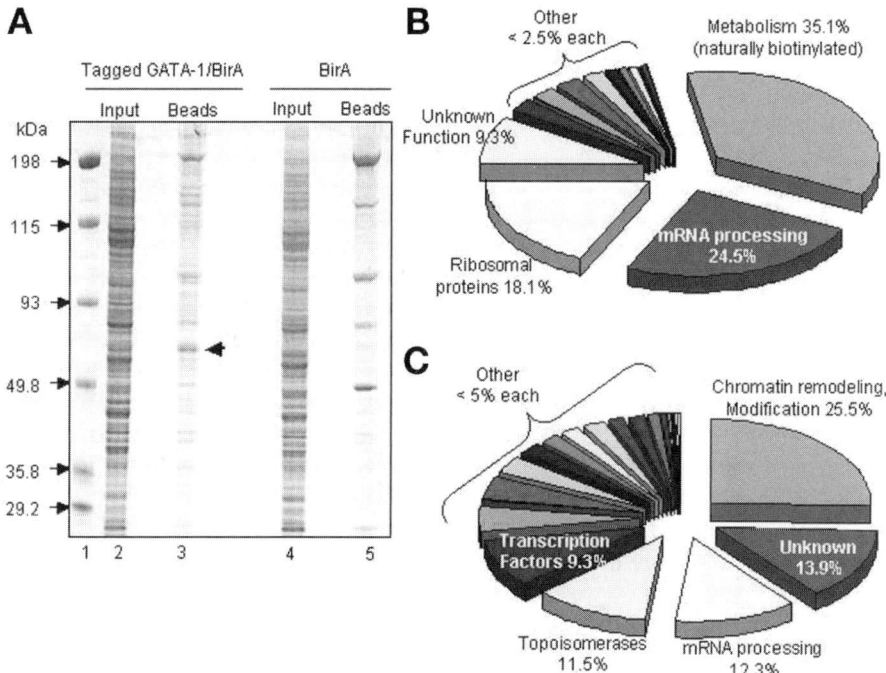

Fig. 4. (**A**) Colloidal blue-stained gel of a binding experiment of crude nuclear extracts to streptavidin beads. Lane 1, Marker (M). Lane 2, input nuclear extract from tagged GATA-1/BirA double-transfected cells. Lane 3, proteins eluted after binding to streptavidin beads. Lane 4, input nuclear extract from BirA-transfected cells. Lane 5, proteins eluted after binding to streptavidin beads. Arrow in lane 3 indicates protein band containing biotinylated GATA-1, as determined by mass spectrometry. (**B**) Classification according to Gene Ontology criteria of proteins identified by mass spectrometry from the control experiment using extracts from cells expressing BirA. (Around 500 peptide sequences were identified.) This represents the background binding. (**C**) Classification according to Gene Ontology criteria of proteins identified by mass spectrometry using extracts from cells expressing biotin-tagged GATA-1. (More than 1000 peptide sequences were identified.) A significant increase in the identification of chromatin remodeling proteins and transcription factors is clearly seen compared with the control experiment shown in (B). (Figure 4A reproduced with permission from **ref. *1***. Copyright [2003] National Academy of Sciences, USA.)

shown to include essential hematopoietic transcription factors such as FOG-1, TAL-1, and Gfi-1b in addition to chromatin remodeling complexes such as MeCP1 and WCRF/ACF (*6*). In addition, analysis of the GATA-1 binding experiment identified abundant chromatin-associated proteins, such as topoisomerases, as background owing to their indirect copurification with GATA-1 by virtue of their association with chromatin. Thus, we have defined background in these experiments as consisting primarily of naturally biotinylated proteins, abundant nuclear proteins associated with RNA metabolism and ribosome bio-genesis, and abundant chromatin-associated proteins that are indirectly copur-ified with chromatin-bound transcription factors.

4. Future Prospects

We have shown that biotinylation tagging is highly efficient in cultured cells (**Fig. 1**) and transgenic mice (*1*), and we have used this approach to identify a number of different complexes formed by the essential hematopoietic transcrip-tion factor GATA-1 (*6*). Owing to its efficiency and ease of application, bio-tinylation tagging offers the prospect of rapidly expanding the characterization of transcription factor complexes. For example, the biotinylation tagging of the hematopoietic transcription factor partners of GATA-1 and the characteriza-tion of their protein complexes will lead to the rapid elucidation of the distinct and overlapping transcriptional networks these factors regulate in hematopoie-sis. Similarly, the biotinylation tagging of chromatin cofactors will lead to a better understanding of their interactions with tissue-specific transcription fac-tors and the molecular basis of their functions (i.e., chromatin remodeling and modification in activation and repression). Furthermore, efforts in reducing the background along the lines described here (i.e., a prepurification steps such as gel filtration or the use of protease cleavage) will help in further expanding the utility of biotinylation tagging, for example, in preserving the native prop-erties of complexes or in determining stoichiometries. The utility of biotinyla-tion tagging will be further increased through the development of additional tools such as the recent derivation of a transgenic mouse strain that expresses BirA ubiquitously in all tissues (*8*), or the construction of a codon-optimized version of BirA for the efficient expression in mammalian cells (*9*). The recent description of the biotinylation of cell surface proteins (*10*) should also serve to expand the utility of this approach. Lastly, it should be noted that in vivo biotinylation tagging can also be employed (e.g., instead of antibodies) in all other applications involving an affinity purification or detection step, such as immunofluorescence (*1*), immunoprecipitation (*1,11*), and chromatin immuno-precipitation (ChIP) assays (*1,12*).

5. Notes

1. We routinely screen 12 to 20 stable transfected MEL cell clones by SDS-PAGE in order to select a clone that expresses the tagged protein at no more than 50% of the expression level of the endogenous protein. This is to ensure that the physio-logical interactions and functions of the protein of interest are not disturbed as a result of the overexpression of the tagged protein.

2. The specific lysis conditions will depend on the make of blender employed. It is recommended that conditions be optimized for cell density, length of lysis time, and speed setting of the blender.

3. The final salt concentration is critical for the extraction of nuclear proteins.

4. We routinely obtain around 100 mg of nuclear extract from 4 L of MEL cell culture at a density of 2×10^6 cells/mL.

5. There are a large variety of column matrices commercially available for gel filtration, with each matrix having different optimal separation ranges and physicochemical properties (e.g., ability to withstand high pressure in the column). Thus, the choice of matrix will depend on the desired range of fractionation and the liquid chromatography operating system available to the user (e.g., FPLC or HPLC).

6. Users must also refer to the manufacturer's instructions and training for use of the column and the FPLC apparatus.

7. The resolution efficiency of new columns, expressed as the number of theoretical plates per meter of column under normal running conditions, should be tested first. This can be done by injecting a sample of acetone (5 mg/mL) in ddH$_2$O water. Indicative efficiency for the analytical grade column is 11,100 theoretical plates/m.

8. While loading the extract, care must be taken that no air bubbles enter the loop. Air bubbles as well as cell debris can damage the column bed.

9. Once a new column is installed, the void (V_0) volume is determined by the peak of elution of dextran blue. To further calibrate the column, a mixture of at least two proteins of known molecular weight should also be injected. Recommended standards are bovine serum albumin (67 kDa), thyroglobulin (669 kDa), and aldolase (158 kDa).

10. If there is any suspicion that the column bed has been damaged, it is best to run the calibration standards again.

11. If the blue color of the sample loading buffer turns yellow, it is because of the protein sample being acidic, which will also affect migration of the sample during SDS-PAGE. A few microliters of Tris-HCl, pH 9.0, are usually sufficient to neutralize the sample.

12. To avoid pressure buildup, the run can be started at a flow rate of 1 mL/min. It is also better to inject the sample with the lower flow rate.

13. The concentration of 150 mM KCl is critical for the efficient binding of biotinylated proteins to streptavidin beads. We have found that even modest increases in salt concentration severely affect binding efficiency.

14. Protease cleavage also works well with shorter incubation times (5–30 min) and a broader temperature range (4–37°C).

15. Avoid handling membrane directly; use gloves and forceps.

16. Under these transfer conditions, the temperature of the buffer can rise significantly, and frothing may occur. This does not affect the transfer.

17. The gel can be stained after blotting in order to visualize residual proteins as a test for the efficiency of transfer as well as an indication of the amount of protein loaded per lane.

18. The primary antibody can be stored and reused. Sodium azide is added to the antibody solution to a 0.02% final concentration and stored at 4°C (sodium azide stock: 10% w/v in ddH$_2$O). **Caution:** sodium azide is highly toxic.
19. To reduce the risk of contaminating the samples for mass spectrometry, particularly with keratins, work is carried out in a hood with double gloves and a lab coat and always using sterile plasticware.
20. The volume of trypsin solution added will depend on the size of the gel slice. The volumes given above are for approx 4 × 2-mm gel slices. At this stage, gel slices should swell, and little solution should remain visible.

Acknowledgments

We are indebted to Dr. Jeroen Krijgsveld (Utrecht University) and Dr. Jeroen Demmers (Erasmus Medical Center) for expert mass spectrometry analysis. Work in our laboratory has been supported by grants from the Dutch Research Organization (NWO), the European Union (grant HRPN-CT-2000-00078), the NIH (grant RO1 HL 073445-01), and the Netherlands Proteomic Center.

References

1. de Boer, E., Rodriguez, P., Bonte, E., et al. (2003) Efficient biotinylation and single-step purification of tagged transcription factors in mammalian cells and transgenic mice. *Proc. Natl. Acad. Sci. USA* **100**, 7480–7485.
2. Schatz, P. J. (1993) Use of peptide libraries to map the substrate specificity of a peptide-modifying enzyme: a 13 residue consensus peptide specifies biotinylation in *Escherichia coli. Biotechnology (NY)* **11**, 1138–1143.
3. Beckett, D., Kovaleva, E., and Schatz, P. J. (1999) A minimal peptide substrate in biotin holoenzyme synthetase-catalyzed biotinylation. *Protein Sci.* **8**, 921–929.
4. Smith, P. A., Tripp, B. C., DiBlasio-Smith, E. A., Lu, Z., LaVallie, E. R., and McCoy, J. M. (1998) A plasmid expression system for quantitative in vivo biotinylation of thioredoxin fusion proteins in *Escherichia coli. Nucleic Acids Res.* **26**, 1414–1420.
5. Cull, M. G. and Schatz, P. J. (2000) Biotinylation of proteins in vivo and in vitro using small peptide tags. *Methods Enzymol.* **326**, 430–440.
6. Rodriguez, P., Bonte, E., Krijgsveld, J., et al. (2005) GATA-1 forms distinct activating and repressive complexes in erythroid cells. *EMBO J.* **24**, 2354–2366.
7. Singer, D., Cooper, M., Maniatis, G. M., Marks, P. A., and Rifkind, R. A. (1974) Erythropoietic differentiation in colonies of cells transformed by Friend virus. *Proc. Natl. Acad. Sci. USA* **71**, 2668–2670.
8. Driegen, S., Ferreira, R., van Zon, A., et al. (2005) A generic tool for biotinylation of tagged proteins in transgenic mice. *Transgenic Res.* **14**, 477–482.
9. Mechold, U., Gilbert, C., and Ogryzko, V. (2005) Codon optimization of the BirA enzyme gene leads to higher expression and an improved efficiency of biotinylation of target proteins in mammalian cells. *J. Biotechnol.* **116**, 245–249.

10. Chen, I., Howarth, M., Lin, W., and Ting, A. Y. (2005) Site-specific labeling of cell surface proteins with biophysical probes using biotin ligase. *Nat. Methods* **2,** 99–104.
11. Koutsodontis, G. and Kardassis, D. (2004) Inhibition of p53-mediated transcriptional responses by mithramycin A. *Oncogene* **23,** 9190–9200.
12. Viens, A., Mechold, U., Lehrmann, H., Harel-Bellan, A., and Ogryzko, V. (2004) Use of protein biotinylation in vivo for chromatin immunoprecipitation. *Anal. Biochem.* **325,** 68–76.

Index